T0281411

Electrical Engineering Fundamentals

Electrical Engineering Fundamentals

S. Bobby Rauf

CRC Press
Taylor & Francis Group
Boca Raton London New York

CRC Press is an imprint of the
Taylor & Francis Group, an **informa** business

First edition published 2020
by CRC Press
6000 Broken Sound Parkway NW, Suite 300, Boca Raton, FL 33487-2742

and by CRC Press
2 Park Square, Milton Park, Abingdon, Oxon, OX14 4RN

© 2021 Taylor & Francis Group, LLC

CRC Press is an imprint of Taylor & Francis Group, LLC

Library of Congress Cataloging-in-Publication Data
Names: Rauf, S. Bobby, 1956- author.
Title: Electrical engineering fundamentals / S. Bobby Rauf.
Description: First edition. | Boca Raton : CRC Press, 2021. |
Includes bibliographical references and index.
Identifiers: LCCN 2020037197 (print) | LCCN 2020037198 (ebook) |
ISBN 9780367376086 (hardback) | ISBN 9780429355233 (ebook)
Subjects: LCSH: Electrical engineering.
Classification: LCC TK145 .R38 2021 (print) | LCC TK145 (ebook) |
DDC 621.3—dc23
LC record available at https://lccn.loc.gov/2020037197
LC ebook record available at https://lccn.loc.gov/2020037198

ISBN: 978-0-367-37608-6 (hbk)
ISBN: 978-0-367-63624-1 (pbk)
ISBN: 978-0-429-35523-3 (ebk)

Typeset in Times
by codeMantra

Dedication

_____

This book is dedicated to my son Leonardo "Leo" Rauf

Contents

Preface

Many, in their quest for knowledge in engineering, find typical textbooks intimidating. Perhaps due to an extensive amount of physics theory, an overwhelming barrage of math and not enough practical applications of the engineering principles, laws, and equations. Therein lies the difference between this text and those voluminous and daunting conventional college engineering textbooks. A brief review of most electrical engineering textbooks will affirm the fact that the study of electrical engineering is replete with concepts, methods, and mathematical techniques that are somewhat more abstract than those employed in non-engineering subjects as well as other engineering disciplines, such as civil, mechanical, environmental, and industrial engineering. This text leads the reader into the more complex and abstract content after explaining the electrical engineering concepts and principles in an easy-to-understand fashion, support by analogies borrowed from day-to-day examples, and other engineering disciplines. Many complex electrical engineering concepts, e.g. low power factor, are examined from multiple perspectives, aided by diagrams, illustrations, and examples that the reader can easily relate to.

This mission to this book is to serve as a resource for exploring and understanding fundamental electrical engineering concepts, principles, analytical, and mathematical strategies.

If your objective as a reader is limited to the acquisition of basic to intermediate knowledge in electrical engineering, then the material in this book should suffice. If, however, the reader wishes to progress their electrical engineering knowledge to advanced level, this book could serve as a useful platform.

As the aphorism goes, "a picture is worth a thousand words"; this book maximizes the utilization of diagrams, graphs, pictures, and flow charts to facilitate quick and effective comprehension of the concepts of electrical engineering.

In this book, the study of electrical engineering concepts, principles, and analysis techniques is made relatively easy for the reader by inclusion of most of the reference data, in the form of excerpts from different parts of the book, within the discussion of each case study, exercise, and self-assessment problem solutions. This is in an effort to facilitate quick study and comprehension of the material without repetitive search for reference data in other parts of the book.

Certain electrical engineering concepts and terms are explained more than once as these concepts appear in different chapters of this text, often with a slightly different perspective. This approach is a deliberate attempt to make the study of some of the more abstract electrical engineering topics more fluid, allowing the reader continuity and precluding the need for pausing and referring to chapters where those specific topics were first introduced.

Due to the level of explanation and detail included for most electrical engineering concepts, principles, computational techniques, and analysis methods, this book is a useful tool for those engineers, energy engineers and non-engineers, who are not current on the subject of electrical engineering.

The solutions for end-of-the-chapter self-assessment problems are explained in just as much detail as the case studies and sample problems in the pertaining chapters. This approach has been adopted so that this book can serve as an electrical engineering skill building and knowledge reinforcement resource. Since all chapters and topics begin with the introduction of important fundamental concepts and principles, this book can serve as a "brush-up," refresher, or review tool for even electrical engineers whose current area of engineering specialty does not afford them the opportunity to keep their electrical engineering knowledge current.

In an effort to clarify some of the electrical engineering concepts effectively for engineers whose engineering education focus does not include electrical engineering, analogies are drawn from non-electrical engineering realms, on certain complex topics, to facilitate comprehension of the relatively abstract electrical engineering concepts and principles.

Each chapter in this book concludes with a list of questions or problems, for self-assessment, skill building, and knowledge affirmation purposes. The reader is encouraged to attempt these problems and questions. The answers and solutions, for the questions and problems, are included under Appendix A of this text.

Most engineers understand the role units play in definition and verification of the engineering concepts, principles, equations, and analytical techniques. Therefore, most electrical engineering concepts, principles, and computational procedures covered in this book are punctuated with proper units. In addition, for the reader's convenience, units for commonly used electrical engineering entities and some conversion factors are listed under Appendix C.

Most electrical engineering concepts, principles, tables, graphs, and computational procedures covered in this book are premised on SI/Metric units. However, US/Imperial units are utilized where appropriate and conventional. When the problems or numerical analyses are based on only one particular unit system, the given data and the final results can be transformed into another desired unit system through the use of unit conversion factors in Appendix B.

Some of the Greek symbols, used in the realm of electrical engineering, are listed in Appendix C, for reference.

WHAT CAN THE READERS GAIN FROM THIS BOOK?

- A clear and strong grasp of electrical engineering fundamentals.
- Better understanding of electrical engineering terms, concepts, principles, laws, analysis methods, solution strategies, and computational techniques.
- The ability to communicate with professional electrical engineers and electricians at their "wavelength."
- Greater confidence in their interactions with electrical engineering design engineers, power engineers, controls engineers, and other electrical engineering experts.
- Certain skills and preparation necessary for succeeding in electrical engineering portion of various certification and licensure exams, i.e. FE, Fundamentals of Engineering (also known as EIT, or Engineer in Training),

PE, Professional Engineering, CEM, Certified Energy Manager, and many other trade certification tests.
- A compact and simplified electrical engineering desk reference instead of a voluminous handbook.
- Better understanding of electrical energy cost and tips on improvement of electrical energy intensity in industrial and commercial environment.
- Better understanding of myriad battery options available in the market; their strengths, weaknesses, opportunities that lie ahead, and potential threats; and how batteries compare with capacitors as energy storage devices.

An epistemic advice to the reader: If you don't understand some of the abstract concepts the first time, don't give up. Read it again! Such is the nature, intrigue, and challenge of engineering, physics, science, and other subjects that require thinking, reflection, and rumination.

Acknowledgments and Credits

Rockwell International®

Many thanks to **Rachel R Schickowski, Hannah M. Schermerhorn, and Stacey Melichar** of Rockwell International® for provision of illustrative electrical power control and monitoring pictures that, I am certain, will be appreciated by the readers.

Fluke Corporation

Many thanks to **Beverly Summers** of Fluke Corporation for granting permission to use material associated with their multi-meter and clamp ammeter.

Control Technologies, LLC

Many thanks to **Wes Lampkin** of Control Technologies, LLC, for his contribution of the PLC relay ladder logic examples.

Author

Mr. S. Bobby Rauf is the president, chief consultant, and a senior instructor at Sem-Train, LLC. Mr. Rauf has over 25 years of experience in teaching under- and post-graduate Engineering, Science, Math, Business Administration, and MBA courses, seminars, and workshops. Mr. Rauf earned his BS in Electrical Engineering, with honors, from NC State University, Raleigh, NC, USA. He earned his Executive MBA degree from Pfeiffer University Misenheimer, NC, USA. He is a registered professional engineer, in the States of Virginia, Wyoming, and North Carolina and is a certified energy manager. He holds a patent in process controls technology.

Mr. Rauf was inducted as "**Legend in Energy**" by AEE, in 2014. Mr. Rauf is a member of **ASEE**, American Society of Engineering Education.

Mr. Rauf is certified to instruct various engineering, ergonomics, and industrial safety courses. He has conducted certification training and trained engineers for Professional Engineering licensure exams in the United States and abroad over the past 15 years.

Mr. Rauf develops and instructs PDH (Professional Development Hour) and, continuing education, engineering skill building courses. He conducts these courses in the form of webinars, live on-site presentations, workshops, pre-recorded audio, and self-study texts. Some of his major clients are **Amazon**, **NYC MTA**, **NYSSPE**, **Patrick Engineering**, and **several universities**. He is also an Adjunct Professor at Gardner-Webb University.

Mr. Rauf has published the following textbooks over the last 10 years:

- *Electrical Engineering for Non-electrical Engineers*
- *Thermodynamics Made Simple for Energy Engineers*
- *Finance and Accounting for Energy Engineers*

He has also developed and published several self-study books, in electronic format, that cater to the continuous professional development needs of engineers, technicians, and technical managers.

Mr. Rauf's last full-time engineering employment, in the corporate world, was at PPG Industries, Inc., where he served as a **senior staff engineer**. During his long career at PPG, his responsibilities included development and management of energy and ergonomics programs for multiple manufacturing plants, in the United States and overseas.

1 Fundamental Electrical Engineering Concepts and Principles

INTRODUCTION

In this first chapter of the *Electrical Engineering Fundamentals* text, we will explore fundamental electrical engineering terms, concepts, principles, and analytical techniques and impart knowledge that is considered elemental in the discipline of electrical engineering. Readers who invest time and effort in studying this text are likely to do so for the key purpose of gaining an introduction into the field of electricity. In this chapter, we will lay the foundations in the electrical engineering realm by covering basic electrical engineering terms, concepts, and principles, without the understanding of which, discussion and study of terms that bear important practical significance, such as power factor, real power, reactive power, apparent power, and load factor, would be untenable.

Most of the material in this chapter pertains to DC, or direct current, electricity. However, some entities discussed in this chapter such as capacitive reactance, inductive reactance, and impedance are fundamentally entrenched in the AC, alternating current, realm.

This text affirms that electrical engineering is rooted in the field of physics and chemistry. Physics, chemistry, and electrical engineering, as most other subject matters in science, depend on empirical proof of principles and theories. Empirical analysis and verification require tools and instruments for measurement of various parameters and entities. Hence, after gaining a better understanding of the basic electrical concepts, we will conclude this chapter with an introduction to three of the most common and basic electrical instruments, namely, multi-meter, clamp-on ammeter, and a scope meter or oscilloscope.

VOLTAGE OR EMF (ELECTROMOTIVE FORCE)

Voltage can be defined as a "force" that moves or pushes electrically charged particles like electrons, holes, negatively charged ions, or positively charged ions by forming an electric field. The term "electromotive" force stems from the early recognition of electrical current as something that consisted, strictly, of the movement of "electrons." Nowadays, however, with the more recent breakthroughs in the renewable and non-traditional electrical power generating methods and systems like microbial fuel cells and hydrocarbon fuel cells, electrical power is being harnessed, more and more, in the form of charged particles that may not be electrons. In batteries,

1

such as those used in automobiles, as we will see in the batteries chapter, the flow of current driven by voltage potential difference consists not only of negatively charged electrons, e⁻, but also types of ions, including H^+ and HSO_4^- ions.[1]

Two, relatively putative, analogies for voltage in the mechanical and civil engineering disciplines are pressure and elevation. In the mechanical realm – or more specifically in the fluid and hydraulic systems – high pressure or pressure differential pushes fluid from one point to another and performs mechanical work. Similarly, voltage – in the form of voltage difference between two points, as with the positive and negative terminals of an automobile battery – moves electrons or charged particles through loads such as motors, coils, resistive elements, wires, or conductors.

As electrons or charged particles are pushed through loads like motors, coils, resistive elements, light filaments, etc., electrical energy is converted into mechanical energy, heat energy, or light energy. In equipment like rechargeable batteries, during the charging process, applied voltage can push ions from one electrode (or terminal) to another and thereby "charge" the battery. Charging of a battery, essentially, amounts to the restoration of battery terminals' or plates' chemical composition to "full strength." So, in essence, the charging of a battery could be viewed as the "charging" of an electrochemical "engine." Once charged, a chemical or electrochemical engine, when presented with an electrical load, initiates and sustains the flow of electrical current, and performs mechanical work through electrical machines. In automobiles, this, essentially, amounts to the turning of the internal combustion motor when we attempt to start them.

Common symbols for voltage are: E, V, V_{DC}, V_{AC}, V_p, V_m, V_{Eff}, and V_{RMS}. Symbols "E" and "V" are synonymous, and both represent voltage. The symbol E stands for electromotive force, while V, simply, denotes voltage. Subscripts in the aforementioned nomenclature are intended to specify the type of voltage. In the absence of a subscript, these symbols can be somewhat ambiguous, in that, they could be construed to represent either AC or DC voltage.

The symbol V_{DC} specifically denotes DC voltage. See Chapter 3 for a detailed comparison between AC and DC, voltage and current. The AC and DC discussion, related to voltage and current, in Chapter 3, is supported by graphs that depict the differences in a graphical fashion.

The symbol V_{AC} represents AC voltage. When dealing with AC voltage, one needs to be specific about whether one is referring to "**peak**" voltage, V_p, or **RMS** voltage, V_{RMS}. Note that peak voltage, V_p, is synonymous with maximum voltage V_m. In addition, V_{Eff}, the effective (AC) voltage, is the same as, RMS voltage, V_{RMS}. The term "RMS" stands for **R**oot **M**ean **S**quare value of AC voltage. The RMS, or effective value of AC voltage, is the work producing portion of the AC voltage. The RMS voltage and current can also be viewed as those values of voltage and current that contribute the same amount of work as the best derived DC. This implies that AC voltage, current, and power, all, have the "work producing" components and the "non-work producing" components. The work producing components, in essence, transform into – or contribute toward the production of – various forms of energy, mechanical work

---

[1] Ions are defined as those atomic or molecular particles that are not neutral and bear a net negative or positive charge.

and break horsepower – or, to be more accurate, break horsepower-hour. The law of conservation of energy must always be obeyed. The concepts of power, break horse-power, and other power-related topics are discussed, in detail, in Chapter 4.

Even though a detailed discussion on the mathematical composition of RMS volt-age is outside the scope of this text, the formula for RMS voltage is as follows (later, in the discussion of current, it will become evident that formulas and nomenclature for voltage and current are quite similar):

$$V_{RMS} = V_{EFF} = \sqrt{\frac{1}{T} \int_0^T V^2(t)\,dt} \qquad (1.1)$$

AC voltage $\mathbf{V_{RMS}}$, $\mathbf{V_{Eff}}$, $\mathbf{V_p}$, and $\mathbf{V_m}$ are inter-related through the following equations:

$$V_p = V_m = \sqrt{2}V_{RMS} = \sqrt{2}V_{EFF} \qquad (1.2)$$

$$V_{RMS} = \frac{V_p}{\sqrt{2}} = \frac{V_m}{\sqrt{2}} \qquad (1.3)$$

Voltage is measured in volts, or V's, named after the Italian physicist Alessandro Volta (1745–1827), who invented the first chemical battery. See details in the chapter on batteries.

CURRENT

Current consists of movement of electrons, ions, or simply charged particles. Movement of electrons can be oscillatory, vibratory, or linear. When DC voltage is applied in an electrical circuit, electrons, ions, or charged particles move in one direction. Such linear, unidirectional movement of charged particles or electrons is DC current. DC electrical current is analogous to fluid flow in mechanical or hydrau-lic systems. Just as pressure, or pressure differential, causes fluid to flow from point A to point B, DC voltage drives electrically charged particles to move from one point to another. Important characteristics of electric current are illustrated in greater detail, in Chapter 3, under the topic of electrodeposition. When electrons vibrate or oscillate, the resulting current is AC current. AC electrical current is established and sustained by AC voltage. A detailed comparison between AC and DC is also illus-trated in the graphical form in Chapter 3.

Common symbols for current are: $\mathbf{I}$, $\mathbf{I_{DC}}$, $\mathbf{I_{AC}}$, $\mathbf{I_p}$, $\mathbf{I_m}$, $\mathbf{I_{Eff}}$, and $\mathbf{I_{RMS}}$. Similar to the voltage symbols, the symbols of current assume a more specific meaning through associated subscripts; therefore, coupling the symbols with subscripts is important.

$\mathbf{I_{AC}}$ represents AC. When dealing with AC, one needs to be specific about whether one is referring to "peak" current, $\mathbf{I_p}$, or RMS current, $\mathbf{I_{RMS}}$. Note that peak current, $\mathbf{I_p}$, is synonymous with maximum current $\mathbf{I_m}$. In addition, $\mathbf{I_{Eff}}$, the effective (AC) cur-rent, is the same as, RMS current, $\mathbf{I_{RMS}}$. The term "RMS current" stands for **R**oot **M**ean **S**quare value of AC. The RMS or effective value of AC is the work produc-ing portion of the AC. AC, like AC voltage and AC power, has a "work producing"

component and a "non-work producing" component. The work producing component of AC contributes toward the production of mechanical work, energy, and break
horsepower.

The formula for RMS current is as follows:

$$I_{RMS} = I_{EFF} = \sqrt{\frac{1}{T}\int_0^T I^2(t)d(t)} \tag{1.4}$$

AC I_{RMS}, I_{Eff}, I_p, and I_m are inter-related through the following equations:

$$I_p = I_m = \sqrt{2}I_{RMS} = \sqrt{2}I_{EFF} \tag{1.5}$$

$$I_{RMS} = \frac{I_p}{\sqrt{2}} = \frac{I_m}{\sqrt{2}} \tag{1.6}$$

The unit for current is Ampere, named after André-Marie Ampère (1775–1836), a
French mathematician and physicist. André-Marie Ampère is revered as the father of
electrodynamics. One amp of current is said to flow when electrical charge is flowing
at the rate of **one Coulomb per second**. This leads to the following mathematical
definition:

$$1 \text{ Amp of Current} = \frac{\text{One Coulomb of Charge}}{\text{Second}}$$

or,

$$1 \text{ Amp of Current} = \frac{1/96,487 \text{ Faraday}}{\text{Second}} \tag{1.7}$$

Example 1.1

In an AC system, a voltage source **V(t) = 156 Sin (377t+0°)** V sets up a current
of **I(t) = 15 Sin (377t+45°)** A. Calculate the RMS values of voltage and current.

Solution:

The peak voltage and the peak current, in accordance with convention, are **156 V
and 15 A**, respectively.
According Eq. 1.3:

$$V_{RMS} = \frac{V_p}{\sqrt{2}} = \frac{V_m}{\sqrt{2}}$$

$$\therefore V_{RMS} = \frac{V_p}{\sqrt{2}} = \frac{156}{\sqrt{2}} = 110 \ V_{rms}$$

Note: This is the voltage indicated by a true RMS voltmeter when measuring the AC voltage at a typical household or workplace wall receptacle in the United States. In other parts of the world, typical receptacle voltage would read closer to 220 V_{rms}. See more discussion toward the end of this chapter.

According Eq. 1.6:

$$I_{RMS} = \frac{I_p}{\sqrt{2}} = \frac{I_m}{\sqrt{2}}$$

$$\therefore I_{RMS} = \frac{15}{\sqrt{2}} = 10.6\ A_{RMS}$$

Note: This is the current indicated by a true RMS clamp-on ammeter when measuring AC. See more discussion toward the end of this chapter.

RESISTANCE

Property of a material that opposes or resists the flow of current is known as electrical resistance, or simply, resistance. Physically, at the atomic level, when electrons, or other types of ions, collide with other "non-current participating" particles (i.e. neutral atoms), the movement of these current constituting particles is impeded. Electrical resistance is analogous to friction in mechanical systems; for instance, friction between the surfaces of two objects that slide against each other. In a fluid flow scenario, electrical resistance is analogous to friction between the fluid and the walls of the pipe. In electrical systems, resistance in conductors (wires) is an undesirable characteristic and results in wasted heat or heat losses. This is similar to frictional head losses in fluid systems – frictional head losses governed by Darcy's and Hazen–Williams' equations. The symbol for electrical resistance is "**R**." Resistance is measured in ohms, or simply, Ω's. The ohm symbol Ω is often prefixed with letters, such as, k for kilo or M for mega. Where, 1 kW would represent 1,000 ohms and 1 MΩ would represent 1,000,000 ohms.

Resistance of any material, component, or conductor can be measured using an ohm-meter or multi-meter, while that particular entity is de-energized. When an ohm-meter or a multi-meter is used to measure resistance, the meter's own power source is used to flow current through the object being tested. Therefore, if the object or component, whose resistance is being measured, is in a "live" or energized electrical circuit, the current already flowing through that circuit is likely to interfere with the resistance measurement by the meter and could result in an inaccurate resistance measurement or, in some cases, cause the meter to be damaged.

$$R = \frac{V}{I}\ \text{or,}\ V = I \cdot R \tag{1.8}$$

The unit of "ohm" can also be defined on the basis of Ohm's law. Ohm's law and its application in AC and DC systems are discussed in greater depth in Chapter 2. At this point, upon examination of Ohm's law, in the form of Eq. 1.8, we can define **1 ohm** as the amount of resistance that would permit the flow of only **1-amp** of current when a voltage of **1-volt** is applied across a specific length of that conductor. In other words:

$$1\,\Omega = \frac{1\text{-volt}}{1\text{-amp}}$$

The Ohm's law equation stated above also stipulates that as the increase in electrical "demand" increases in the form of an increase in current, **I**, with the resistance, **R**, of the conductor or component remaining constant, the voltage – or voltage drop – across the conductor or component increases, resulting in lower voltage at the point of delivery of power or energy to the load. This is similar to the **pressure drop**, or **loss**, experienced in a long compressed air pipe or a long water header due to frictional head loss.

In electrical systems, the term "**source**" means the most immediate source of power or energy, while the term "load" represents the ultimate or immediate downstream entity that is demanding the power or energy.

From physical characteristic and physical composition point of view, we could define resistance as being directly proportional to the **length** of the conductor and inversely proportional to the **area of cross-section** of the conductor. This proportional relationship can be transformed into a mathematical relationship or equation, through the insertion of a constant of proportionality, as follows:

$$R = \rho.\frac{L}{A} \tag{1.9}$$

where
ρ = Resistivity of the conductor, serving as a constant of proportionality
L = Length of the conductor, and
A = Area of cross-section of the conductor.

The resistivity values of two common conductors, copper and aluminum, are as follows:

ρ_{copper} = 17.2 n-Ω.m; read as "nano-ohm-meter," where **n** = nano = 10^{-9}
$\rho_{aluminum}$ = 28.2 n-Ω.m

Example 1.2

A cubic block of electrically conductive material measures 0.02 m on each side. The resistivity of this material is 0.01 Ω-m. What is the resistance between opposite sides of this block?

Solution:

$$R = \frac{\rho L}{A} = \frac{(0.01\,\Omega - \text{m})(0.02\text{ m})}{(0.02\text{ m})^2} = 0.5\,\Omega$$

Example 1.3

A phase conductor of a power distribution line spans, approximately, 500 ft and has a diameter of 1.5 inch. The conductor is composed of copper. Calculate the per-phase electrical resistance of this conductor.

Solution:

Solution strategy: Since the resistivity value of copper, as stated above, is in metric or SI unit system, the length and diameter specifications stated in this problem must be streamlined in metric units before application of Eq. 1.9 for determination of resistance in ohms (Ω's).

L = 500 ft = 152.4 m
Diameter = 1.5 inch = 0.0381 m; $\therefore$ **R** = Radius = D/2 = 0.019 m
A = Area of cross-section = $\pi \cdot R^2$ = (3.14) (0.019)² = 0.00113 m²
ρ_{copper} = 17.2 n Ω-m = 17.2 × 10⁻⁹ Ω-m

$$R = \rho \cdot \frac{L}{A}$$

$$\therefore R = \rho \cdot \frac{L}{A} = 17.2 \times 10^{-9} \left(\frac{152.4}{0.00113} \right) = 0.00232 \, \Omega$$

As described earlier, in electrical systems, resistance in conductors (wires) is an undesirable characteristic and results in wasted heat or heat losses. This energy lost, as heat, as the current flows through a conductor can be quantified through Eq. 1.10.

$$\text{Heat Loss} = P = VI = \frac{V^2}{R} = I^2 R, \text{ in Watts} \tag{1.10}$$

Although the subject of electrical power and energy is discussed, in detail, in Chapter 4, at this point, in an effort to illustrate the substantial impact of resistance on heat losses and design challenges associated with power transmission, let's continue examination of long transmission lines used to transport electrical power from power generating plants to power consumers. On a hot summer afternoon, those power transmission lines that we notice crisscrossing the country side are not only carrying higher currents due to the higher commercial, residential, industrial activity-related loads but also experiencing an increase in resistance due to the rise in resistivity, ρ, of the conductor in accordance with Eq. 1.11. In other words, the resistivities of copper and aluminum, stated above, serve as **constants only at standard temperature of 20°C or 68°F.**

$$\rho = \rho_0 \left(1 + \alpha (T - T_0) \right) = \rho_0 \left(1 + \alpha \Delta T \right) \tag{1.11}$$

In Eq. 1.11:

ρ = Resistivity at current temperature "**T**"
ρ_0 = Resistivity at standard temperature "**T₀**"
T_0 = Standard temperature in °C

T = Current temperature °C
α = Thermal coefficient of resistance, in 1/°C

In addition, it's common knowledge that as temperature rises, most metals or conductors expand. So, with exposure to elevated summer season solar radiation, during those hot afternoons, the transmission lines elongate to a certain extent, resulting in higher "**L**." As those transmission lines elongate, the diameter of the conductors shrinks to some degree, resulting in lower area of cross-section **A**. Hence, the perceptible "sag" in the transmission lines during the hot summer afternoons is not simply an optical illusion.

As we account for the increase in **L**, reduction in **A**, and the rise in ρ collectively, in accordance with Eq. 1.9, we see that all of these factors result in escalation of resistance. Furthermore, if we consider resistive heat loss equation, Eq. 1.10, the **exponential** effect of the rise in load current **I** and the increase in **R** precipitate in a "cascading," unfavorable, physical, and electrical impact on the transmission line conductors. Power transmission system designers have to consider all of these factors in the specification and design of transmission line systems.

RESISTORS IN SERIES

When electrical circuits, AC or DC, consist of multiple resistors, circuit analyses require simplification of such network of resistors into one, **equivalent**, resistor or resistance, R_{EQ}. Often, this equivalent resistance is referred to as a "**total**" resistance. When "**n**" number of resistors are connected in a "daisy chained" or concatenated fashion, as shown in Figure 1.1, they are said to be connected in series. Note the distinct symbol used for the voltage source labeled "V" below. In order to avoid ambiguity, this circular symbol with a "+" or "−" sign enclosed, in electrical diagrams, is dedicated to DC power supplies. Power supplies are devices that convert AC to DC. Although some electrical system designers use this circular symbol to represent batteries, proper symbol for batteries, as shown later, consists of two unequal parallel lines.

When resistors are connected in series, they can be combined in a "linear" fashion, as shown in Eq. 1.12, for "**n**" number of resistors.

$$R_{EQ} = R_T = R_1 + R_2 + R_3 + \cdots + R_n \tag{1.12}$$

When multiple resistors are combined into an equivalent resistor, with resistance value R_{EQ}, the simplified version of the original series circuit would appear as pictured in Figure 1.2.

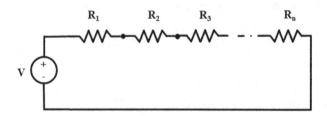

FIGURE 1.1   **n**–Resistors in series.

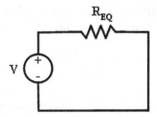

FIGURE 1.2   Equivalent resistance for **n**–resistors in series.

RESISTORS IN PARALLEL

When electrical circuits, AC or DC, consist of multiple resistors, connected in a parallel fashion, as shown in Figure 1.3, circuit analysis would require simplification of the parallel network of resistors into one, equivalent resistor R_{EQ}. Figure 1.3 shows "**n**" number of resistors connected such that the "heads" of all resistors are "bonded" or connected together, with an electrical connection to the anode (positive terminal) of the DC power supply. In addition, the "tails" of all resistors are connected together and to the cathode of the DC power source.

When multiple parallel resistors are combined into an equivalent resistor, with resistance value R_{EQ}, the simplified version of the original series circuit would appear as shown in Figure 1.4. In parallel resistor networks, the calculation of R_{EQ} involves addition of the inverses of all resistors in the parallel network and taking the inverse of the sum as stipulated in Eq. 1.13.

$$R_{EQ} = \cfrac{1}{\cfrac{1}{R_1} + \cfrac{1}{R_2} + \cfrac{1}{R_3} + \cdots + \cfrac{1}{R_n}} \tag{1.13}$$

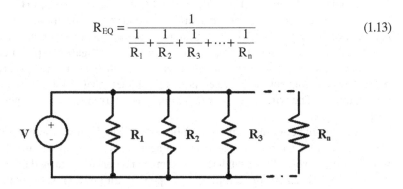

FIGURE 1.3   **n**–Resistors in parallel.

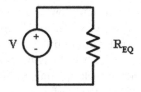

FIGURE 1.4   Equivalent resistance for **n**–resistors in parallel.

In case of a simple two parallel resistor circuit, the equivalent resistance $\mathbf{R_{EQ}}$ could be calculated using the simplified equation, Eq. 1.14.

$$R_{EQ} = \frac{R_1 R_2}{R_1 + R_2} \tag{1.14}$$

As a special case, assume that the parallel resistor network shown in Figure 1.3 consists of "**n**" **equal** parallel resistors. Because the resistors are assumed to be equal, calculation of $\mathbf{R_{EQ}}$, or $\mathbf{R_{EQ\text{-}n}}$, boils down to Eq. 1.15.

$$R_{EQ} = R_{EQ\text{-}n} = \frac{R}{n} \tag{1.15}$$

So, if one were to connect one million 1-Mega Ω resistors in parallel, the net or equivalent resistance, $\mathbf{R_{EQ}}$, in the circuit would be 1-Ω. The reader is encouraged to prove this by using Eq. 1.15.

ELECTRICAL SHORT AND OPEN CIRCUIT

When a conductor, wire or bus bar, is used to pass current between two or more points in an electrical circuit – with the smallest resistance feasible – such a connection is, technically, a "**short**." With that said, note that in electrical engineering realm, normal use of conductors for the delivery of current and voltage to loads (lamps, motors, heaters, etc.) is not commonly referred to as a "**short**." The term "short" is generally reserved for accidental, unintended, or undesirable connection of an electrically energized, higher potential, point to ground or a lower potential point.

When a short segment of wire, assembled with an "**alligator clip**" on each end, is used by electrician and electrical engineers to establish a temporary short circuit, or an electrical connection, between points in an electrical circuit, it is referred to as a "**jumper**." The wire used to form a jumper is referred to as a "jumper lead" in electrical jargon.

A short circuit between two points implies zero or negligible resistance. The opposite of an electrical short, or an electrical connection, is an "**open circuit**." Examples of open circuits would be a switch that is open, a breaker that is turned off, or simply a wire that has been cut or "**clipped**." An open circuit between two points implies an infinite (∞) resistance. The concepts of open and short circuits are illustrated in Figure 1.5. Figure 1.5a and b represents electrical circuit segments between points A and B. When the switch between points (1) and (2) is open, as depicted in Figure 1.5a, we have an open circuit, and no current flows between points **A** and **B**. However, closure of the same switch, as shown in Figure 1.5b, constitutes a short circuit. The closed switch scenario depicted in Figure 1.5b also represents "**continuity**." The term "continuity" is used commonly by electrical engineers and electricians during troubleshooting of equipment. When troubleshooting electrical or electronic equipment failures, engineers and technicians often perform continuity tests on fuses to determine if they have cleared or opened. The instrument used for performing continuity

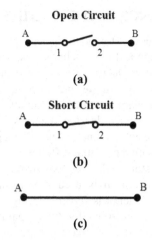

FIGURE 1.5   (a–c) Open, short, and continuous circuits.

checks is a multi-meter. A continuous piece of wire, shown in Figure 1.5c, represents continuity between terminals **A** and **B**.

Example 1.4

Determine the equivalent resistance for the DC circuit shown below from the 12 V battery vantage point. (*Note the symbol on the left side of the circuit used to represent a 12 V battery and not a 12 V power supply.*)

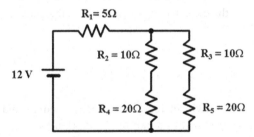

Solution:

The resistances need to be combined in a multi-step process in order to determine R_{eq} for the overall circuit.

Combination of R_2 and $R_4 = R_{2,4} = R_2 + R_4 = 10\,\Omega + 20\,\Omega = 30\,\Omega$

Combination of R_3 and $R_5 = R_{3,5} = R_3 + R_5 = 10\,\Omega + 20\,\Omega = 30\,\Omega$

Combination of $R_{2,4}$ and $R_{3,5}$:

$$R_{2-5} = \frac{(30\,\Omega) \cdot (30\,\Omega)}{(30\,\Omega + 30\,\Omega)} = \frac{900}{60} = 15\,\Omega$$

$$\therefore R_{eq} = 15\,\Omega + 5\,\Omega = 20\,\Omega$$

CAPACITOR, CAPACITANCE, AND CAPACITIVE REACTANCE

A capacitor is a charge and energy storage device capable of storing charge and electrical energy in DC and AC applications. A charged capacitor stores electrical charge on two electrodes; one of the two electrodes is negative, and the other one is positive. The negative electrode is called a **cathode**, and the positive electrode is referred to as an **anode**. This separation of charge and the quantity of charge separated determine the electrical potential – or voltage – developed across the electrodes of the capacitor, as well as the amount of electrical energy stored in a given capacitor. Therefore, as energy storage devices – with a potential difference – capacitors are analogous to mechanical pressure energy storage devices like air receivers and pneumatic cylinders; pressure differential wherein can be used to perform mechanical work.

Construction of a simple capacitor is depicted in Figure 1.6. As shown in the figure, a simple capacitor can be constructed with two parallel square plates, of equal size, separated by a dielectric substance like air, glass, mica, paper, etc. The separation between the two plates (electrodes), "**r**," in conjunction with the area of the plates determines the "**capacitance**" of the capacitor. Capacitance, "**C**," of a capacitor is defined as the charge storage capacity of the capacitor.

Capacitance can be defined, mathematically, through Eq. 1.16.

$$C = \epsilon \frac{A}{r} \tag{1.16}$$

where

    C = Capacitance is quantified or specified in farads.
    A = The area of cross-section – or simply area – of the capacitor electrode plates.
    ϵ = Permittivity of the dielectric medium between the plates.

and

$$\epsilon = \epsilon_r \cdot \epsilon_0$$

where ϵ_r = Relative permittivity of a specific dielectric medium and ϵ_0 = Permittivity in vacuum or in air = 8.854×10^{-12} farads per meter (F·m^{-1}).

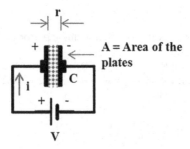

FIGURE 1.6 A simple parallel plate capacitor.

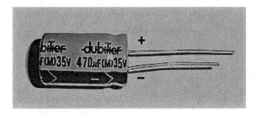

FIGURE 1.7   A cylindrical 470 μF capacitor.

One farad is rather large amount of capacitance for most common capacitor applications. Therefore, many capacitors – especially, at the circuit board level – are specified or labeled in terms of smaller units, such as mF (milli-Farad), μF (micro-Farad),
or nF (nano-Farad). The capacitor shown in Figure 1.7 is rated 470 μF and designed
to operate at a maximum of 35 V.

The mathematical relationship stated as Eq. 1.16 stipulates that capacitance is
directly proportional to the area **A** of the capacitor plates and inversely proportional
to the separation **r** between the plates. In other words, if larger capacitance or **charge
storage capacity** is desired, one must increase the area of the plates and/or decrease
the separation between the capacitor plates. In addition to serving as a "**constant of
proportionality**" for the equation, permittivity $\mathbf{\in}$ injects the property or characteristic
of the dielectric medium into the computation of capacitance through the dielectric
medium's characteristic $\mathbf{\in_r}$ value. Although, for the sake of simplicity, the discussion
on capacitance in this text is limited to flat plate capacitors, many capacitors have
cylindrical construction such as the one shown in Figure 1.7. Some, common, motor
starting capacitors are of cylindrical construction with paper used as the dielectric
medium between the cylindrical electrodes or plates.

Electrical energy stored in a capacitor can be determined through application of
Eq. 1.17:

$$\text{Energy(joules)} = \frac{1}{2}CV^2 \text{ and } Q = CV \qquad (1.17)$$

Even though Eq. 1.17 stipulates that the total charge stored in a capacitor is equal to the
product of capacitance "C" of the capacitor and the voltage "V" applied to the capacitor, charge storage characteristic of the capacitor should not be confused with the
charge storage and power source characteristics of a battery. One difference between
capacitors and batteries is that when capacitors are charging or discharging, charge
flows through a "dielectric" medium, while most batteries consist of electrolytes (i.e.
sulfuric acid) that ionize readily and the ions sustain the flow of current. Capacitors
allow the charge to move between electrode plates through a dielectric medium. Also,
in most cases, capacitors charge and discharge at a faster rate than batteries.

The dynamics of how a capacitor stores and dissipates charge are somewhat different between the DC and AC realms. Unlike resistive circuits, current and voltage associated with capacitors vary in a non-linear fashion. A common, series,
RC (Resistive–Capacitive) circuit is shown in Figure 1.8, consisting of a capacitor,

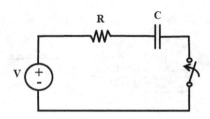

FIGURE 1.8 A series RC circuit.

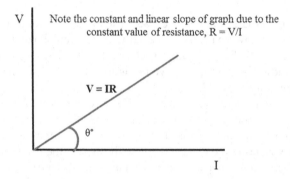

FIGURE 1.9 Linear, voltage versus current, response in a purely resistive circuit.

resistor, and a switch that can be used to control the charging and discharging of the capacitor. The graphs in Figures 1.9–1.11 compare voltage and current responses in circuits that are purely resistive versus a series RC circuit, the type illustrated in Figure 1.8. This "non-linear" charging and discharging of capacitors is referred to as **transient behavior** of RC circuits.

The straight-line graph in Figure 1.9 illustrates and validates Ohm's law. This graph shows that voltage and current are directly proportional, with the "constant" resistance serving as the constant of proportionality and a constant slope of the graph. In a purely resistive circuit, the current would respond, instantaneously and linearly, to the application of voltage across a resistor.

Voltage versus time, transient response in series RC circuit: Contrary to linear and instantaneous response in a purely resistive circuit, the current and voltage response in a common RC circuit is non-linear and non-instantaneous.

$$v_c(t) = v_c(0)\, e^{-\frac{t}{RC}} + V\left(1 - e^{-\frac{t}{RC}}\right) \tag{1.18}$$

The voltage response – or voltage variation – of a capacitor can be predicted through Eq. 1.18, where

R = Resistance in series with the capacitance
C = Capacitance
v_c(0) = Voltage across the capacitor, at time $t = 0$

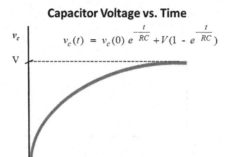

Capacitor Voltage vs. Time

$$v_c(t) = v_c(0)\, e^{-\frac{t}{RC}} + V(1 - e^{-\frac{t}{RC}})$$

FIGURE 1.10   Non-linear, transient, voltage response in a circuit consisting of capacitance and resistance.

$v_c(t)$ = Voltage across the capacitor, at a given time **t**
V = Voltage of the power source
$RC = \tau$ = Time constant of an RC circuit

If the voltage variation or response of a capacitor were graphed, with respect to time, it would resemble the v_c versus **t** graph shown in Figure 1.10.

Current versus time, transient response in series RC circuit: The current response – or current variation – in a capacitor–resistor circuit can be predicted through Eq. 1.19.

$$i_c(t) = \left(\frac{V - v_c(0)}{R} \right) e^{-\frac{t}{RC}} \qquad (1.19)$$

where

R = Resistance in series with the capacitance
C = Capacitance, in farads
$v_c(0)$ = Voltage across the capacitor, at time t = 0
$i_c(t)$ = Current through the capacitor–resistor circuit, at a given time **t**
V = Voltage of the power source

If the current response of a capacitor were graphed, with respect to time, it would resemble the i_c versus **t** graph shown in Figure 1.11. Note that the capacitor current versus time curve in Figure 1.11 validates Eq. 1.19. For instance, the graph shows that at t = ∞, or when steady state is achieved, current $i_c(t)$ through the RC circuit diminishes to zero, and if you substitute t = ∞ in Eq. 1.19, $i_c(t)$ becomes zero. In other words, at steady state, the capacitor transforms into an open circuit and the current ceases to flow. Equation 1.19 also validates the fact that at time t = 0, before the capacitor has had the chance to charge, $v_c(0) = 0$, and the maximum amount current of current would pass through the capacitor and the rest of the electrical circuit, as if the capacitor were an electrical short. The reader is encouraged to prove both t = ∞ and t = 0 states through Eq. 1.19.

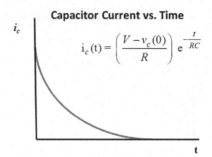

Capacitor Current vs. Time

$$i_c(t) = \left(\frac{V - v_c(0)}{R}\right) e^{-\frac{t}{RC}}$$

FIGURE 1.11 Non-linear, transient, current response in a circuit consisting of capacitance and resistance.

The presence of resistance **R** in capacitive circuits results in what is referred to as a time constant "τ." The relationship between τ, **R**, and **C** is stipulated by Eq. 1.20. The physical significance of time constant τ is that it represents the time it takes to charge a capacitor to **63.2% of the full value** or **63.2% of the full voltage** of the source. Time constant τ also represents the time it takes to discharge a given capacitor to **36.8% of the full voltage** or the voltage of the source.

$$\tau = RC \tag{1.20}$$

Example 1.5

Consider the RC circuit shown in the diagram below. The source voltage is 12 V. The capacitor is in a discharged state before the switch is closed. The switch is closed at $t = 0$. What would the capacitor voltage be at $t = 2\tau$?

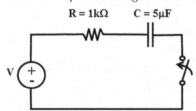

R = 1kΩ   C = 5µF

Solution:

This particular case represents a capacitor charging scenario. Given the values of **R, C, $v_c(0)$,** and the source voltage **V,** Equation 1.18 allows us to calculate the voltage after and elapsed time "**t**" during the capacitor charging phase.

$$v_c(t) = v_c(0) e^{-\frac{t}{RC}} + V\left(1 - e^{-\frac{t}{RC}}\right)$$

In this case,

R = 1 kΩ = 1,000 Ω

C = 5 μF = 5 × 10⁻⁶ F

$v_c(0)$ = 0 V = Voltage across the capacitor at t = 0

$v_c(t)$ = Voltage across the capacitor, at a given time **t** =?

V = Voltage of the power source = 12 V

RC = τ = RC circuit time constant

t = 2τ = 2RC

By substitution of the given values, stated above, Eq. 1.18 can be expanded and simplified as follows:

$$v_c(t) = v_c(0) \cdot e^{-\frac{2\tau}{\tau}} + 12 \cdot \left(1 - e^{-\frac{2\tau}{\tau}}\right)$$

$$= 0 + (12) \cdot \left(1 - e^{-2}\right) = 12(0.865)$$

$$= 10.38 \text{ V}$$

Ancillary Note

The time constant in Example 1.5 is τ = **RC** = (1,000 Ω)·(5 × 10⁻⁶ F) = 5 ms. This implies that the capacitor in this case would be almost fully charged in well below 1 second. This serves as a proof of the statement made earlier in this chapter that capacitors charge and discharge at much faster rate than batteries. To further affirm this point, the reader is encouraged to calculate the voltage across the capacitor in Example 1.5 at t = 5τ.

Capacitors in Series

When a number of capacitors are connected in a "daisy-chained," or concatenated fashion, as shown in Figure 1.12, they are said to be connected in series. In Figure 1.12, "**n**" number of capacitors, C_1 through C_n, are shown connected in series.

When electrical circuits, AC or DC, consist of multiple capacitors, circuit analyses require simplification – or combination – of such network of capacitors into one, **equivalent**, capacitor or capacitance, C_{EQ}. This equivalent capacitance can also be referred to as a "**total**" capacitance.

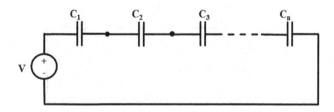

FIGURE 1.12 Capacitors in series.

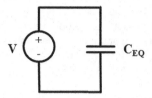

FIGURE 1.13   Equivalent capacitance for capacitors in series.

Unlike series combination of resistors, when capacitors are connected in series, they can be combined in an **"addition of inverses"** format, as stipulated in Eq. 1.21, for **"n"** number of capacitors.

$$C_{EQ} = \frac{1}{\dfrac{1}{C_1} + \dfrac{1}{C_2} + \dfrac{1}{C_3} + \cdots + \dfrac{1}{C_n}} \tag{1.21}$$

When multiple capacitors are combined into an equivalent capacitor, with capacitance value C_{EQ}, the simplified version of the original series circuit can be drawn as shown in Figure 1.13.

As a special case, suppose that the series capacitor network shown in Figure 1.12 consists of "n" series capacitors, with **equal** capacitance. Because the capacitors are assumed to be equal, calculation of C_{EQ}, or C_{EQ-n}, and application of Eq. 1.21 can be simplified to Eq. 1.22.

$$C_{EQ} = C_{EQ\text{-}n} = \frac{C}{n} \tag{1.22}$$

If a series capacitive circuit consists of only three capacitors, as shown in Figure 1.14, Eq. 1.21 can be reduced to Eq.1.23. Further simplification of Eq. 1.23 would result in a, simplified, three capacitor series equivalent capacitance equation, Eq. 1.24.

$$C_{EQ} = \frac{1}{\dfrac{1}{C_1} + \dfrac{1}{C_2} + \dfrac{1}{C_3}} \tag{1.23}$$

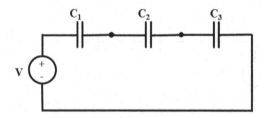

FIGURE 1.14   Equivalent capacitance for three capacitors in series.

$$C_{EQ} = \frac{C_1 C_2 C_3}{C_1 C_2 + C_2 C_3 + C_1 C_3} \tag{1.24}$$

Example 1.6

Determine the equivalent capacitance for the DC circuit shown in Figure 1.14 if $C_1 = 5\ \mu F$ and $C_2 = C_3 = 10\ \mu F$.

Solution:

Application of Eq. 1.24 to the three series capacitor circuit shown in Figure 1.14 yields:

$$C_{EQ} = \frac{(5\times10^{-6})(10\times10^{-6})(10\times10^{-6})}{(5\times10^{-6})(10\times10^{-6})+(5\times10^{-6})(10\times10^{-6})+(10\times10^{-6})(10\times10^{-6})}$$

$$C_{EQ} = 2.5\ \mu F$$

CAPACITORS IN PARALLEL

When electrical circuits consist of capacitors connected in parallel, as shown in Figure 1.15, circuit analyses – as with a network of series connected capacitors – would require simplification of the parallel network of capacitors into one equivalent capacitor C_{EQ}. Figure 1.15 shows "**n**" number of capacitors connected such that the "heads" of all capacitors are (electrically) "bonded" or are connected together, with an electrical connection to the anode (or positive terminal) of the DC power supply and the "tails" of all capacitors are connected together to the cathode (or negative terminal) of the power source.

Simplification of a network of capacitors in parallel is similar to the approach utilized in the combination of **resistors in series**. When multiple **parallel** capacitors are combined into an equivalent capacitor, C_{EQ}, the simplified or condensed equivalent of the original parallel circuit could also be represented by the equivalent circuit diagram in Figure 1.13.

For a parallel capacitor circuit consisting of "**n**" number of parallel capacitors, equivalent capacitance can be calculated by applying Eq. 1.18

$$C_{EQ} = C_1 + C_2 + C_3 + \cdots + C_n \tag{1.25}$$

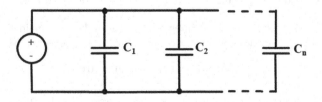

FIGURE 1.15   Equivalent capacitance for "n" capacitors in parallel.

Example 1.7

Determine the equivalent capacitance for the DC circuit shown below if $C_1 = C_2 = 5\ \mu F$ and $C_3 = 10\ \mu F$.

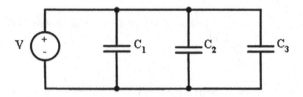

Solution:

Applying Eq. 1.25 to the three parallel capacitor circuit shown in the diagram above yields:

$$C_{EQ} = C_1 + C_2 + C_3$$

$$C_{EQ} = (5\times10^{-6}) + (5\times10^{-6}) + (10\times10^{-6})$$

$$= 20\times10^{-6}\ F,\text{ or } 20\ \mu F$$

Example 1.8

Determine the equivalent capacitance in series and parallel combination circuit shown below. The capacitance values are: $C_1 = C_2 = 5\ \mu F$ and $C_3 = C_4 = 10\ \mu F$.

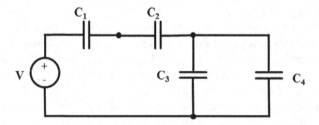

Solution:

Similar to the parallel and series combination approach described in the resistor section, our aim would be to simplify the circuit by first combining the linearly additive segment (or segments), followed by reciprocal combination of the remaining capacitors.

The capacitors in this circuit that lend themselves to linear combination are C_3 and C_4. Therefore, the combined capacitance, C_{3-4}, would be as follows:

$$C_{3-4} = C_3 + C_4 = 10\ \mu F + 10\ \mu F = 20\ \mu F$$

Then, by applying Eq. 1.24 to this special hybrid capacitor combination case, with the third capacitance being C_{3-4} instead of C_3:

$$C_{EQ} = \frac{C_1 C_2 C_{3-4}}{C_1 C_2 + C_2 C_{3-4} + C_1 C_{3-4}}$$

$$C_{EQ} = \frac{(5\times10^{-6})(5\times10^{-6})(20\times10^{-6})}{(5\times10^{-6})(5\times10^{-6})+(5\times10^{-6})(20\times10^{-6})+(5\times10^{-6})(20\times10^{-6})}$$

$$C_{EQ} = 2.22\,\mu F$$

CAPACITIVE REACTANCE

When a capacitor is incorporated into an AC circuit, its impact in that circuit is quantified through an entity referred to as the capacitive reactance. The symbol for capacitive reactance is X_c. Capacitive reactance can be defined, mathematically, as follows:

$$X_c = \frac{1}{\omega C} = \frac{1}{2\pi f C} \qquad (1.26)$$

where
f = Frequency of the AC power source, i.e., 60 Hz in the United States and 50 Hz in some other parts of the world.
ω = Rotational or angular speed, in radians per second = $2\cdot\pi\cdot f$.
C = Capacitance in farads.

Capacitive reactance is measured in ohms, or Ω's. It is important to note that X_c is often misrepresented as Z_c. To the contrary, as explained in the impedance section, Z_c is the **impedance contribution** by the capacitor and is represented as follows:

$$Z_c = -jX_c$$

$\therefore$　$Z_c \neq X_c$ (Note : The notion that $Z_c = X_c$ is a common mistake)

Example 1.9

Assume that the circuit in Example 1.8 is powered by a 60 Hz AC source instead of the DC source. Determine the total capacitive reactance, X_c, seen by the AC source. (**Note the symbol used to represent an AC power source.**)

Solution:

If the DC source is replaced by an AC source, the circuit would appear as follows:

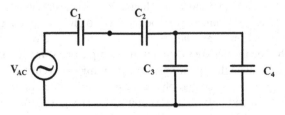

As computed in Example 1.4, the combined or net capacitance contributed to the circuit by the parallel and series network of capacitors is $C_{EQ} = 2.22$ µF. Then, by applying Eq.1.26:

$$X_c = \frac{1}{\omega C} = \frac{1}{2\pi fC} = \frac{1}{2(3.14)(60 \text{ Hz})(2.22 \times 10^{-6} \text{ F})} = 1,194 \, \Omega$$

INDUCTOR, INDUCTANCE, AND INDUCTIVE REACTANCE

Similar to the capacitor, an inductor can be viewed as an energy storage device, which can be applied in DC or AC applications. Unlike a capacitor – where energy is stored in the form of separation of charges, resulting in a potential difference – in an inductor, the energy is stored in the magnetic field that is produced through the flow of electric current. This phenomenon was first discovered by Michael Faraday in 1831. The magnetic field – which can be referred to as magnetic flux – is **not** established instantaneously upon flow of current through an inductor. Instead, much like a "time constant, τ" based charge build-up in a capacitor, the current change and magnetic field build up in an inductor ramps up, or down, at a non-linear rate. This non-linear rate is a function of the inductor's time constant, τ. Inductor's time constant concept will be explained later in this chapter.

Since the energy stored in the magnetic field of an inductor is due to the flow of current – and, ultimately, due to the movement of electrons – we could view an inductor as being analogous to a rotating "flywheel," in the mechanical realm – where the energy is stored in a rotating mass. As with a flywheel – where any attempt to stop the rotation of the flywheel is opposed by momentum and kinetic energy in the rotating mass of the flywheel – **any attempt to change the flow of current in an inductor is opposed by the "inductance"** of the inductor. It is due to the inductance of an inductor that if you, for instance, have 10 A flowing in an electrical circuit, and you try to "break" the circuit by opening a switch, current flow is maintained briefly through an **electric arc**, where permitted, across the opening switch, where the electric arc – which is, in essence, plasma or ionized air – serves as a temporary "channel" for the flow of current. Conversely, when a switch – in a de-energized inductive circuit – is closed to connect a power source to an inductive device, no current flows through the circuit, initially, as **inductance** of the inductor in the electrical circuit opposes the **change** in the flow of current from **"zero"** to some measurable **"non-zero"** level. Therefore, inductance is defined as the capacity or tendency of an inductor to resist the **change in the flow of current**.

In addition, it is worth noting that just like the kinetic energy stored in a rotating flywheel can be "tapped" to perform mechanical work, the energy stored in the **magnetic field** of a "charged" inductor can be released to push a ferromagnetic cylindrical core, in one direction or another, to open or close a valve or to open or close an electrical switch. The former application is an example of a **solenoid operated valve**, while the later represents the operation of a **contactor or a relay**.

The principle of inductance and physical aspects of inductors (or coils) are illustrated in Figure 1.16a and b. Basically, if you take a straight piece of wire, as shown in Figure 1.16a and wind it around a cylindrical core, the final product would be a coil

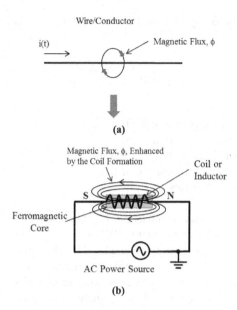

FIGURE 1.16 (a) Straight current-carrying conductor and (b) "coiled" current-carrying conductor.

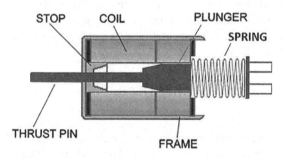

FIGURE 1.17 Construction of a solenoid.

or an inductor. As shown in Figure 1.16a, current flowing through a straight wire produces a "weak" magnetic field. While flowing through a "coiled" conductor, current produces a stronger and denser magnetic field, capable for conducting "work," such as pushing of a "plunger" against the restraint of a spring to open or close a valve or to open or close an electrical circuit in a relay or a contactor.

Figure 1.16b represents a conceptual view of a solenoid or a coil, illustrating the fundamental principle of electromagnetism. On the other hand, physical construction of a simple solenoid or inductor is depicted in Figure 1.17, in a diametrical cross-sectional view. When the coil of the solenoid is energized, the plunger – or core – responds to the magnetic flux by moving to the left. As the plunger moves to the left – against the spring – with a force that is proportional to the magnetic flux, the "pin" attached to the tip of the plunger pushes mechanical devices such as relay contacts or valves to change their state from open to closed or vice versa.

Inductance is denoted by "L," and it can be defined, mathematically, through Eq. 1.27.

$$L = \frac{\mu \cdot N^2 \cdot A}{l} \qquad (1.27)$$

where
μ = Permeability of the medium, in H/m
N = Number of turns of coil (unit-less)
A = Cross-sectional area of the core (in m²)
l = Mean length through the core (in m)
Unit for inductance: H (henry)

The mathematical relationship stated as Eq. 1.27 stipulates that inductance is directly proportional to the area of cross-section "A" of the core. This equation also states that the inductance is directly proportional to "N^2," the "square" of the number of turns in the coil, and is inversely proportional to the mean length "l" of the core. In other words, if larger inductance or **energy storage capacity** is desired, one must increase the area of cross-section of the core, increase the number of turns, or reduce the mean length of the core and the coil. Of course, the values of these variables can be increased or decreased, simultaneously, to achieve the desired results. In addition to serving as a "**constant of proportionality**" for the equation, permittivity "μ" injects the physical characteristics of the core into the computation of inductance. So, if a material with higher relative permeability, μ_r, is chosen, the inductance of a coil would be greater. The permeability μ of a specific medium can be defined, mathematically, as follows:

$$\mu = \mu_r \cdot \mu_o$$

where
μ_r = Relative permeability of the core material. Relative permeability of steel is 100 and that of insulating materials like wood and Teflon is 1.0, which is the same as free space, vacuum, or air. Therefore, in the design of an inductor, if stronger magnetic field is desired, or greater amount work needs to be performed, a ferromagnetic substance like iron should be used instead of a dielectric substance like air. The reader is encouraged to prove this statement on the basis of Eqs. 1.27 and 1.28.
μ_o = Permeability of free space or vacuum = $4\pi \times 10^{-7} = 1.257 \times 10^{-6}$H/m

Electrical energy stored in an inductor can be determined through Eq. 1.28:

$$E_{stored} = \frac{1}{2}LI^2 \qquad (1.28)$$

where the energy is measured in joules (or N-m), inductance L in H (henry), and current I in Amp.

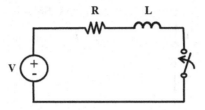

FIGURE 1.18 Series RL circuit.

Inductive electrical systems – that is electrical circuits with inductance – and inductive devices like motors and transformers contain inductance **L**, as well as resistance, **R**. In order to understand the current and voltage response in inductive circuits – or to understand current and voltage variation in inductive systems – we will examine a simple series RL circuit diagram in Figure 1.18.

In a typical inductive-resistive circuit, as the one shown in Figure 1.18, the voltage and current relationships, as a function of time, are governed by Eqs. 1.29–1.31.

$$v_L(t) = L \frac{di_L}{dt} \tag{1.29}$$

$$v_L(t) = -i(0)\,Re^{-\frac{R}{L}t} + Ve^{-\frac{R}{L}t} \tag{1.30}$$

$$i_L(t) = i_R(t) = i(0)e^{-\frac{R}{L}t} + \frac{V}{R}\left(1 - e^{-\frac{R}{L}t}\right) \tag{1.31}$$

Equations 1.30 and 1.31 hold greater practical significance, in that, they can be used to predict the changes in voltage and current – with respect to time – for given values of source voltage **V**, series resistance **R**, and series inductance **L**. The relationship between resistance **R** and inductance **L**, or the **relative** size of **R** and L, determines the charging and discharging rate of the inductor via the time constant "τ." The time constant τ can be defined, mathematically, in terms of **R** and **L**, in the form of Eq. 1.32.

$$\tau = \frac{L}{R} \tag{1.32}$$

It is ostensible from examination of Eqs. 1.30 and 1.31 that in circuits that consist of inductance and resistance, unlike purely resistive circuits – but similar to RC circuits – current and voltage associated with inductors vary in a non-linear fashion. This "non-linear" charging and discharging of an inductor in an RL circuit – similar to RC circuits – is referred to as **transient behavior**. The graphs in Figures 1.19 and 1.20 compare voltage and current responses – on the basis of Eqs. 1.30 and 1.31, respectively.

Voltage versus time, transient response in series RL circuit: Examination of Eq. 1.30 and Figures 1.18 and 1.19 reveals the following facts:

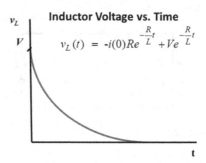

FIGURE 1.19 Non-linear **voltage** response in a circuit consisting of inductance and resistance.

   a. At time, t = 0, which is when the switch in the circuit is closed, the inductor resists the change in flow of current, or, $i(0) = 0$; so, the first segment of Eq. 1.30 becomes zero and drops out. Also, at t = 0, the exponent of "e" becomes zero, which makes $e^{-Rt/L} = 1$. This results in $v_L(t) = V$ (i.e. the source voltage) at t = 0. In other words:

       $v_L(t) = V$ at the instant the switch is closed in the series RL circuit.

   b. On the other end of the time spectrum, where t = ∞, or when steady-state condition has been achieved:

$$v_L(\infty) = -i(0)\,Re^{-\infty} + Ve^{-\infty}$$

$$= -(0)\cdot\left(\frac{1}{\infty}\right) + V\left(\frac{1}{\infty}\right) = 0$$

The analyses above support the following basic tenets of series RL inductive circuits:

   i. The voltage across the inductor, at the instant the switch in a series RL circuit is closed, is the same as the source voltage, implying that no current flows through the inductor.

   ii. The voltage across the inductor, after a large amount of time has elapsed, diminishes to zero. And, with voltage drop across the inductor zero, the inductor acts as a short in a series RL circuit under steady-state conditions – with steady-state current, V/R, flowing through it.

Current versus time, transient response in series RL circuit: Examination of Eq. 1.31 and Figures 1.18 and 1.20 reveals the following facts associated with current versus time response in a series RL circuit:

   a. At time, t = 0, which is when the switch in the circuit is closed, $i(0) = 0$; so, the first segment of Eq. 1.31 becomes zero and drops out. Also, at t = 0, the exponent of "e" becomes **zero**, which makes $e^{-Rt/L} = 1$. This results in $i_L(t) = 0 + V/R(1 - e^0) = 0$, this means that at **t = 0**, or at the instant the switch is closed, the inductance of the inductor resists the initiation of current flow; hence, no current flows through the inductor at **t = 0**.

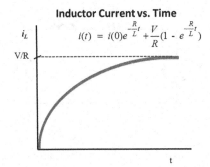

Inductor Current vs. Time

$$i(t) = i(0)e^{-\frac{R}{L}t} + \frac{V}{R}(1 - e^{-\frac{R}{L}t})$$

FIGURE 1.20 Non-linear **current** response in a circuit consisting of inductance and resistance.

b. On the other end of the time spectrum, when $t = \infty$, or when steady-state condition has been achieved:

$$i_L(\infty) = i_R(\infty) = i(0)e^{-\frac{R}{L}t} + \frac{V}{R}\left(1 - e^{-\frac{R}{L}t}\right)$$

$$i_L(\infty) = i_R(\infty) = (0)(e^{-\infty}) + \frac{V}{R}(1 - e^{-\infty})$$

$$i_L(\infty) = i_R(\infty) = \frac{V}{R}(1 - 0) = \frac{V}{R}$$

The analyses stated above support the following basic characteristics of inductive circuits:

i. No current flows through the inductor and resistor combination at the instant the switch is closed, or at t = 0, which is congruent with one of the fundamental characteristics of an inductor described earlier. In other words, at the outset, the inductance of the inductor, successfully, resists the rise of the current to a non-zero value.

ii. Current through the inductor and resistor combination develops to the maximum level after a long span of time. The maximum level of current in the inductor and resistor combination is equal to **V/R**. This also implies that the inductor acts as a **"short"** when steady-state condition is achieved or after 10 τ amount of time.

Example 1.10

Consider the series RL circuit shown in the diagram below. The source voltage is 12 V and R = 10 Ω. The switch is closed at t = 0. What would be the magnitude of current flowing through this circuit at t = τ?

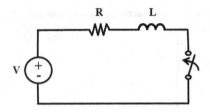

Solution:

In most series RL cases, the current value at a certain time "t" can be predicted through Eq. 1.31.

$$i_L(t) = i_R(t) = i(0)e^{-\frac{R}{L}t} + \frac{V}{R}\left(1 - e^{-\frac{R}{L}t}\right)$$

Note: In this case, the value of **L** is not given, but the elapsed time is given as a function of time constant as, "**1·τ**."

Since τ = **L/R**, Eq. 1.31 can be rewritten, in τ form as follows:

$$i_L(t) = i_R(t) = i(0)e^{-\frac{t}{\tau}} + \frac{V}{R}\left(1 - e^{-\frac{t}{\tau}}\right)$$

Then, by substituting t = τ, and given the fact that i(0) = 0, the **i_L(t)** equation simplifies into the following form:

$$i_L(t) = i_R(t) = i(0)e^{-\frac{\tau}{\tau}} + \frac{V}{R}\left(1 - e^{-\frac{\tau}{\tau}}\right)$$

$$= \frac{V}{R}(1 - e^{-1})$$

$$= \frac{V}{R}(0.632) = \frac{12}{10}(0.632) = (1.2)(0.632) = 0.759 \text{ A}$$

This analysis of current response validates a characteristic fact about inductors: current develops to **63.2%** of its full potential in "one time constant" or **1 τ** worth of time.

Example 1.11

Consider the series RL circuit given in Example 1.9, in discharge mode, with voltage source removed. Inductor L = 10 mH. The switch has been closed for a long period of time, such that the current has developed to the maximum or steady-state level 1.2 A. How much time would need to elapse for the current to drop to 0.8 A after the switch is opened?

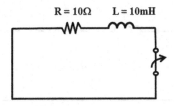

$R = 10\Omega \qquad L = 10\text{mH}$

Solution:

Apply series RL current equation, Eq. 1.31.

$$i_L(t) = i_R(t) = i(0)e^{-\frac{R}{L}t} + \frac{V}{R}\left(1 - e^{-\frac{R}{L}t}\right)$$

Given:
$t = ?$
$L = 10 \times 10^{-3}$ H
$R = 10\ \Omega$
$V = 0$
$i(0) = 1.2$ A
$i_L(t) = 0.8$ A

$$i_L(t) = (0.8) = (1.2)e^{-\frac{10}{0.01}t} + (0)\left[1 - e^{-\frac{10}{0.01}t}\right]$$

$$0.8 = (1.2)e^{-\frac{10}{0.01}t}$$

$$0.667 = e^{-\frac{10}{0.01}t}$$

$$\ln(0.667) = \ln\left(e^{-\frac{10}{0.01}t}\right)$$

$$-0.4055 = -1,000t$$

$$t = 0.00041 \text{ s or } 0.41 \text{ ms}$$

SERIES INDUCTOR COMBINATION

When analyzing DC circuits with inductors connected in series, derivation of equivalent inductance, L_{eq}, can accomplished by, simply, adding the inductance values linearly as represented by Eq. 1.33. Figure 1.21 depicts "n" inductors, L_1 through L_n, connected in series.

$$L_{EQ} = L_1 + L_2 + L_3 \cdots + L_n \qquad (1.33)$$

Derivation of the combined equivalent inductance L_{EQ} permits us to represent Figure 1.21 in the form of a condensed version depicted in Figure 1.22.

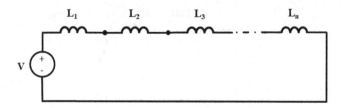

FIGURE 1.21   Series combination of "n" inductors.

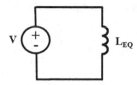

FIGURE 1.22 L_{EQ}, equivalent inductance representing series combination of "n" inductors.

Example 1.12

Determine the equivalent inductance for three inductors connected in the series combination circuit shown below. The inductance values of the three inductors are: $L_1 = 5$ mH, $L_2 = 5$ mH, and $L_3 = 10$ mH.

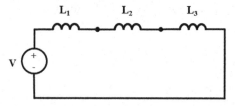

Solution:

Apply Eq. 1.33:

$$L_{EQ} = L_1 + L_2 + L_3$$

$$= 5\,\text{mH} + 5\,\text{mH} + 10\,\text{mH}$$

$$= 20\,\text{mH}$$

PARALLEL COMBINATION OF INDUCTORS

The circuit depicted in Figure 1.23 shows "n" number of inductors connected in parallel. The formula for determining the equivalent inductance L_{eq} for this circuit is represented by Eq. 1.34.

$$L_{EQ} = \frac{1}{\dfrac{1}{L_1} + \dfrac{1}{L_2} + \dfrac{1}{L_3} + \cdots + \dfrac{1}{L_n}} \tag{1.34}$$

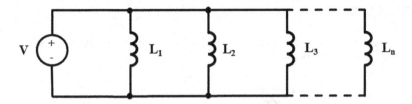

FIGURE 1.23 Parallel combination of "**n**" inductors.

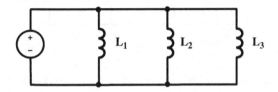

FIGURE 1.24 Combination of three parallel inductors.

If a parallel resistor circuit or network consists of only three inductors, the circuit would appear as shown in Figure 1.24 and the L_{eq} equation for this circuit would reduce to Eq. 1.36.

$$L_{EQ} = \frac{1}{\dfrac{1}{L_1} + \dfrac{1}{L_2} + \dfrac{1}{L_3}} \tag{1.35}$$

$$L_{EQ} = \frac{L_1 L_2 L_3}{L_1 L_2 + L_2 L_3 + L_1 L_3} \tag{1.36}$$

Example 1.13

Determine the equivalent inductance L_{EQ} for three parallel inductor DC circuit shown in Figure 1.24 if $L_1 = 1$ H, and $L_2 = 5$ H, and $L_3 = 10$ H.

Solution:

Apply Eq. 1.36 to compute L_{EQ} for the three parallel inductor circuit shown in Figure 1.24:

$$L_{EQ} = \frac{L_1 L_2 L_3}{L_1 L_2 + L_2 L_3 + L_1 L_3}$$

$$= \frac{(1\ H)(5\ H)(10\ H)}{(1\ H)(5\ H) + (5\ H)(10\ H) + (1\ H)(10\ H)}$$

$$= 0.77\ H$$

Example 1.14

Calculate the net or total inductance as seen from the 24 V source vantage point in the circuit shown below.

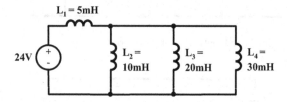

Solution:

We need to focus on the parallel combination of L_2, L_3, and L_4 first. Apply Eq. 1.36 to calculate the equivalent inductance L_{234} for the three parallel inductors:

$$L_{234} = \frac{L_2 L_3 L_4}{L_2 L_3 + L_3 L_4 + L_1 L_4}$$

$$= \frac{(10 \text{ mH})(20 \text{ mH})(30 \text{ mH})}{(10 \text{ mH})(20 \text{ mH}) + (20 \text{ mH})(30 \text{ mH}) + (10 \text{ mH})(30 \text{ mH})}$$

$$= 5.45 \text{ mH}$$

This reduces the circuit as shown below:

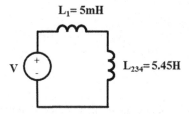

Inductors L_1 and L_{234}, in this reduced circuit, lend themselves to a linear combination. Therefore, the equivalent inductance L_{EQ} for the entire parallel and series inductor hybrid circuit would be as follows:

$$L_{EQ} = L_1 + L_{234} = 5 \text{ mH} + 5.45 \text{ mH} = 10.45 \text{ mH}$$

INDUCTIVE REACTANCE

When an inductor is incorporated into an AC circuit, its impact in that circuit is quantified through an entity referred to as the **inductive reactance**. The symbol for inductive reactance is X_L. Inductive reactance can be defined, mathematically, as follows:

$$X_L = \omega L = 2\pi f L \qquad (1.37)$$

where

f = Frequency of the AC power source, i.e., 60 Hz in the United States and 50 Hz in some other parts of the world.

ω = Rotational speed, in radians per second.

L = Inductance in henry, or H.

Inductive reactance is measured in ohms, or Ω's. It is important to note that X_L is often misconstrued as Z_L. To the contrary, as explained in the impedance section, Z_L is the **impedance contribution by the inductor** and is represented as follows:

$Z_L = jX_L$

∴   **$Z_L \neq X_L$**(Note : The notion that $Z_L = X_L$ is a common mistake.)

Example 1.15

Assume that the circuit in Example 1.14 is powered by a 60 Hz AC source. Calculate the inductive reactance, X_L, as seen by the AC voltage source.

Solution:

If the DC source is replaced by an AC source, the circuit would appear as follows:

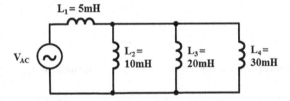

L_{EQ}, as seen by the AC voltage source, is shown in the simplified equivalent circuit below:

As computed in Example 1.13, the combined or net inductance contributed to the circuit by the parallel and series network of inductors is $L_{EQ} = 10.45$ mH. Then, by applying Eq.1.37, the inductive reactance, **X_{L-EQ}** as seen by the AC voltage source V_{AC}, would be as follows:

$$X_{L-EQ} = \omega \cdot L = (2\pi f) \cdot L_{EQ} = 2(3.14)(60 \text{ Hz})(10.45 \text{ mH}) = 3.94 \ \Omega$$

IMPEDANCE

The narrative definition of impedance would be that it is the current resisting and impeding characteristic of load or conductor in an AC circuit. As implied in the definition of this term, impedance is an AC entity. While in DC circuits the factor that opposes the flow of DC is resistance, the entity that influences the flow of AC, in AC circuits, is impedance. Impedance, like AC, voltage, and power, is a vector entity. By definition, a vector can be completely defined by two key characteristics, namely, the **magnitude** and the **direction**. Therefore, impedance and other entities in AC circuits, such as current, voltage, and power, can be defined fully, only, through specification of both magnitude and direction.

Symbol for impedance is $\mathbf{Z}$. When typewritten, as with most AC entities, the symbol $\mathbf{Z}$ for impedance is represented in bold font. When handwritten, most AC entities are denoted by the respective symbol with a half arrow. So, impedance would be denoted by "$\vec{\mathbf{Z}}$." The unit for impedance is ohm, or Ω; similar to the unit for resistance $\mathbf{R}$, capacitive reactance $\mathbf{X_c}$, and inductive reactance $\mathbf{X_L}$. Consistency of units between $\mathbf{R}$, $\mathbf{X_L}$, $\mathbf{X_C}$, and $\mathbf{Z}$ is one justification for the following mathematical definition for $\mathbf{Z}$:

$$\mathbf{Z} = \mathbf{R} + \mathbf{jX_l} - \mathbf{jX_c} \qquad (1.38)$$

where

$$\mathbf{jX_L} = \mathbf{Z_L} \qquad (1.39)$$

where $\mathbf{Z_L}$ is vectoral representation of impedance contribution by the inductance in the circuit.

And,

$$-\mathbf{jX_c} = \mathbf{Z_c} \qquad (1.40)$$

where $\mathbf{Z_C}$ is vectoral representation of impedance contribution by the capacitance in the circuit.

The dimensions, DC resistance, AC resistance, and impedance values for various commercially available, standard, conductors can be found in Chapter 9 of the NEC®. A sample of such data is shown in Table 1.1.

The following footnotes, from NEC® 2017, supplement the information in NEC® Table 9, Chapter 9.[2,3]

In-depth discussion of impedance and impedance-related calculations can be found in later chapters of this book. At this juncture, illustration of how total or

---

[2] These values (the resistances, reactances, and impedance values) are based on the following constants: UL-Type RHH wires with Class B stranding, in cradled configuration. Wire conductivities are 100% IACS copper and 61% IACS aluminum, and aluminum conduit is 45% IACS. Capacitive reactance is ignored, since it is negligible at these voltages. These resistance values are valid only at 75°C (167°F) and for the parameters as given but are representative for 600 V wire types operating at 60 Hz.

[3] Effective Z is defined as R Cos(θ) + X Sin(θ), where θ is the power factor angle of the circuit. Multiplying current by effective impedance gives a good approximation for line-to-neutral voltage drop. Effective impedance values shown in this table are valid only at 0.85 power factor. For another circuit power factor (PF), effective impedance (Z_e) can be calculated from R and X_L values given in this table as follows:

$$Z_e = R \times PF + X_L \, Sin[arcCos(PF)].$$

TABLE 1.1

Sample Impedance Data; NEC® Table 9, Chapter 9, Alternating Current Resistance and Reactance for 600 V Cables, Three Phase, 60 Hz

Ohms to Neutral per Kilometer
Ohms to Neutral per 1,000 ft

Size (AWG or kcmil)	X_L (Reactance) for All Wires		Alternating-Current Resistance for Uncoated Copper Wires			Alternating-Current Resistance for Aluminum Wires			Effective Z at 0.85 PF for Uncoated Copper Wires			Effective Z at 0.85 PF for Aluminum Wires		
	PVC, Aluminum Conduits	Steel Conduit	PVC Conduit	Aluminum Conduit	Steel Conduit	PVC Conduit	Aluminum Conduit	Steel Conduit	PVC Conduit	Aluminum Conduit	Steel Conduit	PVC Conduit	Aluminum Conduit	Steel Conduit
12	0.177	0.223	6.6	6.6	6.6	10.5	10.5	10.5	5.6	5.6	5.6	9.2	9.2	9.2
	0.054	0.068	2.0	2.0	2.0	3.2	3.2	3.2	1.7	1.7	1.7	2.8	2.8	2.8
1/0	0.144	0.18	0.39	0.43	0.39	0.66	0.69	0.66	0.43	0.43	0.43	0.62	0.66	0.66
	0.044	0.055	0.12	0.13	0.12	0.2	0.21	0.2	0.13	0.13	0.13	0.19	0.2	0.2
250	0.135	0.171	0.171	0.187	0.177	0.279	0.295	0.282	0.217	0.23	0.24	0.308	0.322	0.33
	0.041	0.052	0.052	0.057	0.054	0.085	0.09	0.086	0.066	0.07	0.073	0.094	0.098	0.1
500	0.128	0.157	0.089	0.105	0.095	0.141	0.157	0.148	0.141	0.157	0.164	0.187	0.2	0.21
	0.039	0.048	0.027	0.032	0.029	0.043	0.048	0.045	0.043	0.048	0.05	0.057	0.061	0.064

equivalent impedance is derived by combining resistance and reactance, vectorially, is accomplished through Example Problem 1.16.

Example 1.16

Find the RMS current flowing through the circuit below. Assume the voltage source is a 60 Hz AC.

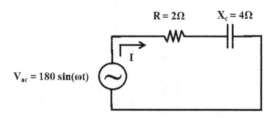

Solution:

To determine the comprehensive, vectoral, value of the AC RMS current flowing through the given circuit, we must use Ohm's law. According to Ohm's law, $I_{RMS} = V_{RMS}/Z_e = V_{RMS}/Z_t$. Note that the use of bold fonts is deliberate, because the voltage, current, and total impedance are all vectors or complex entities.

Also, as a rule of thumb, when vectoral or complex values are being divided or multiplied, they are converted into polar or phasor form. Addition and subtraction operations are conducted in rectangular form. Of course, if you are performing all mathematical operations using a scientific calculator, values of various entities can be entered in either rectangular or polar/phasor form. The mathematical operations in this scenario would be as follows:

$$V_{RMS} = \frac{V_p}{\sqrt{2}} = \frac{180}{\sqrt{2}} = 127 \text{ V}_{RMS}, \text{ and since the angle is not stated,}$$

$$\bar{V}_{RMS} = 127 \angle 0° \text{ V}$$

Total or equivalent impedance in the circuit $= Z_e = 2 + (-j4)$

Then, by using pythagorean theorem or a scientific calculator,

$$Z_e = Z_t = 2 - j4 \ \Omega = 4.47 \angle -63.4° \ \Omega$$

and,

$$I_{RMS} = \frac{127 \angle 0° \text{ V}}{4.47 \angle -63.4° \ \Omega} = 28.41 \angle 63.4 \text{ A rms}$$

The last, division, operation can be accomplished by using a scientific calculator.

Alternatively, you simply divide 127 by 4.47, yielding the magnitude of 28.41.

The angle in the denominator is "conjugated," meaning the sign of the angle is reversed

and is added to the numerator yielding the final current angle of 63.4°.

Note that the given sinusoidal or trigonometric specification of the voltage source is converted to RMS phasor vectoral form at the very outset.

Concept of impedance, other AC entities such as AC, AC voltage, AC power, and associated computational methods are discussed in greater depth in Chapter 3.

MAGNETIC CIRCUITS VERSUS ELECTRICAL CIRCUITS

The two diagrams shown in Figure 1.26 illustrate similarities and differences between a basic electrical circuit and a magnetic circuit. In-depth discussion on the subject of magnetism is beyond the scope of this text. Nevertheless, a contrast between the basic magnetic and electric circuits below will afford the reader an opportunity to gain basic understanding of magnetism, or, more specifically, electromagnetism.

The circuit shown in Figure 1.26a represents a basic DC electrical circuit. This circuit consists of a DC voltage source labeled "**V**." As explained earlier, voltage is synonymous to the term "electromotive force." So, we can explain the phenomenon in the electrical circuit as the electromotive force "**V**" driving DC "**I**" through the load or resistor "**R**." Note the direction of the current is from the left to the right, in a clockwise loop, emerging from the positive electrode of the DC voltage source and terminating into the negative electrode of the voltage source. This clockwise current flow is assigned on the basis of an electrical convention which stipulates that the current flow consists of "**holes**," or positively charged particles, being repelled or driven out of the positive terminal.

See Figure 1.25 for a better understanding of the formation of a positively charged hole from a neutral hydrogen atom by removal of the only electron "**e⁻**" from the "1s" orbital of the atom.

The "hole"-based convention also affirms the fact that electrical current is not necessarily, always, due to the flow or movement of electrons. Electrical current, as explained in the section on the topic of current, can be due the flow or movement of

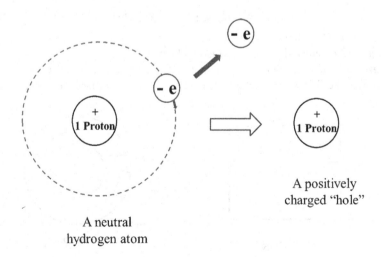

A neutral
hydrogen atom

A positively
charged "hole"

FIGURE 1.25 Illustration of the formation of a positively charged hole.

negatively or positively charged particles. The positively or negatively charged particles at the atomic or molecular level are referred to as ions.

The relationship between V, I, and R in the electrical circuit is governed by Ohm's law. Ohm's law, in conjunction with other basic electrical laws – used to analyze electrical circuits – will be explained in more detail in Chapter 2. For now, note that Ohm's law is stated, mathematically, in the form of Eq. 1.41. In other words, according to Ohm's law, electromotive force is equal to the product of current and resistance.

$$\textbf{Electromotive Force, } V = I \cdot R = (\textbf{Current}) \times (\textbf{Resistance}) \qquad (1.41)$$

The circuit shown in Figure 1.26b represents a basic electromagnetic circuit. This circuit consists of a toroid or donut-shaped core – typically constructed out of iron. In this magnetic circuit, a conductor, or wire, is wrapped in four turns around the left side of the toroid core. When current is passed through wound conductor, magnetic field is established in the core as represented by the dashed circular line, with an arrow pointing in clockwise direction. This magnetic field is referred to as magnetic flux, ϕ. Magnetic flux is measured in weber. The unit weber is named for the German physicist **Wilhelm Eduard Weber** (1804–1891). In the magnetic realm, the flux serves as a counterpart to the current, **I**, from the electrical realm. Just like the electromotive force, **EMF**, or voltage, drives the current through the resistor, **R**, the magnetomotive force (**MMF**), $\mathcal{F}$, drives the magnetic flux, ϕ, through the toroid magnetic core. Magnetomotive force is measured in ampere-turns. In electrical systems, load is represented by the resistor **R**. In the magnetic circuit, the flow of magnetic flux is opposed by reluctance $\mathcal{R}$. Just as Ohm's law, represented by Eq. 1.41, governs the relationship between electromotive force (voltage), current, and resistance in the electrical realm, Eq. 1.42 represents the relationship between the magnetomotive force, $\mathcal{F}$, the magnetic flux, ϕ, and the reluctance $\mathcal{R}$, in the magnetic domain.

$$\mathcal{F} = \Phi \cdot \mathcal{R} = (\text{Magnetic Flux}) \times (\text{Reluctance}) \qquad (1.42)$$

Equation 1.42 is referred to as **Hopkinson's Law.**

Magnetic reluctance can also be perceived as magnetic resistance; a resistance that opposes the flow of magnetic flux. Like resistance, reluctance is a scalar entity,

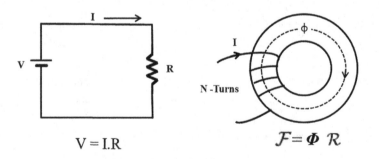

FIGURE 1.26   (a and b) Magnetic and electrical circuit comparison.

but unlike electric resistance, it stores magnetic energy instead of dissipating it. Reluctance is measured in ampere-turns per weber or turns per henry. Ferromagnetic substances such as iron have low reluctance while dielectric substances like air and vacuum offer high reluctance to magnetic flux. That is the reason why transformers, contactors, relays, and other similar electromagnetic devices utilize iron – or iron alloy – cores.

Analogous to Eq. 1.1, which represents the relationship between resistance, resistivity, length, and area of cross-section, the reluctance of a uniform magnetic circuit can be calculated as follows:

$$\mathcal{R} = \left(\frac{1}{\mu_r \mu_o}\right) \cdot \frac{l}{A} \tag{1.43}$$

or

$$\mathcal{R} = \left(\frac{1}{\mu}\right) \cdot \frac{1}{A} \tag{1.44}$$

where

l is the mean length of the circuit or core, in meters.

μ_o is the permeability of free space or vacuum = $4 \cdot \pi \cdot 10^{-7}$ henry per meter.

μ_r is the relative magnetic permeability of the core material. This is a dimensionless number.

μ is the permeability of the core material in henry per meter.

A is the area of cross-section of the core or the magnetic circuit defined in m^2.

BASIC ELECTRICAL INSTRUMENTS

Three of the most common instruments used to make electrical measurements and perform electrical system troubleshooting are as follows:

    I. Multi-meter, or a VOM, volt-ohm-meter

    II. Clamp-on ammeter

    III. Oscilloscope or scope meter

MULTI-METER

The modern multi-meter, sometimes just written as "multimeter," has evolved from its basic predecessor, the **Ohm-meter**. The original Ohm-meters were designed to measure resistance of electrical components and to verify continuity and integrity of electrical or electronic circuits. Voltage measuring feature was later added to the basic resistance measuring function of the Ohm-meter in the form of a more versatile instrument called the **Volt-Ohm-Meter**, or **VOM**. Due to the miniaturization of electronic components, additional functions were added to the basic VOM resulting in the contemporary multi-meter that transitioned from analog to digital format. See

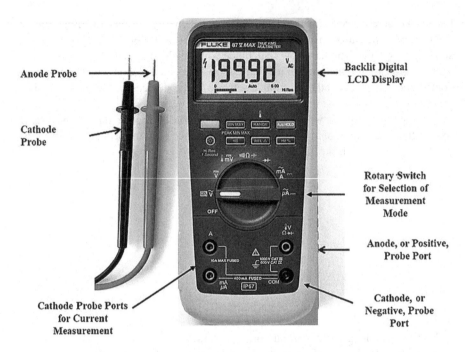

Anode Probe

Backlit Digital
LCD Display

Cathode
Probe

Rotary Switch
for Selection of
Measurement
Mode

Anode, or Positive,
Probe Port

Cathode Probe Ports
for Current
Measurement

Cathode, or
Negative, Probe
Port

FIGURE 1.27 Fluke® 87 V Max (printed herein and reproduced with permission of Fluke Corporation) True RMS Digital Multi-meter.

the diagram of a digital multi-meter in Figure 1.27. Some of the following features and functions are common among most multi-meters available on the market today:

a. Voltage measurement, AC and DC
b. Resistance measurement
c. Current measurement
d. Temperature measurement
e. Capacitor testing
f. Diode testing
g. Transistor testing

Standard accuracy of a portable digital multi-meter is, approximately, 0.3%. Bench-top and lab-grade multi-meters are known to offer accuracy that is better than ±0.1%. Many multi-meters include the "**Peak-Hold**" feature, which allows one to capture peak reading when the measured parameter is not steady. Some of the more sophisticated multi-meters have the ability to interface with PCs for direct data transfer, plotting, and storage.

When used for measuring voltage, multi-meters operate in "high impedance" mode. In other words, when multi-meters are used to measure voltage, the circuit being tested "sees' high impedance between the probes. This allows for the application of a multi-meter to be a **non-invasive** to the circuit being tested. This means that for most voltage measurements, a multi-meter can be applied to an electrical circuit without turning off the power to the system being tested. On the other hand, if a multi-meter is being used for measuring current, the circuit must be turned off, interrupted and severed at the point of measurement, and the meter must be injected in series to route the current through the multi-meter for measurement.

One must consider certain important safety measures associated with the application of multi-meters. Multi-meters, similar to other electrical instrumentation, are designed to operate within specified voltage range. Most, commonly used, multi-meters are rated for **low-voltage** applications. Low-voltage category ranges from 0 to 600 VAC or VDC. Application of a low-voltage instrument on higher voltages can result in a potentially catastrophic failure of the instrument and a potentially lethal arc flash incident. In addition, from safety point of view, it is important to inspect the instrument, periodically, for signs of mechanical or electrical damage. The probes must always be inspected for signs of wear, fraying, and signs of "**mechanical** stress or strain." The outermost surface of the cathode and anode probes is, essentially, insulation. Therefore, if the insulation is damaged or atrophied, person handling the probes could be exposed to electrical shock hazard. Note that even minor cracks or fissures in the "plastic" insulation of the probes can allow human perspiration to seep through and expose the user to hazardous potentials being tested with the probes. In most cases, prudent and safe course of action would be to **decommission, destroy, and discard** faulty probes and replace them with, suitable, manufacturer-recommended replacements.

CLAMP-ON AMMETER OR CLAMP METER

When current in an electrical or electronic circuit must be measured without interruption of an electrical or electronic circuit, a clamp-on ammeter can be used. Note that clamp-on ammeters are sometimes just referred to as clamp meters because their functions extend beyond the measurement of just current. Current measurement with a clamp-on ammeter, unlike a multi-meter, requires no probes. See Figure 1.28. When measuring current, the rotary selector switch on the face of the ammeter is switched to AC measurement setting. The spring-loaded "clamping" current transformer is opened by pressing the knob on one side of the ammeter. The current-carrying conductor is surrounded by the open current transformer clamp and the current transformer clamp is allowed to close around the conductor. As the current transformer closes or loops around the current-carrying conductor, it develops magnetic flux that is proportional to the flow of current through the conductor. This magnetic flux is **transduced** into voltage that is subsequently scaled and displayed, digitally. The current indicated on the display of a typical AC ammeter is an RMS value. In single-phase AC applications, the clamp-on ammeter should be clamped around the ungrounded or energized conductor, but not both. If the neutral and the energized conductors are clamped together, the currents through the two conductors cancel each other resulting in an erroneous null or zero readout on the display. Similarly, while measuring line currents in multiphase AC systems, **only one phase** must be enclosed in the clamp-on ammeter current transformer. If all three phases of a balanced three-phase circuit are enclosed by an AC ammeter, the instrument will erroneously indicate "zero" current even though the load is

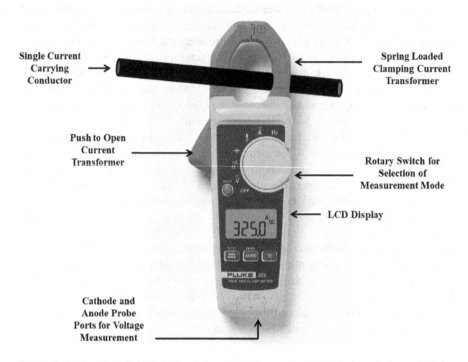

Single Current
Carrying →
Conductor

Spring Loaded
Clamping Current
Transformer

Push to Open
Current
Transformer

Rotary Switch for
Selection of
Measurement Mode

LCD Display

Cathode and
Anode Probe
Ports for Voltage
Measurement

FIGURE 1.28 Fluke® 325 (printed herein and reproduced with permission of Fluke Corporation) True RMS Clamp Meter.

operating and drawing current through each of the three-phase conductors. This is because the current flowing through each of the three phases is equal in magnitude but 120° apart. The vector sum of three vectors equal in magnitude but 120° apart is "zero." Some clamp-on ammeters are equipped with auxiliary features like voltage measurement; such is the case with the clamp-on ammeter shown in Figure 1.28. The ports for connecting the voltage measuring anode and cathode probe leads are located at the bottom of the ammeter.

The voltage rating-related safety precaution, advised in the use of multi-meters, applies to clamp meters or clamp-on ammeters, as well. The clamp-on ammeter depicted in Figure 1.28 is a low-voltage (0–600 VAC) device. Use of low-voltage clamp-on ammeters on **Medium- or High-Voltage** electrical circuits is unsafe and can result in catastrophic failures, such as **arc flash incidents**. Proper precaution must also be exercised concerning the maximum current rating of a clamp-on meter. Current and voltage ratings of a clamp-on meter must **not** be exceeded. Mis-application of test instrumentation can result in catastrophic faults or arc flash incidents.

As noted earlier, most clamp-on ammeters used routinely by electrical engineers and electricians are designed to detect and measure AC. Although uncommon, DC clamp-on ammeters, operating on **Hall Effect Principle**, are available for "non-invasive" DC measurement. The clamp-on ammeter depicted in Figure 1.28 includes the DC measurement function.

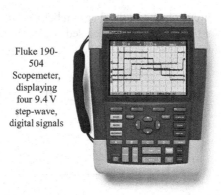

Fluke 190-504 Scopemeter, displaying four 9.4 V step-wave, digital signals

FIGURE 1.29 Fluke® 190-504 (printed herein and reproduced with permission of Fluke Corporation) Oscilloscope or Scope Meter.

OSCILLOSCOPE/SCOPE METER

The instrument depicted in Figure 1.29 is called an oscilloscope, and its more contemporary version – because of its compactness – is referred to as a "Scope Meter." An oscilloscope is a sophisticated and versatile instrument. It has myriad functions and features. Some of those functions and features are as follows:

• Measurement of voltage, current, resistance, temperature, and phase difference between signals such as voltage and current to assess the power factor.
• Display of signal waveforms, with magnitude of the signals plotted as a function of time.
• Side by side graphical comparison of various signals.
• Depiction, detection, and quantification of imperfections and noise on signals.
• Determination of the response of electronic devices such amplifiers by injecting a wavefunction or signal that is generated by the scope.

SELF-ASSESSMENT PROBLEMS AND QUESTIONS – CHAPTER 1

1. In an AC system, a voltage source $V(t) = 120Sin(377t + 0°)$ V rms, sets up a current of $I(t) = 5Sin(377t + 45°)$ A rms. Calculate the maximum values of voltage and current in this case.
2. A phase conductor of a transmission line is 1 mile long and has a diameter of 1.5 inch. The conductor is composed of aluminum. Calculate the electrical resistance of this conductor.
3. What is the resistance of the following circuit as seen from the battery?

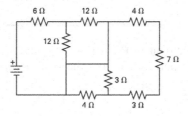

4. Consider the RC circuit shown in the diagram below. The source voltage is 12 V. The capacitor before the switch is closed is 2 V. The switch is closed at $t = 0$. What would the capacitor voltage be at $t = 5$ seconds?

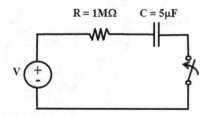

5. Determine the equivalent capacitance for the DC circuit shown in the circuit diagram below if $C_1 = 5\ \mu F$ and $C_2 = 10\ \mu F$.

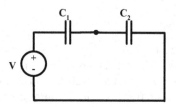

6. Determine the equivalent capacitance for the DC circuit shown below if this circuit consists of twenty 100 μF capacitors in series.

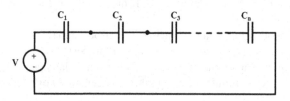

7. Determine the equivalent capacitance in series and parallel combination circuit shown below. The capacitance values are: $C_1 = 10\ \mu F$, $C_2 = 10\ \mu F$, $C_3 = 20\ \mu F$, and $C_4 = 20\ \mu F$.

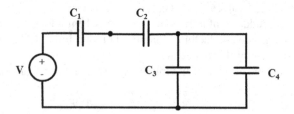

8. Assume that the circuit in Problem 4 is powered by a 60 Hz AC source instead of the DC source. Determine the total capacitive reactance, X_c, seen by the AC source.

9. Consider the series RL circuit shown in the diagram below. The source voltage is 12 V, R = 10 Ω, and L = 10 mH. The switch is closed at t = 0. What would be the magnitude of current flowing through this circuit at t = 2 ms?

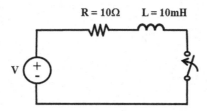

10. Consider the series RL circuit given in Problem 9, in discharge mode, with voltage source removed. Parameters such as R = 10 Ω and L = 10 mH are the same. The switch has been closed for a long period of time, such that the current has developed to the maximum or steady-state level 1.04 A. How much time would need to elapse for the current to drop to 0.5 A after the switch is opened?

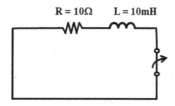

11. Determine the equivalent inductance L_{EQ} for three parallel inductor DC circuit shown in the diagram below if L_1 = 2 mH, L_2 = 5 mH, and L_3= 20 mH.

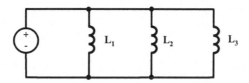

12. Calculate the net or total inductance as seen from the 24 V source vantage point in the circuit shown below.

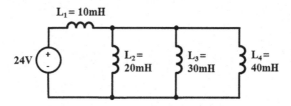

13. Assume that the circuit in Problem 12 is powered by a 60 Hz AC source. Calculate the inductive reactance, X_L, as seen by the AC voltage source.

2 DC Circuit Analysis, Diodes, and Transistors – BJT, MOSFET, and IGBT

INTRODUCTION

In this chapter, we will begin with the review of three fundamental laws of physics and electrical engineering, namely, (1) Ohm's law, (2) Kirchhoff's voltage law (KVL), and (3) Kirchhoff's current law (KCL). These three laws are instrumental in performing basic circuit analysis. Laws serve a vital purpose in electrical engineering analysis and problem-solving. Most of these laws can be applied in direct current (DC) as well as alternating current (AC) domains. Most engineering problems entail determination of unknown values of key parameters, under certain known circumstances or conditions. Laws allow us the opportunity to formulate equations that can be solved for values of unknown parameters. Solving for **one unknown** variable requires a minimum of **one equation**. Solution for determination of the values of **two unknown variables** requires **two equations**, and so on. Therefore, laws afford us the opportunity to model an engineering problem or scenario in the form of set of equations that can be solved to adequately define and identify unknowns. Not unlike other laws of physics, the KVL and the KCL are used to solve for important electrical parameters such as current, voltage, resistance, power, energy, reactance, impedance, and reactance. While the later three parameters were introduced in Chapter 1, they are discussed in greater detail in Chapter 3.

Important electrical circuit principles and circuit simplification techniques, such as voltage division, current division, superposition theorem, Thevenin Equivalent, and Norton Equivalent, will be discussed in this chapter in addition to the introduction to basic laws. The application of the laws will be illustrated through sample problems in this chapter. Note that there are many other principles and methods that are at electrical engineer's disposal to solve or analyze complex circuits such as the conversion of **Y** (or Star) load (or resistor) configuration to **Delta** "**Δ**" load configuration or vice and versa and two port networks. However, in-depth discussion of these and many other advanced circuit analysis methods is outside the scope of this text.

We will introduce the reader to two basic electronic devices, namely, a diode and a transistor, their basic characteristics, and some of their applications. We will conclude this chapter with a pictorial "tour" of a typical electronic printed circuit board (PCB) to allow the reader an opportunity to gain a measure of familiarity with basic electronic devices.

OHM'S LAW

Ohm's law was introduced, briefly, in Chapter 1. Ohm's law stipulates that voltage, or voltage drop, in a DC (or an AC) circuit is equal to the product of current flowing in the circuit and the resistance (or impedance) in the electrical circuit. In other words, voltage or voltage drop in an electrical circuit is directly proportional to the resistance (or impedance) of the circuit and the current flowing through it. Ohm's law can, therefore, be stated mathematically as follows:

$$V = IR, \ I = \frac{V}{R}, \text{ or } R = \frac{V}{I} \text{ for DC circuits} \tag{2.1}$$

$$\vec{V} = \vec{I}\vec{Z}, \ \vec{I} = \frac{\vec{V}}{\vec{Z}}, \text{ or } \vec{Z} = \frac{\vec{V}}{\vec{I}} \text{ for AC circuits} \tag{2.2}$$

The first mathematical representation of Ohm's law, in the form of Eq. 2.1, pertains to DC circuits, and in this statement of Ohm's law, all three parameters – **V**, **I**, and **R** – are scaler. On the other hand, mathematical representation of Ohm's law, in the form of Eq. 2.2 applies to AC circuits, where voltage "**V**" and current "**I**" are vectors or complex entities and **R**, as always, is scalar. Note that the symbols with "half-arrows" above them denote complex AC, or vector, entities. These half arrows are shown here to introduce the reader to this method for representing vector or complex entities. Later in this text, vector or complex entities will be represented mostly in bold fonts. As one examines the statement of Ohm's law in the form of Eqs. 2.1 and 2.2, it becomes obvious that Ohm's law can be interpreted and applied in several ways. This fundamental aspect of Ohm's law, its versatility, and wide application will become more evident through various circuit analysis problems in this chapter, and others in this text, beginning with Example 2.1.

Example 2.1

The DC circuit shown below consists of a hybrid, parallel–series, network or resistors: $R_1 = 10 \ \Omega$, $R_2 = 5 \ \Omega$, $R_3 = 1 \ \Omega$, and $R_4 = 10 \ \Omega$. Calculate the following parameters in this circuit: (a) R_{eq} or R_{total} for the entire circuit. (b) The amount of current "i" flowing through resistor R_1.

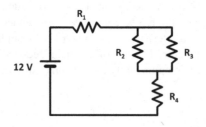

Solution:

a. R_{eq} or R_{total} for the entire circuit:
$$R_{eq} = R_1 + R_4 + (R_2 \cdot R_3)/(R_2 + R_3)$$
$$= 10\ \Omega + 10\ \Omega + (5/6\ \Omega) = 20.833\ \Omega$$
b. The amount of current "i" flowing through resistor R_1:
By Ohm's law: $V = I \cdot R$, or, $I = V/R$.
Therefore, $i = 12\ V/R_{eq}$
$$= 12\ V/20.833\ \Omega = 0.576\ A.$$

KIRCHHOFF'S VOLTAGE LAW (KVL)

KVL stipulates that the algebraic sum of voltage drops around any closed path, within a circuit, is equal to the sum of voltages presented by all of the voltage sources. The mathematical representation of KVL is as follows:

$$\sum V_{Drops} = \sum V_{Source} \qquad (2.3)$$

KVL can also be stated as follows:
Sum of ALL voltages in a circuit loop $= 0$ or:

$$\sum V = 0 \qquad (2.3a)$$

Some electrical engineers find the later representation of KVL somewhat easier to apply when performing circuit analyses because, with this version, once the voltages and respective polarities have been identified, you simply sum up all the voltage values with appropriate signs, as observed, while going around the loop in the chosen direction. This importance of this approach, and alternative approaches, is illustrated through Example 2.2.

Example 2.2 – KVL

Variation of current in the circuit shown below needs to be studied as a function of the three resistors and the voltage source, V_s. Using Ohm's law and KVL, develop an equation that can be used to compute the value of current **I** for various values of R_1, R_2, R_3, and V_s.

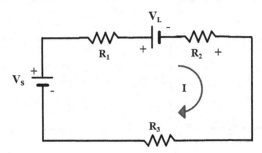

Solution:

Similar to other engineering disciplines, in most electrical engineering problems, multiple methods can be employed to derive the solution. The suitability of one method over another depends on the known parameters and the complexity of the circuit.

APPROACH I

Reduce or simplify the given circuit to a "net" voltage source and equivalent resistance R_{eq}. Since R_1, R_2, and R_3 are in series:

$$R_{eq} = R_1 + R_2 + R_3$$

Based on the assumption that V_s is indeed the source driving this circuit, by electrical convention, the current in this circuit would flow "out" of the positive terminal, or anode, of the voltage source. Hence, the current would flow in the clockwise direction as shown in the diagram below.

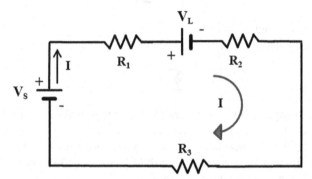

By the same token, application of the electrical current convention to the voltage device, or load voltage, V_L would mean that it would **try** to set up current in the counterclockwise direction. However, because we assumed that voltage source V_s is driving the net flow of current through the circuit, its dominance over V_L is implied, and the net voltage in the circuit would be as follows:

$$V_{Net} = V_s - V_L$$

This results in the simplification of the circuit as depicted below:

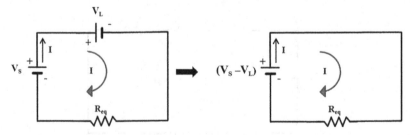

Then, application of Ohm's law yields:

$$I = \frac{(V_s - V_L)}{R_{eq}} \text{ or, } I = \frac{V_s - V_L}{R_1 + R_2 + R_3}$$

Approach II

This approach is premised on the application of KVL to the given circuit after the circuit has been annotated with voltage designations, voltage polarities, and current direction.

According to another electrical convention, voltage polarities are assigned such that the current enters the resistances (or loads, in general) on the **positive** side and exits from the **negative** side. The voltage sources or existing "voltage loads" retain their stated polarities. The aforementioned steps result in the transformation of the original (given) circuit as follows:

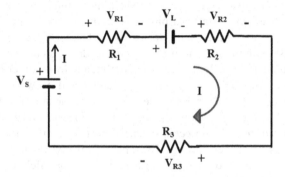

Apply Ohm's law to define the voltages, or voltage drops, across the three resistors.

$$V_{R1} = IR_1 \quad V_{R2} = IR_2 \quad V_{R3} = IR_3$$

With all voltages – voltage source, voltage load, and voltage drops across the resistors – identified and their polarities noted, apply KVL by "walking" the annotated circuit beginning at the cathode or negative electrode of the voltage source, V_s. Note the voltages and respective polarities as you make a complete loop around the circuit in the clockwise direction of the current. This results in the following equation:

$$\Sigma V = 0$$

$$-V_s + V_{R1} + V_L + V_{R2} + V_{R3} = 0$$

Expansion of this KVL-based equation through substitution of the resistor voltage drop formulas, derived earlier, yields:

$$-V_s + IR_1 + V_L + IR_2 + IR_3 = 0$$

Further rearrangement and simplification result in:

$$I(R_1 + R_2 + R_3) = (V_s - V_L)$$

$$I = \frac{V_s - V_L}{R_1 + R_2 + R_3}$$

KIRCHHOFF'S CURRENT LAW (KCL)

According to KCL, total current flowing **into** a **node** is equal to the total current that flows **out** of the **node**. The mathematical representation of KCL is as follows:

$$\sum i_{in} = \sum i_{out} \qquad\qquad (2.4)$$

Proper identification of a **node** before application of KCL is pivotal in the application of KCL. The concept of a node is not unique to electrical circuit. Intersections and junctions in fluid piping systems are sometimes referred to as nodes, by mechanical engineers and technicians. The term finds its use in disciplines as diverse as human anatomical "lymph node" system and the algorithm nodes in computer systems. In the electrical realm, a node is sometimes construed as a point where two conductors merge or get connected. However, as illustrated through application of KCL in Example 2.3, a more meaningful definition of a node in electrical circuits is that it is a point where ***three or more conductors are electrically terminated or connected together.*** Just as significance and effectiveness of KVL was illustrated through Example 2.2, we will demonstrate the utility of KCL and selection of a meaningful node in a "parallel" electrical circuit through Example 2.3.

Example 2.3

Determine the value of voltage source **current** in the parallel circuit below.

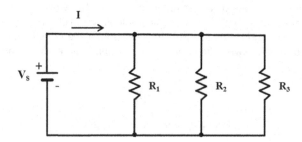

Solution:

Similar to Example 2.2, we will present two different approaches for determining the value of unknown source current. The first approach simply utilizes Ohm's law and the parallel circuit simplification method. The second approach, on the other hand, utilizes KCL and "**nodal**" analysis technique.

APPROACH I

Reduce or simplify the given circuit to a voltage source and equivalent resistance R_{eq}. Since R_1, R_2, and R_3 are in **parallel**, application of Eq. 1.13 yields:

$$R_{eq} = \frac{1}{\dfrac{1}{R_1} + \dfrac{1}{R_2} + \dfrac{1}{R_3}}$$

$$R_{eq} = \frac{R_1 R_2 R_3}{R_1 R_2 + R_2 R_3 + R_1 R_3}$$

This simplifies the given parallel DC circuit as follows:

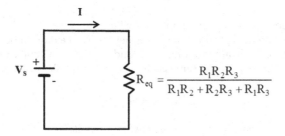

$$R_{eq} = \frac{R_1 R_2 R_3}{R_1 R_2 + R_2 R_3 + R_1 R_3}$$

Next, Ohm's law is applied to determine the source current:

$$I = V_s \left(\frac{R_1 R_2 + R_2 R_3 + R_1 R_3}{R_1 R_2 R_3} \right)$$

APPROACH II

This approach is premised on the application of KCL to the given circuit after the node had been identified and circuit has been annotated with voltage designation, voltage polarity, branch currents, and current directions. See the circuit diagram below:

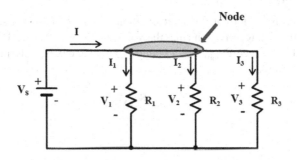

Subscribing to the definition of a node as a point where three or more conductors merge, the shaded segment in the diagram above is designated as the node for this circuit. Next, before KCL can be applied to determine the source current, the individual currents, through each of the resistors, need to be defined – using Ohm's law – in terms of the specific resistance values and the voltages around them:

$$I_1 = \frac{V_1}{R_1} \quad I_2 = \frac{V_2}{R_2} \quad I_3 = \frac{V_3}{R_3}$$

Then, application of KCL at the designated node yields the following equation:

$$I = I_1 + I_2 + I_3$$

Substitution of the values of branch currents, as defined earlier, yields:

$$I = \frac{V_1}{R_1} + \frac{V_2}{R_2} + \frac{V_3}{R_3}$$

At this juncture, it is important to note that when circuit elements are in parallel – as is the case with R_1, R_2 and R_3 – their voltages (or voltage drops around them) are equal. In fact, not only are the voltages around the parallel circuit elements equal to each other but they are the same as the source voltage, V_s. In other words:

$$V_s = V_1 = V_2 = V_3$$

Therefore, the current equation can be rewritten as follows:

$$I = \frac{V_s}{R_1} + \frac{V_s}{R_2} + \frac{V_s}{R_3} = V_s \left(\frac{1}{R_1} + \frac{1}{R_2} + \frac{1}{R_3} \right)$$

And the source current would be as follows:

$$I = V_s \left(\frac{R_1 R_2 + R_1 R_3 + R_2 R_3}{R_1 R_2 R_3} \right)$$

which is the same as the answer derived through Approach 1.

VOLTAGE DIVISION

Voltage division is a shortcut for determination of voltage across a series resistor. According to the voltage division rule, the voltage across resistance R, in a DC circuit, with total resistance R_{total}, and a voltage source V, can be determined through the following formula:

$$V_R = \frac{R}{R_{total}} V \qquad\qquad (2.5)$$

For AC circuits, the voltage on impedance Z_i, in a loop with total impedance Z_{total}, with a voltage source V, would be as follows:

$$V_i = \left(\frac{Z_i}{Z_{total}}\right) \cdot V \qquad (2.6)$$

Example 2.4

Determine the following for the DC circuit shown below:

    a. Equivalent resistance for the entire circuit, if $R_1 = 5\ \Omega$, $R_2 = R_3 = 10\ \Omega$, and $R_4 = R_5 = 20\ \Omega$
    b. Current flowing through resistor R_1
    c. Voltage across resistor R_5

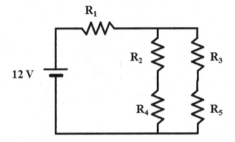

Solution:

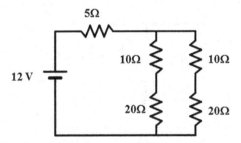

    a. Combination of R_2 and $R_4 = R_{2,4} = R_2 + R_4 = 10\ \Omega + 20\ \Omega = 30\ \Omega$
    Combination of R_3 and $R_5 = R_{3,5} = R_3 + R_5 = 10\ \Omega + 20\ \Omega = 30\ \Omega$
    Combination of $R_{2,4}$ and $R_{3,5} =$

$$R_{2-5} = \frac{(30\ \Omega) \cdot (30\ \Omega)}{(30\ \Omega + 30\ \Omega)} = \frac{900}{60} = 15\ \Omega$$

$$R_{eq} = R_1 + R_{2-5} = 5\ \Omega + 15\ \Omega = 20\ \Omega$$

b. Current through R_1 would be the same as the current through the 12 V supply:

$$I = \frac{V}{R_{eq}} = \frac{12\ V}{20\ \Omega} = 0.6\ A$$

c. One method for determining V_{R5}, voltage across R_5, is to first calculate V_{R2-5}, the voltage across the combined resistance of resistances R_2, R_3, R_4, and R_5. Then, by applying voltage division, calculate V_{R5} as follows:

According to Ohm's law,

$$V_{R2-5} = I \cdot (R_{2-5}) = (0.6\ A) \cdot (15\ \Omega) = 9\ V$$

Then, by applying the voltage division rule:

$$V_{R5} = (9\ V) \cdot \left(\frac{R_5}{R_5 + R_3} \right)$$

$$= (9\ V) \cdot \left(\frac{20\ \Omega}{20\ \Omega + 10\ \Omega} \right)$$

$$= (9\ V) \cdot (0.67) = 6\ V$$

Example 2.5

What is the voltage across the 6 Ω resistor?

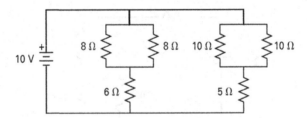

Solution:

The right-hand side of the circuit, consisting of the two 10 Ω resistors and the 5 Ω resistor, is irrelevant insofar as the determination of voltage across the 6 Ω resistor is concerned.

Parallel combination of the two 8 Ω resistors results in an equivalent resistance of 4 Ω as follows:

$$R_{8\Omega//8\Omega} = \frac{(8\ \Omega) \cdot (8\ \Omega)}{8\ \Omega + 8\ \Omega}$$

Using voltage division, the voltage across the 6 Ω resistor would be as follows:

$$V_{6\Omega} = (10\ \text{V}) \cdot \left(\frac{6\,\Omega}{4\,\Omega + 6\,\Omega} \right)$$

$$= 6\ \text{V}$$

CURRENT DIVISION

The current through a resistor **R** in parallel, or in **shunt**, with another resistance $R_{parallel}$ and a current into the node of **I** is:

$$I_R = \left(\frac{R_{parallel}}{R_{total}} \right) \cdot I \tag{2.7}$$

where
R_{total} = The sum of the resistances in parallel (and **not** the parallel combination R_{EQ})
$R_{parallel}$ = Resistance value of the resistor opposite the "subject" resistor
I = Current through the source

When current division is applied in **AC** circuits, the formula for current through an impedance **Z**, in parallel with another impedance $Z_{parallel}$, would be as follows:

$$I_R = \left(\frac{Z_{parallel}}{Z_{total}} \right) \cdot I \tag{2.8}$$

where
I = Current, in its complex AC form, flowing into the node formed by the parallel impedances
Z_{total} = The sum of the impedances in parallel (and **not** the parallel combination Z_{EQ})
$Z_{parallel}$ = The impedance of the load opposite to the subject impedance
I = AC through the source

Example 2.6

Determine the current flowing through the 10 Ω resistance in the circuit shown below.

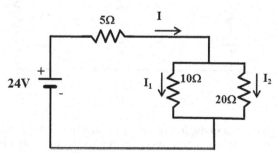

Solution:

We must determine the value of source current **I**, first. In order to determine the value of current **I** flowing through the source and the 5 Ω resistor, we must consolidate all resistors into an equivalent resistance **R$_{EQ}$** and then apply Ohm's law.

$$R_{EQ} = 5\,\Omega + \left(\frac{(10\,\Omega)\cdot(20\,\Omega)}{(10\,\Omega)+(20\,\Omega)} \right) = 5\,\Omega + 6.67\,\Omega = 11.67\,\Omega$$

$$I = \frac{V}{R_{EQ}} = \frac{24\text{ V}}{11.67\,\Omega} = 2.06\text{ A}$$

Apply current division formula in the form of Eq. 2.7:

$$I_{10\,\Omega} = I_1 = \left(\frac{R_{parallel}}{R_{total}} \right) I = \left(\frac{20\,\Omega}{10\,\Omega + 20\,\Omega} \right) 2.06\text{ A} = 1.36\text{ A}$$

MULTI-LOOP CIRCUIT ANALYSIS

As expected, analyses of circuits that consist of more than one current loop tend to be more complex, and require formulation of multiple equations and utilization of additional conventions and principles. Similar to single-loop circuit analysis covered earlier, in most cases, there are multiple strategies and approaches available for solving multi-loop circuits. We will illustrate one approach through Example 2.7.

Example 2.7

Using the KVL method, determine the current flows, I$_1$ and I$_2$, in the circuit below.

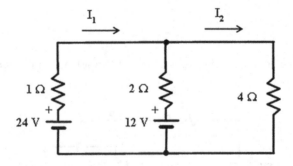

Solution:

Current I$_1$ and I$_2$ are two of the five unknown parameters in the circuit above. The other unknown parameters are V$_1$ Ω, V$_2$ Ω, and V$_4$ Ω. One approach for determining the values of I_1 and I_2 would be to formulate two equations, using the KVL, such that each equation includes the same two unknown variables, I_1 and I_2. Then, by applying the simultaneous equation technique to the two-equation system, with two unknowns, we can determine the values of I_1 and I_2.

Application of KVL requires that voltage around each circuit element be defined in terms of known values and the unknown variables. In addition, the sign or polarity of each voltage must be assigned. See the diagram below.

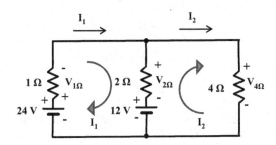

Rules for assuming the current direction and assigning voltage polarities to various loads (resistors) and sources (voltage sources) are as follows:

1. As shown in the figure above, by convention, the currents are assumed to be emanating from the positive pole (positive electrode or anode) of the voltage source and are assumed to be terminating into the negative pole (negative electrode or cathode) of the voltage source.
2. As shown in the figure above, the end or side of the resistor or load that the current enters from is labeled as positive.
3. The polarity for a voltage source is assumed as encountered in the direction of current flow.

Examination of the circuit above reveals that there are three loops in the given circuit. The left loop will be referred to as loop 1, the loop on the right segment of the circuit is loop 2. The third loop in this circuit is formed by the outer perimeter. We will focus on the first two loops to derive two equations for the determination of the two unknown currents. Assume that I_1 is greater than I_2. Conventionally, it is acceptable to make such assumptions as long as the assumptions are, strictly, adhered to in deriving all equations necessary for the solution.

"Walking" loop 1, beginning at the negative terminal of the 24 V DC source, yields the following equation:

$$-24\text{ V} + V_{1\Omega} + V_{2\Omega} + 12\text{ V} = 0 \tag{2.9}$$

"Walking" loop 2, beginning at the negative terminal of the 12 V DC source, yields the following equation:

$$-12\text{ V} - V_{2\Omega} + V_{4\Omega} = 0 \tag{2.10}$$

Based on Ohm's law:

$$V_{1\Omega} = (I_1) \cdot (1\,\Omega) = I_1 \tag{2.11}$$

$$V_{4\Omega} = (\mathbf{I}_2) \cdot (4\ \Omega) = 4\mathbf{I}_2 \tag{2.12}$$

$$V_{2\Omega} = (\mathbf{I}_1 - \mathbf{I}_2) \cdot (2\ \Omega) = 2 \cdot (\mathbf{I}_1 - \mathbf{I}_2) \tag{2.13}$$

Then, by substituting Eqs. 2.11–2.13 into Eqs. 2.9 and 2.10, we get:

$$-24 + \mathbf{I}_1 + 2 \cdot (\mathbf{I}_1 - \mathbf{I}_2) + 12 = 0$$

$$3\mathbf{I}_1 - 2\mathbf{I}_2 = 12 \tag{2.14}$$

$$-12 - 2 \cdot (\mathbf{I}_1 - \mathbf{I}_2) + 4\mathbf{I}_2 = 0$$

$$-2\mathbf{I}_1 + 6\mathbf{I}_2 = 12 \tag{2.15}$$

Equations 2.14 and 2.15 represent the two simultaneous equations that were needed to solve for currents $\mathbf{I}_1$ and $\mathbf{I}_2$. These equations will be solved simultaneously to determine the values of $\mathbf{I}_1$ and $\mathbf{I}_2$.

$$3\mathbf{I}_1 - 2\mathbf{I}_2 = 12 \tag{2.14}$$

$$-2\mathbf{I}_1 + 6\mathbf{I}_2 = 12 \tag{2.15}$$

For simultaneous equation solution, multiply left- and right-hand sides of Eq. 2.14 by 3 and add it to Eq. 2.15:

$$9\mathbf{I}_1 - 6\mathbf{I}_2 = 36$$

$$\underline{-2\mathbf{I}_1 + 6\mathbf{I}_2 = 12}$$

$$7\mathbf{I}_1 = 48$$

$$\therefore\ \mathbf{I}_1 = 6.86\ \text{A}$$

Then, by substituting this value of $\mathbf{I}_1$ into Eq. 2.15 yields:

$$-2(6.86) + 6\mathbf{I}_2 = 12$$

Or

$$\mathbf{I}_2 = 4.29\ \text{A}$$

Note: The values of unknown currents $\mathbf{I}_1$ and $\mathbf{I}_2$ can also be determined by applying Cramer's Rule to Eqs. 2.14 and 2.15 in matrix format and linear algebra.

CIRCUIT ANALYSIS USING CRAMER'S RULE AND LINEAR ALGEBRA

Cramer's Rule can be applied to solve for unknowns, in lieu of simultaneous equations or substitution methods, after a set of equations have been formulated using the loop analysis method described above. Since the Cramer's Rule involves matrices and linear algebra, it is the method that electrical circuit analysis computer software is premised on. The application of Cramer's Rule is illustrated through Example 2.8.

Example 2.8

The values of all known parameters for the following multi-loop circuit are listed in the table below. Find the values of currents I_1, I_2, and I_3.

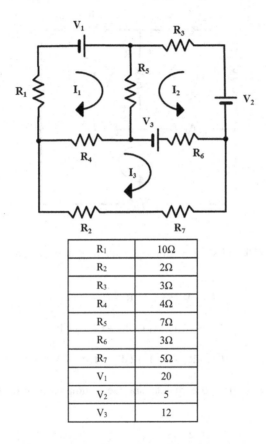

R_1	10Ω
R_2	2Ω
R_3	3Ω
R_4	4Ω
R_5	7Ω
R_6	3Ω
R_7	5Ω
V_1	20
V_2	5
V_3	12

Solution:

In this example, the fundamental strategy, conventions, and principles needed to calculate the three unknown currents I_1, I_2, and I_3 would be the same as the ones described in Example 2.4. The obvious difference is that this circuit has a total of four loops – including the outermost loop – and has three unknown variables in the form of currents I_1, I_2, and I_3. Therefore, we will need a minimum of three equations. Those three equations, as before, are derived by applying the KVL to each of the three inner loops. The voltage drops across each of the load components (or resistors) are defined on the basis of Ohm's law, $V = I \cdot R$.

Before we embark on the formulation of current computation equations, let's ensure that the circuit is in its most simplified form. In that vane, by inspection, we notice that the two series resistors in the bottom loop, R_2 and R_7, can be added together or combined as follows:

$$R_{2-7} = R_2 + R_7$$

This simplification and assignment of voltage drop polarities results in the following circuit schematic:

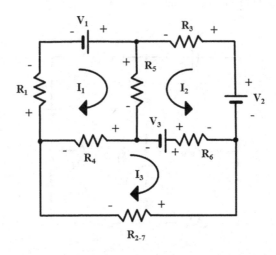

Application of KVL in loop 1, with the assumptions that $I_1 > I_2$ and I_3, and $I_2 > I_3$ yields:

$$I_1 R_1 - V_1 + (I_1 + I_2)R_5 + (I_1 - I_3)R_4 = 0$$

$$I_1 R_1 - V_1 + I_1 R_5 + I_2 R_5 + I_1 R_4 - I_3 R_4 = 0$$

$$(R_1 + R_5 + R_4)I_1 + I_2 R_5 - I_3 R_4 = V_1 \qquad (2.16)$$

Application of KVL in loop 2, maintaining the assumption that $I_1 > I_2$ and I_3, and $I_2 > I_3$ yields:

$$-V_2 + I_2 R_3 + (I_1 + I_2)R_5 - V_3 + (I_2 + I_3)R_6 = 0$$

$$-V_2 + I_2 R_3 + I_1 R_5 + I_2 R_5 - V_3 + I_2 R_6 + I_3 R_6 = 0$$

$$R_5 I_1 + (R_3 + R_5 + R_6)I_2 + I_3 R_6 = V_2 + V_3 \qquad (2.17)$$

Application of KVL in loop 3, maintaining the assumption that $I_1 > I_2$ and I_3, and $I_2 > I_3$ yields:

$$I_3 R_{2-7} - (I_1 - I_3)R_4 - V_3 + (I_2 + I_3)R_6 = 0$$

$$I_3 R_{2-7} - I_1 R_4 + I_3 R_4 - V_3 + I_2 R_6 + I_3 R_6 = 0$$

$$-R_4I_1 + R_6I_2 + (R_4 + R_6 + R_{2-7})I_3 = V_3 \qquad (2.18)$$

The three simultaneous equations thus derived are as follows:

$$(R_1 + R_5 + R_4)I_1 + R_5I_2 - R_4I_3 = V_1 \qquad (2.16)$$

$$R_5I_1 + (R_3 + R_5 + R_6)I_2 + R_6I_3 = V_2 + V_3 \qquad (2.17)$$

$$-R_4I_1 + R_6I_2 + (R_4 + R_6 + R_{2-7})I_3 = V_3 \qquad (2.18)$$

Substitution of the given resistor and voltage source values into Eqs. 2.16–2.18 yields the following simultaneous equations:

$$21I_1 + 7I_2 - 4I_3 = 20$$

$$7I_1 + 13I_2 + 3I_3 = 17$$

$$-4I_1 + 3I_2 + 14I_3 = 12$$

Apply Cramer's Rule to solve for the three unknown currents I_1, I_2, and I_3. The augmented matrix thus developed would be as follows:

$$\begin{vmatrix} 21 & 7 & -4 & | & 20 \\ 7 & 13 & 3 & | & 17 \\ -4 & 3 & 14 & | & 12 \end{vmatrix}$$

The coefficient matrix, denoted as **A**, would be as follows:

$$\begin{vmatrix} 21 & 7 & -4 \\ 7 & 13 & 3 \\ -4 & 3 & 14 \end{vmatrix}$$

The determinant of the coefficient matrix, denoted as $|\mathbf{A}|$, would be as follows:

$$|\mathbf{A}| = 21\{(13 \times 14) - (3 \times 3)\} - 7\{(7 \times 14) - (-4 \times 3)\} - 4\{(7 \times 3) - (-4 \times 13)\} = 2,571$$

The determinant of the substitutional matrix, $\mathbf{A_1}$, for determining the value of I_1, is denoted as $|\mathbf{A_1}|$, and

$$\mathbf{A_1} = \begin{vmatrix} 20 & 7 & -4 \\ 17 & 13 & 3 \\ 12 & 3 & 14 \end{vmatrix}$$

$$|\mathbf{A_1}| = 20\{(13 \times 14) - (3 \times 3)\} - 7\{(17 \times 14) - (12 \times 3)\} - 4\{(17 \times 3) - (12 \times 13)\} = 2,466$$

The determinant of the substitutional matrix, A_2, for determining the value of I_2, is denoted as $|A_2|$, and

$$A_2 = \begin{vmatrix} 21 & 20 & -4 \\ 7 & 17 & 3 \\ -4 & 12 & 14 \end{vmatrix}$$

$$|A_2| = 21\{(17 \times 14) - (12 \times 3)\} - 20\{(7 \times 14) - (-4 \times 3)\} - 4\{(7 \times 12) - (-4 \times 17)\} = 1{,}434$$

The determinant of the substitutional matrix, A_3, for determining the value of I_3, is denoted as $|A_3|$, and

$$A_3 = \begin{vmatrix} 21 & 7 & 20 \\ 7 & 13 & 17 \\ -4 & 3 & 12 \end{vmatrix}$$

$$|A_3| = 21\{(13 \times 12) - (3 \times 17)\} - 7\{(7 \times 12) - (-4 \times 17)\} + 20\{(7 \times 3) - (-4 \times 13)\} = 2{,}601$$

Applying Cramer's Rule, the unknown variables, currents I_1, I_2, and I_3, can be calculated by dividing the determinants of substitutional matrices A_1, A_2, and A_3, respectively, by the determinant of the coefficient matrix A.

Therefore,

$$I_1 = \frac{|A_1|}{|A|} = \frac{2{,}466}{2{,}571} = 0.959 \text{ A}$$

$$I_2 = \frac{|A_2|}{|A|} = \frac{1{,}434}{2{,}571} = 0.558 \text{ A}$$

$$I_3 = \frac{|A_3|}{|A|} = \frac{2{,}601}{2{,}571} = 1.012 \text{ A}$$

SUPERPOSITION THEOREM

Superposition theorem provides us a method for analyzing complex circuits, with multiple power sources and multiple loads or resistors, in a simplified fashion, with fewer steps than typical loop analysis methods. The superposition theorem states that for a linear system the voltage or current response in any branch of a bilateral linear circuit – consisting of more than one independent source – equals the algebraic sum of the responses caused by each independent source acting alone.

The superposition method of circuit simplification and analysis is implemented as follows:

1. Net current is determined by summing the currents caused by each current source, with the voltage sources shorted.

2. Net voltage is determined by summing the voltages caused by each voltage source, with the current sources open-circuited.
3. Short all voltage sources, open all current sources, and then turn on only one source at a time.
4. Simplify the circuit to get the current or voltage of interest.
5. Repeat until the responses of all sources have been evaluated.
6. Add the results to get the answer.

The superposition method is illustrated through Example 2.9.

Example 2.9

Using the superposition theorem, determine the current flowing through the center leg of the bilateral circuits below.

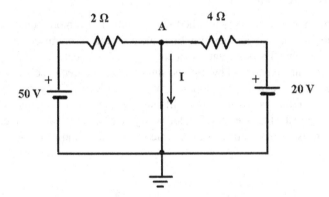

Solution:

First "turn off" the 20 V voltage source in the right-side loop and calculate the current contributed to the center leg by the 50 V source. This scenario and the current calculation are depicted below.

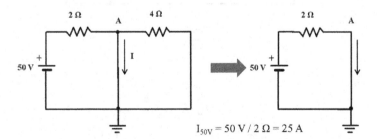

$$I_{50V} = 50 \text{ V} / 2 \text{ } \Omega = 25 \text{ A}$$

Next, "turn off" the 50 V voltage source in the left-side loop and calculate the current contributed to the center leg by the 20 V source. This scenario and the associated current calculation are depicted below.

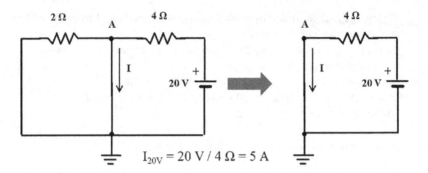

$$I_{20V} = 20\ V\ /\ 4\ \Omega = 5\ A$$

Therefore, in accordance with the superposition theorem, the total current through the center leg would be 25 A + 5 A = 30 A.

THEVENIN EQUIVALENT

The Thevenin Equivalent is a simplified representation of a more complex electrical circuit. Determination of a Thevenin Equivalent circuit is premised on Thevenin's Theorem. Thevenin's Theorem states that any linear circuit containing several voltages and resistances can be replaced by just one single voltage source in series with a single resistance connected across or in "shunt" form to the load. In essence, according to Thevenin's Theorem, it is possible to simplify any electrical circuit, no matter how complex, into an equivalent two-terminal circuit with just a single constant voltage source in series with a resistance or impedance connected to a load, as shown in Figure 2.1.

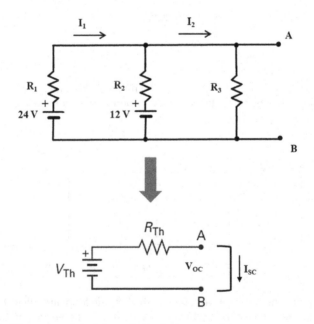

FIGURE 2.1   Illustration of the Thevenin Equivalent principle.

The Thevenin Equivalence principle can be translated to mathematical form as follows:

$$R_{Th} = \frac{V_{OC}}{I_{SC}}$$

where

R_{Th} = The Thevenin Equivalent resistance

V_{OC} = The open circuit voltage or the voltage one would measure with a voltmeter with points A and B open-circuited as shown in Figure 2.1.

I_{Sc} = The short-circuit current or the current one would measure with an ammeter between points A and B, with the points short-circuited as shown with a "jumper" or "] " in Figure 2.1 Thevenin Equivalent circuit.

Example 2.10

Derive the Thevenin Equivalent for the circuit given below.

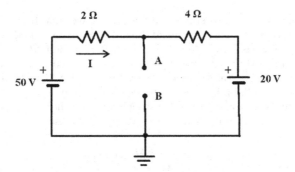

Solution:

The requirement "Thevenin Equivalent" implies the determination of V_{Th} and R_{Th}. Combination of these two key components, as depicted in diagram below, constitutes the Thevenin Equivalent for the given circuit.

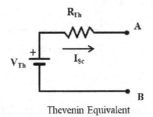

Thevenin Equivalent

First, let's focus on the calculation of V_{Th}, the voltage shown in the diagram below by applying Ohm's law.

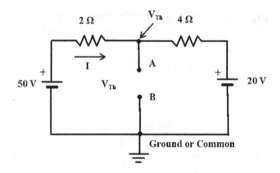

$$I = \frac{50\ V - V_{Th}}{2\ \Omega}\{Ohm's\ Law\};\ or,\ V_{Th} = 50\ V - (2\ \Omega)\cdot I$$

And, since there is an open circuit between point A and Ground,

$$I = \frac{50\ V - 20\ V}{6\ \Omega}\{Ohm's\ Law\} = 5\ A$$

Then, $V_{Th} = 50\ V - (2\ \Omega)\cdot(5\ A) = 40\ V$

Next, we focus on R_{Th}. To determine the value of R_{Th}, we view the circuit from the vantage point of A and Ground, inward. Although some might find this observation less than obvious, as you look into the A – B (A to Ground) plane, you find that 2 Ω and the 4 Ω resistors are in parallel. You can visualize this by imagining that you crease the page along the line formed by points A, B, and Ground. Also, in accordance with the principles of Thevenin's Theorem, when computing R_{Th}, you neutralize the 50 V and 20 V voltage sources. Hence, the parallel combination of the 2 Ω and the 4 Ω resistors, as shown below, would result in R_{Th} of 1.33 Ω.

$$R_{Th} = \frac{(2\ \Omega)(4\ \Omega)}{2\ \Omega + 4\ \Omega} = 1.33\ \Omega$$

The Thevenin Equivalent would then appear as follows:

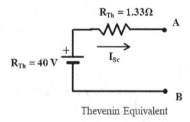

Thevenin Equivalent

NORTON EQUIVALENT

Norton Equivalent is a simplified representation of a more complex electrical circuit consisting of current sources, voltage sources, and resistors or loads. Determination of a Norton Equivalent circuit is premised on Norton's Theorem. Norton's Theorem states that any linear circuit containing several current sources, voltage sources, and

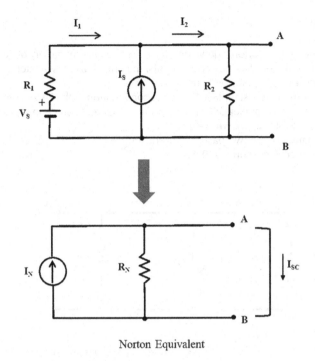

Norton Equivalent

FIGURE 2.2   Illustration of the Norton Equivalent principle.

resistances or loads can be replaced by just one single current source in parallel with a single resistance connected across or, in "shunt," form to the load. In essence, according to Norton's Theorem, it is possible to simplify any electrical circuit, no matter how complex, to an equivalent two-terminal circuit with just a single constant current source in shunt or parallel with a resistance or impedance connected to a load, as shown in Figure 2.2. Note that Norton current, I_N, is also the current that would flow through a jumper or short across terminals A and B of the Norton Equivalent circuit as shown in Figure 2.2. In other words, $I_N = I_{Sc}$.

Example 2.11

Develop the Norton Equivalent for the circuit given below.

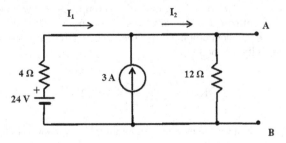

Solution:

The requirement "Norton Equivalent" implies the determination of I_N and R_N. Combination of these two key components, in parallel, as shown later in the solution will represent the Norton Equivalent.

Using Ohm's law, first replace the series combination of the 24 V source and the 4 Ω resistor with a parallel combination of a 6 A current source in shunt with the original 4 Ω resistor as shown below.

Combination of the two parallel current sources, 6 A and 3 A, results in a net, total, Norton current source of 9 A.

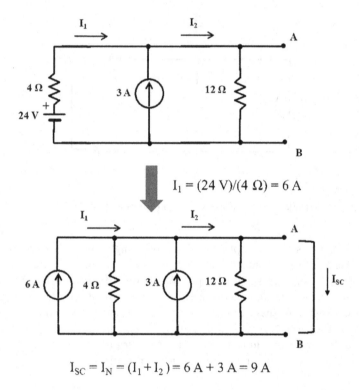

$$I_1 = (24\ \text{V})/(4\ \Omega) = 6\ \text{A}$$

$$I_{SC} = I_N = (I_1 + I_2) = 6\ \text{A} + 3\ \text{A} = 9\ \text{A}$$

Next, we focus on R_N. To determine the value of R_N, we view the circuit from the vantage point of A and B, or ground/common. As you look to the left from points A and B, you find that 12 Ω and the 4 Ω resistors are in parallel. Also, in accordance with the principles of Norton's Theorem, when computing R_N, you neutralize the 6 A and 3 A current sources. This amounts to "open-circuiting" the two current sources. Then, the parallel combination of the 12 Ω and the 4 Ω resistors, as shown below, would result in R_N of 3 Ω.

$$R_N = \frac{(12\ \Omega)(4\ \Omega)}{12\ \Omega + 4\ \Omega} = 3\ \Omega$$

The Norton Equivalent for the given circuit would then appear as follows:

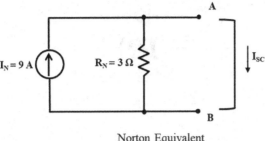

Norton Equivalent

DIODES

A diode, unlike its "functional" predecessor – the vacuum tube rectifier – is constructed out of **semiconductor** materials such as silicone, germanium, and gallium arsenide. While a semiconductor is not a good conductor at room temperature, it doesn't fall distinctly in the category of insulators, such as, glass, ceramics, urethanes, plastics, and polyvinyl chlorides (PVCs). A semiconductor can, however, be transformed into a "partially" or "selectively" conductive substance through a process called "**doping**." The term doping implies addition of "impurities" into a pure substance like silicone. If these impurities are added to create a region with a predominant concentration of negative charge carriers, or electrons, the end result would be the formation of a region called **n-type** semiconductor. At the same time, if impurities are added, adjacently, to create a region with predominant positive charge carriers, or holes, a **p-type** semiconductor is formed. The plane where the **p-** and **n**-doped materials interface with each other is called the **p–n junction**. A p–n junction is where the essential function of a diode takes place in response to the application of proper voltage.

A diode can be viewed as a device that functions as an electronic "check valve." As we know, the function of a check valve, in mechanical or hydraulic systems, is to permit the flow of fluids in one specific direction. In other words, an attempt by the fluid to move in the reverse direction is blocked by a check valve. A diode performs the same function in the flow of current. A diode permits the flow of current only from a higher voltage (or electrical potential) point in an electrical circuit to a lower voltage or ground potential point. This unidirectional behavior is called **rectification**, and this function of a diode finds a common application in the conversion of AC to DC. Common applications of diodes in rectifiers and other equipment are discussed later in this chapter.

The symbol and drawing of common circuit board-type diode are shown in Figure 2.3.

The left side of the diode, labeled "anode" is normally connected to the positive or higher voltage point in the circuit. The right side of the diode, labeled "cathode," on the other hand, is normally connected to the ground, negative, or lower potential point in the circuit. When a diode is connected in this manner, it is said to be **forward biased**. As apparent from the diagram of a typical diode in Figure 2.1, a band on one side of the diode denotes the cathode side of the diode. If, however, the voltage

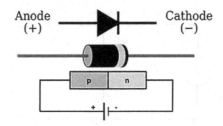

FIGURE 2.3   Symbol, diagram, and schematic of a basic diode circuit.

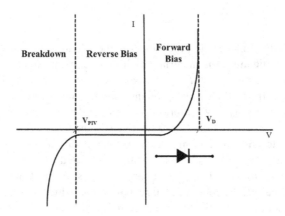

FIGURE 2.4   Current response of a diode in forward and reverse bias modes.

is reversed such that the anode of the diode is connected to the negative voltage potential and the cathode is connected to the positive potential, the diode is said to be **reverse biased**. The current response of a diode in forward and reverse bias modes of operation is depicted in Figure 2.4.

In Figure 2.4, V_D represents the forward bias voltage. Forward bias voltage of a diode is the voltage at which the diode begins to conduct current or is the voltage where the diode is said to be "turned on." A diode can be perceived as a "self-actuating" electronic switch, as well. The forward bias voltage, V_D, is approximately **0.7 V** for silicone-type diodes and **0.3 V** for germanium diodes.

As depicted in Figure 2.4, when a diode is reverse biased, the amount of current (or reverse current) it conducts is negligible, until the magnitude of the reverse bias voltage approach approaches V_{BR}, breakdown voltage. The breakdown voltage is also called the V_{PIV}, peak inverse voltage, or V_{PRV}, peak reverse voltage. The reverse voltage is considerably higher than the forward bias voltage; approximately **ten** times the magnitude of V_D. Although, technically, all diodes are capable of rectifying AC, the term **"rectifier"** is essentially reserved for diodes designed to operate at currents in excess of 1 A and, therefore, utilized frequently to convert AC into DC. Conventionally, the term diode refers to applications involving currents less than or equal to 1 A.

SPECIAL PURPOSE DIODES

Three of the most common, special purpose, diodes that find applications in our daily lives and mainstream electrical or electronic equipment are LEDs, SCRs, and Zener diodes.

LED: LEDs, or light-emitting diodes, are the leading sources of non-natural light in contemporary buildings of all types. LEDs are substantially more efficient than the traditional sources of light. For instance, LED-based lamps – for equivalent number of lumens – are almost ten (10) times more energy efficient than the traditional incandescent bulbs. Symbol for an LED is shown in Figure 2.5. LEDs can emit light in different colors. Unlike the regular diodes, the LEDs turn on at higher forward bias voltages. For example, a red LED, typically, turns on 2.0 V, yellow at 2.1 V, green at 2.2 V, blue at 3.5 V, and white at 3.2 V. LEDs emit light when they are turned on by the application of higher potential at the anode relative to the cathode or the forward bias voltage. When the LEDs turn on, electrons within the LED material fall from higher energy orbitals to lower energy orbitals; in doing so, energy is released by the LEDs in the form of photons or light.

The SCRs, also known as thyristors, are applied heavily in variable frequency drives (VFDs), inverters, converters, UPS, uninterruptible power supply systems, while Zener diodes are key operating elements in DC voltage regulators.

SCR: The acronym "SCR" stands for "Silicone Controlled Rectifier" or "Semiconductor Controlled Rectifier." These specialty diodes are also referred to as "thyristors." Schematic of an SCR is shown in Figure 2.5.

As apparent in Figure 2.5, an SCR differs from a regular diode, mainly, due to an added feature called the "gate." The gate serves as "trigger" or "firing" mechanism for an SCR. Specific voltage application at the gate triggers or "pulses" the SCR on and allows the current to flow. The diode portion of the SCR continues to conduct after the gate voltage dissipates. The SCR, or diode portion of the SCR, stops conducting once the forward bias voltage drops below the threshold voltage, V_D.

SCRs are mainly used in devices associated with the control of high power and high voltage. Their innate characteristic and mode of operation makes them suitable for use in medium- to high-voltage AC power control applications, such as lamp dimming, regulators, and motor control. Thyristors are also commonly used for rectification of high-power AC in high-voltage DC power transmission applications. They are also used in the control of welding machines.

Zener Diode: Zener diodes are similar, in construction, to basic diodes. The key difference is that Zener diodes, unlike regular diodes, are capable of recovering from avalanche reverse bias breakdown mode when the reverse bias is removed. Therefore, as a mechanical analogy, one could compare Zener diodes to "relief valves" that open under abnormally high pressure and reinstate their normal blocking function once

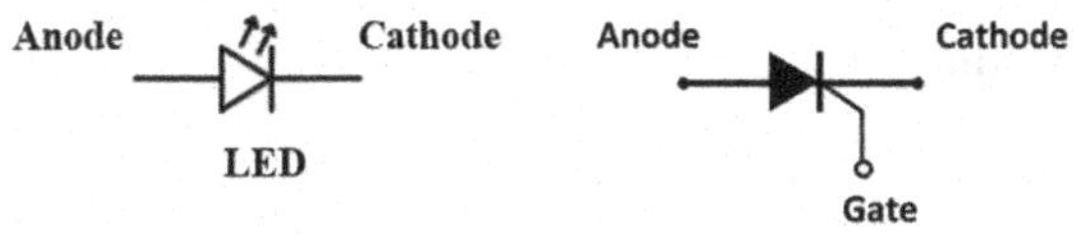

Anode   Cathode   Anode   Cathode

LED   Gate

FIGURE 2.5 Symbols or diagrams for LED and SCR or thyristor.

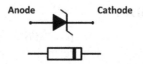

FIGURE 2.6   Symbol and diagram for Zener diode.

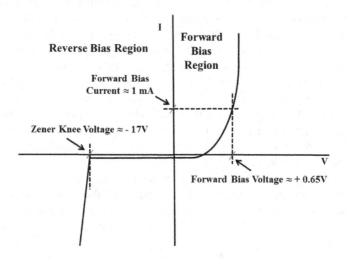

FIGURE 2.7   Current response of a Zener diode.

the pressure subsides to a nominal level. The Zener diode is named after Clarence Zener, who discovered its electrical property. The symbol or electronic representation of a Zener diode is depicted in Figure 2.6.

A Zener diode allows current to flow in the forward direction in the same manner as an ideal diode but will also permit it to flow in the reverse direction when the voltage is above the breakdown voltage. The breakdown voltage is also referred to as the "Zener knee voltage," "Zener voltage," or "avalanche point". Because of this basic characteristic of a Zener diode, it is commonly used to provide a reference voltage for voltage regulators or to protect other semiconductor devices from momentary voltage pulses or excessive voltage "spikes." The current response of a Zener diode is shown in Figure 2.7.

COMMON APPLICATIONS OF DIODES

Some common applications of diodes are as follows:

1. Half-wave rectifier
2. Full-wave rectifier
3. Clamping circuit
4. Base clipper
5. Peak clipper

Among the diode applications listed above, the two most common ones are (1) half-wave rectifier and (2) full-wave rectifier. These two applications are explored below.

HALF-WAVE RECTIFIER

Half-wave rectifier is a circuit consisting of two core components, a diode and resistor. See Figure 2.8a. As shown in Figure 2.8a, the sinusoidal AC waveform is applied to the input side of the diode–resistor circuit. The diode, as shown in Figure 2.8b, is the core "active" component. A half-wave rectifier circuit banks on the innate characteristic of a diode to allow current to flow only "one way," in the forward-biased direction. In other words, the diode permits the current to flow only during the positive half of the AC cycle. When the AC voltage "dives" into the negative realm – acting as a "check valve" – the diode shuts off the flow of current. This response of the diode is plotted graphically in Figure 2.8c, in the form of a series of positive wave crests, average of which represents the DC voltage produced.

Note, however, that half-wave rectifiers only produce **one** positive DC crest per AC cycle. The DC output of a half-wave rectifier can be computed through Eq. 2.19.

$$V_{dc} = \frac{V_p}{\pi} \tag{2.19}$$

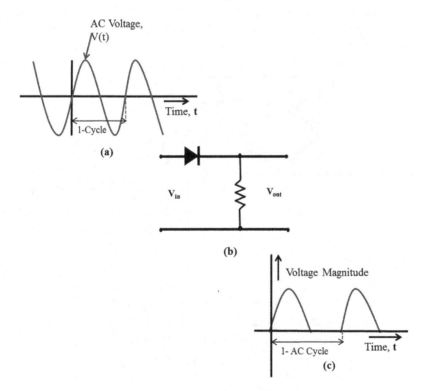

FIGURE 2.8   (a–c) Half-wave rectifier.

FULL-WAVE RECTIFIER

Common full-wave rectifier is a circuit consisting of four diodes and a resistor. See Figure 2.9. The sinusoidal AC waveform applied to the input side of the **four-diode bridge** and resistor circuit is shown in Figure 2.9a. The four-diode bridge configuration shown in Figure 2.9b is the essential power conversion segment of the overall full-wave bridge rectifier circuit. When AC voltage is applied between terminals **A** and **B** on the input side of the diode bridge, as V_{in}, the positive crest of the sinusoidal waveform drives current through terminal **A**, diode CR_1, resistor **R**, diode CR_4 to terminal **B**. During this positive crest-initiated flow of current, diodes CR_1 and CR_4 are forward biased. As the current assumes this course, it "drops" a positive "half" wave across the resistor or terminals **C** and **D**. This positive DC crest is the first crest from the left, in Figure 2.9c. During the positive half of the AC cycle, diodes CR_2 and CR_3 are reverse biased; therefore, they do not conduct, and **no** current flows through the CR_2, **R**, and CR_3 paths.

During the negative half of the AC cycle, diodes CR_1 and CR_4 are reverse biased and no current flows through them. However, because terminal **A** is negative during

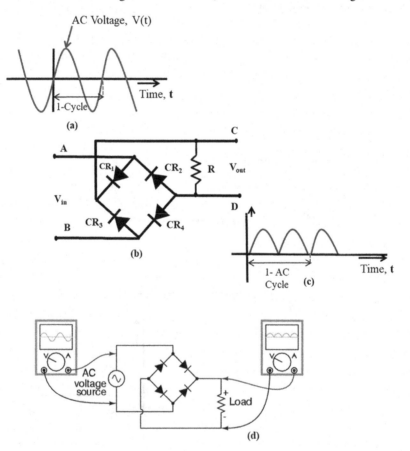

FIGURE 2.9   (a–d) Full-wave rectifier.

the negative half of the AC cycle, diodes CR_2 and CR_3 are forward biased and the current flow is driven from terminal **B**, through diode CR_3, resistor **R**, and diode CR_2. This path of current flow also results in a positive voltage drop or positive voltage crest across terminals **C** and **D**. This positive DC crest is represented by the **second** crest in Figure 2.9c. Therefore, **one** AC cycle on the V_{in} side results in **two** positive crests on the V_{out} side. The average value of the positive crests shown in Figure 2.9c represents the DC output of this full-wave bridge rectifier circuit. Of course, the DC output in most DC power supplies is refined or corrected into a straight-line form through application of resistors, capacitors, and inductors. The formula for full-wave rectified DC voltage is represented by Eq. 2.20.

$$V_{DC} = 2 \cdot \left(\frac{V_p}{\pi} \right) \tag{2.20}$$

Figure 2.9d shows how the signals would appear if an oscilloscope or scope meters were connected to the full-wave rectification circuit on the input and the output sides. Notice the zero average – zero DC – complete sinusoidal voltage signal on the scope meter connected on the input side (left-hand side) of the circuit. On the right-hand side – or the output side – the scope meter shows the fully rectified signal – with a finite average – in the form of two voltage crests in the positive voltage realm. In essence, the average value of this double crest – full-wave rectified – waveform would be as stated in Eq. 2.20.

TRANSISTORS

Similar to diodes, transistors are semiconductor devices. The approach to constructing transistors is similar to the approach used for fabrication of diodes. The "n" and "p" doping approach employed with the construction of transistors is shown in Figure 2.10. As somewhat evident from Figure 2.10, a transistor – functionally, and from construction point of view – appears as set of two diodes connected "back to back." There are three major types of transistors: the BJT, or bipolar junction transistor, MOSFET, or metal-oxide-semiconductor field-effect transistor, and the IGBT, or insulated gate bipolar transistor. Although, in this text we will delve deeper into the operations and application of BJT transistors, we will discuss a few important facts associated with the MOSFET and IGBT transistors, and compare these three transistors, to a degree.

When transistors are used in "saturation" mode, they act like digital switches that generate 1's and 0's. Note that 1's and 0's are the only digits in a binary number system. The brain of a typical computer is a microchip referred to as a microprocessor, and a microprocessor performs all computation and performs all decision-making in binary number system and binary logic. A typical microprocessor consists of about two billion transistors. These billions of transistors are fundamental building blocks of logic gates such as "And Gates," "OR Gates," "Nand Gates," and "Flip Flops," and combination of these gates results in the formation of a microprocessor.

BJT Transistors: Within the BJT category, there are two subcategories: The **"npn"** and the **"pnp"** transistors. We will limit our discussion to the introduction of **npn** BJT-type transistors in this text.

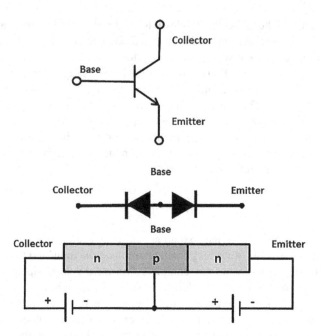

FIGURE 2.10   Bipolar (FET) junction, npn transistor.

Unlike the diode, an npn transistor consists of three "doped" zones: (1) An n-doped segment that is known as the "collector," (2) a center p-doped segment referred to as the "base," and (3) a second n-doped segment known as the "emitter." See the bottom section of Figure 2.10 for the relative location and the construction of an npn transistor. The interface lines where the collector–base and the base to emitter regions meet are referred to as the n–p, collector to base junction, and the p–n, base to emitter junction, respectively. As shown in Figure 2.10, in a normal npn transistor application, the collector to base junction is reverse biased and the base to emitter junction is forward biased. As shown in Figure 2.10, common configuration and application of a transistor resembles a back-to-back connection of two diodes; the one on the left being reverse biased and the one on the right, forward biased.

By energizing the base of a BJT transistor, the small amount of base current ends up catalyzing and amplifying the flow of a large amount of current from the emitter to the collector. When you increase the base current, the collector current increases significantly. The symbol used to represent a BJT transistor in an electrical or electronic circuit is depicted in the top segment of Figure 2.10, with proper labeling of the three terminals of the transistors, in terms of collector, base, and emitter.

A common application of BJT transistors in a typical "output module" of a PLC: A BJT transistor can be applied as an "electronic switch" to execute a Programmable Logic Controller (PLC) microprocessor command in the field. This is illustrated in the form of a transistor-based PLC output, operating a solenoid valve, as shown in Figure 2.11. Applying ample positive bias to the left of R_1 will turn "on" the npn transistor, causing a "short" to ground. This permits the current to flow through the solenoid coil. Flow of current through the solenoid coil generates magnetic flux

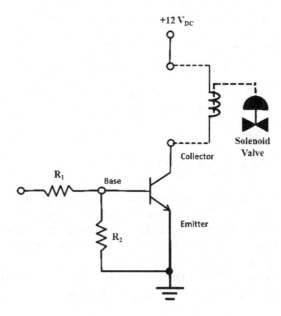

FIGURE 2.11 Bipolar (FET) junction application in a PLC output module.

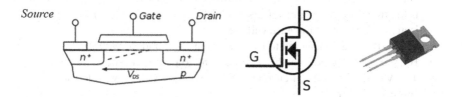

FIGURE 2.12 MOSFET construction, symbol, and picture.

through the core of the coil, thus applying magnetic force on the "plunger" or "pin" in the core of the solenoid coil. Depending on the design of the solenoid, the magnetic force can either open or close the solenoid valve.

MOSFET or Metal-Oxide-Semiconductor Field-Effect Transistor: The MOSFET is a type of field-effect transistor (FET) and has three terminals, **source**, **gate**, and **drain**. The gate is used to bias or control the MOSFET operation. Figure 2.12 shows the cross-section, or construction, the symbol, and a picture of a typical MOSFET transistor. The gate voltage determines the conductivity of the device. This ability to change conductivity with the amount of voltage applied at the gate can be used for amplifying or switching electronic signals. The MOSFET was invented by Mohamed Atalla and Dawon Kahng at Bell Labs in 1959 and is believed to be the most widely used semiconductor device in the world. The construction or cross-section, the symbol, and a picture of a typical MOSFET are depicted in Figure 2.12.

IGBT or Insulated Gate Bipolar Transistor: An IGBT is a three-terminal power semiconductor device primarily used as an electronic switch which, as it was developed, came to combine high efficiency and fast switching. Figure 2.13 shows the cross-section, or construction, the symbol, and a picture of a typical IGBT. Note

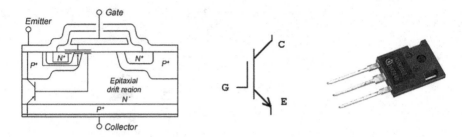

FIGURE 2.13   IGBT construction, symbol, and picture.

that physical shapes of transistors vary based on their ratings and types. An IGBT consists of four alternating layers (P-N-P-N) that are controlled by a metal-oxide-semiconductor (MOS) gate structure. IGBT transistors are used in switching power supplies in high-power applications: VFDs, electric cars, trains, variable speed refrigerators, lamp ballasts, and air-conditioners.

Comparison between BJT, MOSFET, and IGBT: Some comparisons and similarities between BJT, MOSFET and IGBT are as follows:

- While in BJTs it is the **base current** that determines the flow of collector to emitter current, in MOSFETs, the current between the source and the drain is the function of the **gate voltage**. In IGBTs, the collector to emitter current is also a function of the **gate voltage**.
- The IGBTs are considered to be a hybrid of BJTs and the MOSFETs.
- BJTs are considered to be high-voltage and low-current type of devices.
- The MOSFETs are high-current and low-voltage devices, with low drain to source resistance.
- The MOSFETs are capable of operating at high speeds, upwards of around 500 kHz.
- IGBTs are high-voltage and high-current devices. Therefore, they're used for higher power drives. The switching speed of IGBTs is relatively low, at around 50 kHz.
- Ideal applications of MOSFETs are power management, welders, DC power supplies, and inverters.
- IGBTs are transistors of choice for UPS and three-phase motor control.
- Ideal applications for BJTs are: audio amplifiers, single-phase control, and stepper motors.
- IGBT and MOSFET tend to be more expensive than BJTs.

ELECTRONIC DEVICE APPLICATIONS ON PRINTED CIRCUIT BOARDS

This section is designed to provide a brief introduction to a common PCB and an assortment of common electronic devices typically installed on a PCB to perform control, monitoring, and data collection functions. Some of the electronic and electrical devices identified on this board are explained in greater detail in the preceding and subsequent chapters.

FIGURE 2.14 Electrical and electronic devices on a PCB-I.

The devices pin pointed in Figure 2.14 are listed below, in clockwise order, beginning from the top right corner:

1. A toroid-type **inductor or coil** applied for DC power refining and filtering purpose.
2. DC **voltage regulator** applied for voltage regulation purposes. See Chapter 3 for a comprehensive discussion on regulation.
3. Solid-state logic gate **integrated circuit** (IC) micro-chip applied to perform logic, algorithm, and computations for control purposes. This type of IC consists of logic gates such as OR gates, NOR gates, Exclusive OR gates, AND gates, NAND gates, and Flip Flops.
4. A typical, low-voltage electrical **fuse** for protection of the board and upstream power source against faults and shorts.
5. A **transistor**, with collector, base, and emitter pins visible to the right.
6. A **diode**, labeled using the conventional "CR" prefix-based nomenclature.
7. A typical "**proprietary**" **IC** device. While the functional specifications of such devices are made available to electronic control engineers for application purposes, the contents and design of such devices are, typically, kept confidential and are not published.

Figure 2.15 depicts another segment of the same printed circuit control board. The devices noted in this segment of the control board, in clockwise order, beginning from the top right corner are as follows:

1. Power transistor, npn (or NPN); with collector current rating of 30 A at 40 V, capable of switching at a maximum frequency 200 kHz. Maximum power rating of 150 W DC. The black heat-dissipating fins surrounding the power transistor are designed to radiate waste heat and to protect the transistor against overheating.

FIGURE 2.15   Electrical and electronic devices on a PCB-II.

2. A low-wattage resistor with visible color-coded bands identifying the resistance value and the tolerance specification of the resistor.
3. Higher wattage ceramic resistor. Notice the remarkable difference in the physical size and construction of this higher wattage and higher operating temperature ceramic resistor and its lower wattage counterpart described above. The ceramic resistor in Figure 2.15 is encased in a ceramic enclosure to withstand higher temperature and heat.
4. A capacitor.
5. A row of LEDs designed to annunciate the state of various inputs received and outputs transmitted to electrical and electronic equipment in the field. Each of the output terminals represents a hardware-based command from on-board controls out to the field.
6. Four diodes being used in a full-wave bridge formation to convert AC to DC to meet on-board DC needs.

An important feature that is instrumental in minimizing the time it takes to replace a control board, in case of board malfunction, is the "quick connect/disconnect" type terminal strip – labeled as (1) – shown on the right side to the control board (PCB), as shown in Figure 2.16. Prior to the advent of this quick connect-/disconnect-type terminal strip, board replacement required meticulous examination and care in reading the electrical drawings, identification of each wire and terminal, and finally, exercise of proper craftsmanship in termination of wires to ensure correct and reliable electrical connections. The lack of such diligence resulted in miswiring, electrical faults, extensive troubleshooting period, prolonged commissioning time, etc.

When disconnecting a control board, the quick connect/disconnect terminal strip, shown in Figure 2.16, allows technicians and engineers to simply pull off the

FIGURE 2.16 Electronic devices on a PCB-III.

connector, with reasonable certainty that wires will remain securely intact for quick reinstallation. Since all wires normally remain terminated in the connector, once the new (replacement) board is physically secured in place, the connector is pushed/plugged onto the connecting edge of the new board. Note that the circuit boards and respective connectors are, typically, equipped with an interlocking feature to prevent incorrect orientation of the connector. As obvious, incorrect insertion of the connector onto the board can result in electrical faults and damage to the electrical devices.

The picture of the circuit board in Figure 2.16 also shows a set of two LEDs, labeled as item (2). These LEDs serve as indicators of certain control conditions or signals.

Three devices pointed out on the circuit board pictured in Figure 2.17 are as follows:

1. A large size circuit board capacitor.
2. Higher wattage ceramic resistor. Notice the legend on the right side of the resistor. Typically, this legend includes the resistance value and power capacity specifications.
3. Metal-oxide varistor, rated 100 J (100 Joules) and 1 kVA. Metal oxide varistor, also referred to as MOV, serves as stray energy arresting device on electronic circuits and electrical systems, in general. In this capacity, an MOV serves to absorb voltage spikes and rogue energy that might otherwise spread around various parts of a circuit board or electrical control system. Stray energy or voltage spikes, left unchecked, can damage ICs, or IC semiconductor chips. ICs are relatively sensitive. They operate at low voltages, typically, at around 5 V DC. The MOV in this particular case, as visible in the picture, is designed to absorb a maximum of 100 J of electrical energy. The "rate of absorption of this energy," or the power absorption capacity, is labeled on the MOV as 1 kVA; to be exact, the apparent power (**S**) rating of this MOV is 1,000 VA.

FIGURE 2.17   Electrical and electronic devices on a PCB-IV.

SELF-ASSESSMENT PROBLEMS AND QUESTIONS – CHAPTER 2

1. Determine the following for the DC circuit shown below if $R_1 = 5\ \Omega$, $R_2 = R_3 = 10\ \Omega$, and $R_4 = R_5 = 20\ \Omega$:
   a.   Current flowing through resistor R_1
   b.   Voltage across resistor R_5

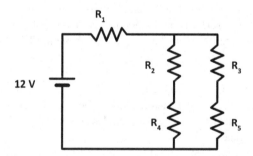

2. What is the current through the 6 Ω resistor in the circuit shown below?

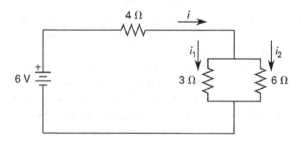

3. Find the current through the 0.5 Ω resistor in the circuit shown below.

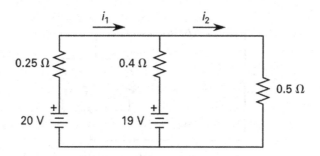

4. Determine the value of currents I_1, I_2, and I_3 in the circuit shown below if the voltage source V_3 fails in short-circuit mode. The specifications of all components are listed in the table below:

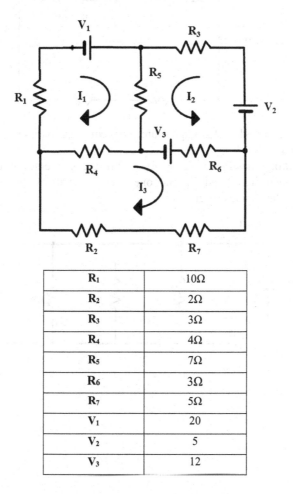

R_1	10Ω
R_2	2Ω
R_3	3Ω
R_4	4Ω
R_5	7Ω
R_6	3Ω
R_7	5Ω
V_1	20
V_2	5
V_3	12

5. Use current division to determine the value of current I_1 in the circuit below:

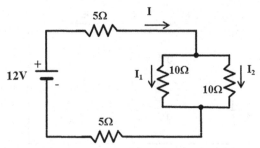

6. Using KVL, calculate the current circulating in the series resistor network below:

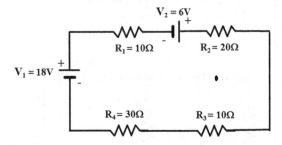

7. Determine the value of voltage source **current** in the parallel circuit below using KCL.

   Ancillary question: If one of the 5 Ω resistors is removed (or replaced with an open circuit) and the other one is replaced with a short circuit, what would be the source current?

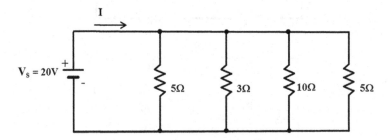

3 Alternating Current

INTRODUCTION

Even though the concepts AC, alternating current, and AC impedance were introduced very briefly in Chapter 1, this chapter builds on that early introduction to AC and segues into a comparison between AC and direct current (DC). This comparison accentuates the complexity of AC as compared with DC due to the use of complex numbers and vectors for complete representation of AC entities and parameters. In this chapter, we get an opportunity to appreciate how the complexity of AC is amplified by three-phase AC consideration in more common, industrial and commercial applications. In an effort to allow readers ample opportunity to explore and learn basic, single- and three-phase AC principles and mathematical computations, we introduce various single- and three-phase AC functions, equations, and mathematical techniques, reinforced by examples and self-assessments problems. The depth to which the readers may endeavor to explore the AC topics in this chapter depends on their need, appetite, and aptitude.

Due to the practical applications of the process of electrodeposition, in the coatings industry –and the innate relationship between electrodeposition and electricity – this process will be defined and illustrated through equations and practical numerical examples. This chapter provides the reader an introduction to AC electrical transformers of various types; single-phase and three-phase. Of course, like other topics in this text, the concepts, principles, equations, and applications in this chapter are illustrated with, analogies, numerical examples and end-of-the-chapter problems. All in all, this chapter provides an adequate introduction to AC and prepares the reader for intermediate level study of AC concepts.

ALTERNATING CURRENT (AC) VERSUS DIRECT CURRENT (DC)

The contrast between AC and DC isn't just rooted in physics but goes as far back as the 1800s when, for a period of time, both vied for the residential, commercial, and industrial markets. Electricity was first discovered and harnessed, mostly, in the DC form. However, due to the physical constraints associated with transmission, distribution, and application of DC in industrial, commercial, and residential applications, AC vanquished DC in the power distribution arena. The early competition between DC and AC is often remembered through early "rivalry" between Edison, the proponent of DC, and Tesla, the pioneer behind AC. Nevertheless, today, DC holds its own in low-voltage control and digital logic applications. In fact, albeit relatively uncommon, DC is being reintroduced in the ultra-high voltage realm in certain special situations. One advantage of DC power transmission lies in the fact that DC resistance in conductors is less than AC resistance, due mainly to the reason that the AC current flow results in skin effect. The 900,000 V DC transmission line system project, in India, is a more contemporary example. There has been some interest in 900 V DC transmission line systems in the United States, as well. The ultra-high voltage, 900 V DC systems find greater appeal in long, direct, power transmission applications, primarily because DC would always need to be converted to AC before it can be applied in most common applications.

If we could attribute the acceptance of AC for residential, commercial, and industrial applications to one pivotal entity, it would be the **transformer** and concept of **voltage transformation**. In essence, when DC was first distributed and transmitted to end users, it was at a fixed voltage level because DC voltage transformation was not feasible. Due to resistance in the transmission lines – as stipulated by Ohm's law: $\mathbf{V = IR}$ – the DC power generators had to be located in the vicinity of the consumers to mitigate the undesirable voltage regulation issues. Longer transmission runs resulted in higher resistance, or "$\mathbf{R}$," which, in turn, resulted in larger voltage drop, "$\mathbf{V}$." In the mechanical fluids' realm, this would be analogous to unacceptable pressure drop – due to frictional head losses – if fluid is distributed over extended distances without some sort of booster pumps.

Introduction of AC accompanied practical means for "stepping up" and "stepping down" of voltage through the use of transformers. So, AC could be generated at, for instance, 4,000 V, stepped up to 100,000 V – through the application of transformers – for transmission purposes, and miles away, it could be stepped down to usable levels, such as 480 V, 240 V, or 120 V. That is, mostly, the way AC is generated, transmitted, and distributed to consumers today.

We learned earlier that movement of electrons, or other charged particles, constitutes electrical current. When the charged particles move in one specific direction – such that there is a net displacement in their position or when the charged particles travel a net distance – **direct current** or **DC** is said to exist. On the other hand, when the electrons or charged particles oscillate or vibrate about a point – similar to a pendulum – such motion of charged particles constitutes **alternating current** or **AC**.

If voltage and current were plotted on Cartesian coordinates, the resulting graphs would be as depicted in Figures 3.2 and 3.3, respectively. In fact, if AC voltage, DC voltage, and current were compared using an **oscilloscope**, the graph displayed on the screen would be similar to voltage and current graphs shown in Figures 3.2 and 3.3,

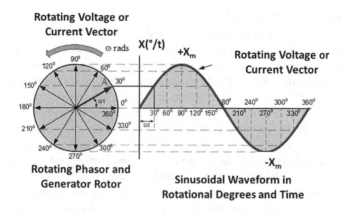

FIGURE 3.1 AC generation, a plot of amplitude versus degrees of rotation.

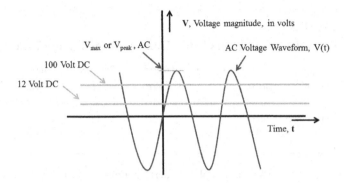

FIGURE 3.2   Voltage comparison, AC versus DC.

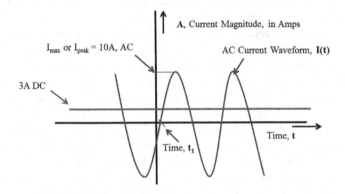

FIGURE 3.3   Current comparison, AC versus DC.

respectively. However, the inception of AC voltage or current is better depicted in Figure 3.1, where **X** represents voltage or current, X(°/t) simply denotes a voltage or current signal varying as a function of time, $+X_m$ represents the positive **maximum** or peak value of voltage or current signal, and $-X_m$ represents the negative **minimum** value of voltage or current signal.

Examination of Figures 3.2 and 3.3 shows that the Y-axis (or abscissa) represents time, **t**, in seconds. However, even though it is the parameter time "t" that is plotted along the X-axis in Figures 3.2 and 3.3, these graphs would look almost identical if time were replaced by rotation expressed in degrees, radians, or cycles, similar to the representation of X-axis in Figure 3.1. The left-hand side of Figure 3.1 represents multiple entities. The counterclockwise rotation not only represents the rotation of AC voltage and current vectors but also depicts the rotation of the generator rotor. The graphical depiction in Figure 3.1 implies that the voltage or current signal magnitude, in this generation of AC, would peak when the generator rotor coil's tangential velocity is at 90° relative to the magnetic field. Note that since Sin 90° is equal to "1," a "sine" function (sin θ) would be maximum when the angle is 90°. Also note that the reason behind the equivalence of time, degrees, radians, and cycles, along the X-axis, is that once we anchor the power or electrical utility frequency to 60 IIz,

TABLE 3.1

Equivalence between Cycles, Degrees, Radians, and Time

	Cycles	Degrees	Radians	Time (ms)
60 Hz power systems	1	360°	2π	16.67
	1/2	180°	π	8.33
50 Hz power systems	1	360°	2π	20
	1/2	180°	π	10

in the United States, as shown in Table 3.1, certain amount of time is always going to be equal to certain number of degrees, radians, or cycles.

As shown in Table 3.1, when an electrical engineer or electrician proclaims that a certain breaker, in the United States or in any other 60 Hz country, will trip in half a cycle, they are implying that that particular breaker is set to trip in 16.67 milliseconds (ms). If a similar proclamation is made by an engineer in the United Kingdom, half a cycle would be equivalent to 20 ms. Also, if voltage and current signals in a predominantly inductive or reactive system are compared graphically, the X-axis of such a graph could represent time, degrees, radians, or cycles. This means that if such a graph shows that the current waveform is lagging behind the voltage waveform by 30°, it would also imply that the two waveforms are separated by 30°/360°, or 1/12th of cycle, $(1/12) \cdot 2\pi$, or $\pi/6$ radians, or $(1/12) \cdot (16.67$ ms), or 1.39 ms. Even though a thorough investigation of power factor will be undertaken later, in Chapter 5, the power factor for this scenario would be Cos 30°, or 0.866 or 86.66%.

As we examine Figure 3.2 further, we see that the two horizontal flat lines represent DC voltage, plotted on two-dimensional X–Y coordinates. The lower flat line represents a typical 12 V automotive battery. The upper flat line, labeled 100 V DC, represents the DC equivalent of 110 V_{RMS} or 156 V_{peak} – i.e. the typical US electrical outlet voltage. The 100 V DC equivalent of 110 V_{RMS} applies to the DC power derived through full-wave rectification of AC input power. The conversion of AC to DC and associated computations are illustrated through Example 3.1 and equations introduced in the electrodeposition section.

The Y-axis (or ordinate) represents the voltage magnitude, **V**, in volts. The sine waveform, oscillating about the time axis, represents the AC voltage, varying as a function of time, **t**.

DC and AC **currents** can also be contrasted using graphical representations, in a manner similar to the DC and AC **voltage** comparison conducted above. The DC and AC **current** functions are depicted in Figure 3.3 in a form that is similar to their voltage counterparts. The horizontal flat line represents the DC current set up by virtue of the DC voltage of the DC power source; which can be a battery or a DC power supply. The Y-axis of this **current versus time** plot represents the current magnitude, **I**, in Amp. The DC current, in this case, is assumed to be a constant 3 A. The sine waveform, oscillating about the time axis, represents the AC current, I(t), varying as a function of time, t. The peak or maximum value of the AC current, in this case, is assumed to be 10 A. Note that in this illustration, the AC current appears to be surfacing into the positive current territory at time t_1. If one were to assume that this current was produced or driven by the voltage depicted in Figure 3.2, where the voltage broaches through the X-axis at time t = 0, one could say that the current is *lagging behind the voltage* because the current waveform is intersecting the horizontal time axis at a finite time to the right of the origin. As we will explore later, in greater depth, in the power factor section, such a situation where current lags behind the voltage is said to cause a **lagging power factor**.

ELECTRODEPOSITION – DC AND THE AVERAGE VALUE OF AC

Electrodeposition, as mentioned earlier, plays an important role in the coatings' operations. Simply put, it involves the use of electrical charge for the deposition of paint pigments to metallic surfaces. When electrical charge is used to transport paint pigments to a metallic surface, the pigments tend to penetrate and adhere to the metal surface more strongly, uniformly, and durably than if the surface were painted through the traditional spray or brush painting methods. Since electrodeposition process involves net displacement of relocation of the pigments, DC electrical current and voltage have to be employed. AC voltage, due to the periodic oscillation or reversal of voltage in the positive and negative realms, would **not** deliver the net displacement of the pigment particles.

In electrodeposition or electroplating scenarios, often the amount of electroplating required is specified in the form of the amount of charge that is needed to be transferred in order to achieve a certain thickness of electroplating or electrodeposition. The basic equation that is useful in determining how **long** a known amount of DC must be passed in order to achieve the required charge transfer is as follows:

$$I_{DC} = \frac{q}{t}, \text{ or, } t = \frac{q}{I_{DC}} \qquad (3.1)$$

where
q = Charge in Coulombs
I_{DC} = DC current and t = time in seconds. Note that symbols "q" and "Q" may
be used interchangeably to denote charge.

Conversion of AC voltage into DC voltage, accomplished through the full-wave rectification, can be quantified using the following equation:

$$V_{DC} = 2 \cdot \left(\frac{V_{max}}{\pi} \right) \qquad (3.2)$$

where

$$V_{max} = \sqrt{2} \cdot V_{rms} \qquad (3.3)$$

In order to derive the DC needed to compute the duration of electrodeposition, the effective resistance of the plating tank, with paint – or liquid pigment solution – is needed. This effective resistance can be measured using an ohm-meter or a multi-meter. With the effective resistance known, the DC can be calculated using Ohm's law:

$$I_{DC} = \frac{V_{DC}}{R} \tag{3.4}$$

A common computation associated with application of AC and DC in the electrode-position process is illustrated in Example 3.1.

Example 3.1

A plating tank with an effective resistance of 150 Ω is connected to the output of a full-wave rectifier. The AC supply voltage is 120 V_{rms}. Determine the amount of time it would take to perform 0.01 faradays worth of electroplating?

Solution:

Background/Theory: The amount or coating or electroplating being accomplished in this example is specified in terms of the electrical charge that must be moved or transferred; which in this case is specified in Faradays. **Faraday** is a unit for quantifying electrical charge, similar to **Coulomb.**

1 amp = 1 Coulomb/sec

or

1(Amp)·(s) = 1 C

and

96,487 Coulomb = 1 Faraday

Therefore, the amount of charge transfer, in Coulombs, in this electrodeposition case would be as follows:

$$q = \left(\frac{96{,}487 \text{ Coulombs}}{1 \text{ Faraday}} \right) \cdot (0.01 \text{ Faraday}) = 964.87 \text{ Coulombs}$$

Since we are interested in the amount of time it takes to transfer a known amount of charge, rearrangement of Eq. 3.1 results in:

$$t = \frac{q}{I}$$

The next step entails determination of the DC produced by the full-wave rectification of 120 V_{max} AC. The net effect of full-wave rectification of AC into DC is illustrated through the graphs depicted in Figure 3.4a and b. The graph in Figure 3.4a shows the AC waveform on the **input** side of a full-wave rectifier.

The graph in Figure 3.4b shows the **DC output** side of a full-wave rectifier; the net effect being the "flipping" or reversing of the negative troughs of the AC waveform such that the resulting output waveform has a definite **average value.** Note that the average of the sinusoidal AC waveform is **zero** because the AC voltage or current dwells the same amount of time in the positive realm – or above the X-axis – as it does in the negative realm or below the X-axis. In other words, on the AC side, since the total **positive** area portended by the voltage or current

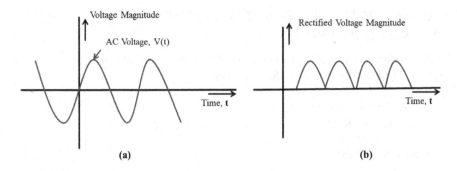

FIGURE 3.4 (a and b) Graphical depiction of AC voltage rectification.

graphs equals the corresponding **negative** areas on the **AC input** side of the rectifier, the average value is zero, and the DC content – voltage or current – on the
AC input side is **zero**.

Based on the data provided, we must first convert the given V_{rms} voltage into
V_{max} using Eq. 3.3 and then compute V_{DC} by applying Eq. 3.2. As shown below,
these two steps are followed by the application of Ohm's law, in the form of Eq.
3.4, to compute the DC, I_{DC}.

$$V_{max} = \sqrt{2} \cdot V_{rms} = (1.414) \cdot (120 \text{ V}) = 170 \text{ V} \tag{3.3}$$

$$V_{DC} = 2 \cdot \left(\frac{V_{max}}{\pi}\right) = 2 \cdot \left(\frac{170}{3.14}\right) = 108 \text{ V} \tag{3.2}$$

and

$$I_{DC} = \frac{V_{DC}}{R} = \frac{V_{Ave}}{R} = \frac{108}{150} = 0.72 \text{ A} \tag{3.4}$$

Then, application of Eq. 3.1 yields:

$$t = \frac{964.87 \text{ Coulombs}}{0.72 \text{ Coulombs/second}} = 1,340 \text{ seconds or } 22.33 \text{ minutes}$$

ALTERNATING CURRENT AND IMPEDANCE

As explained briefly in Chapter 1, impedance is the current opposing or current
impeding characteristic of a load (or conductor) in an AC circuit. As implied in the
definition of this term, impedance is an AC entity. While in DC circuits the factor
that opposes the flow of current is resistance, the entity that influences the flow of AC
is impedance. Also note that, in the practical realm, there is no such thing as "DC
Impedance." In AC systems, impedance, like AC, voltage, and power, is a vector
entity, and therefore, it can be fully defined by two key characteristics, the **magnitude** and the **direction**.

MATHEMATICAL AND GEOMETRIC REPRESENTATION OF COMPLEX AC ENTITIES

Vector, complex number-based entities, such as impedance, AC, and voltage can be expressed or quantified in the following forms:

a. Polar or phasor form
b. Rectangular form
c. Sinusoidal form
d. Exponential form

POLAR OR PHASOR FORM

The terms "polar" and "phasor" are used synonymously. Phasor or polar representation of AC entities such as impedance, current, voltage, and power requires definition of the magnitude of the respective entity and the direction in degrees (°). The polar/phasor representation of **AC current** would be represented as follows:

$$\mathbf{I} = I_{rms}\angle\theta°$$

(3.5)

where $\mathbf{I}$ is the vector or complex value of AC. The magnitude of the overall AC, in polar/phasor representations, unless otherwise specified, is the rms (root mean square) value and is denoted by, simply I or $\mathbf{I_{rms}}$. In the phasor representation of AC above, $\theta°$ is the angle of the AC.

As an example of phasor/polar representation of AC, consider a current $\mathbf{I} = 10\angle30°$ A. In this phasor representation of AC, 10 represents the rms **magnitude** of AC in Amperes, and 30° represents the **angle** of the current. This is illustrated in Figure 3.5. This tion or understanding of AC, as a vector, is analogous to the role of a **force** vector in mechanical or civil engineering realms; such as, in the study

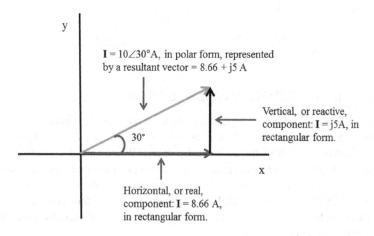

FIGURE 3.5   Vector representation of phasor value of AC.

of statics, where, the magnitude of a vector entity such as force may be defined in Newtons (N) or pounds force (lbf) and the direction of the force in degrees.

In most practical applications and computations, the rms value of current, or I_{rms}, is used instead of the maximum/peak magnitude of current I_{max}. For instance, the overcurrent protection devices, i.e. fuses, breakers – and circuit isolation apparatus like the disconnect switches – to name a few, are specified in rms terms. The same is true for AC voltages.

RECTANGULAR FORM

Rectangular representation of AC entities such as impedance, current, voltage, and power entails numerical definition of those entities in the form of horizontal and vertical, vector, components. An AC current of **10∠30° A** rms, represented in the polar or phasor form, can be translated to the corresponding rectangular form as: **8.66 + j5 A** rms, where the horizontal component of 8.66 A represents the "**real**" component of the overall AC and 5 A represents the "**imaginary**" or "**reactive**" component of the overall AC. This conversion of the AC from its phasor to rectangular form can be accomplished through a scientific calculator or by performing trigonometric calculations. The trigonometric approach would involve the computation of horizontal and vertical components of the rectangular form, as follows:

$$I_{real} = 10 \, Cos \, 30° = 10(0.866) = 8.66 \, A$$

and

$$I_{reactive} = 10 \, Sin \, 30° = 10(0.5) = 5 \, A$$

In the mechanical engineering or kinetics realm, the rectangular representation of AC would be similar to the representation of a 10∠30° mph velocity vector in the form of its horizontal (0°) and vertical (90°) components as 8.66 and 5 mph, respectively.

SINUSOIDAL OR TRIGONOMETRIC FORM

The sinusoidal representation of AC parameters or entities involves the application of trigonometry. Sinusoidal representation of an AC entity, such as AC voltage, would be as follows:

$$V = V(t) = V_m \, Sin(\omega t + \theta) \tag{3.6}$$

where "**V**" and V(t) denote sinusoidal AC voltage, V_m is the maximum or peak voltage, ω represents the angular frequency in rad/s, and θ represents the angle of the AC voltage in degrees. Figure 3.6 depicts this AC voltage function in the graphical form. Note that the "Sine" function in Eq. 3.6 can be replaced by a Cosine function by adding 90° to the angle θ. In other words, $V(t) = V_m Cos(\omega t + \theta + 90)$.

Note that the AC **I**, or **I(t)**, would be represented by a graph, very similar to the voltage **V(t)** graph, with the exception of the fact that the **I(t)** wave form would be

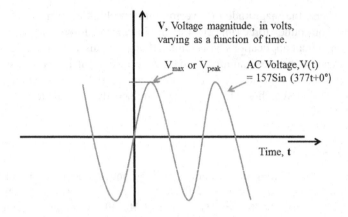

FIGURE 3.6   Graphical representation of sinusoidal voltage.

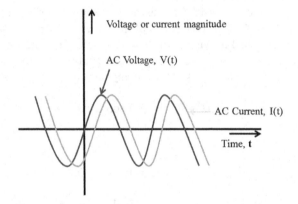

FIGURE 3.7   Graphical contrast between AC voltage and current wave forms.

shifted to the left or right of the voltage wave form, depending upon whether the reactance of the AC load is predominantly inductive or capacitive. This is illustrated in Figure 3.7 by a sinusoidal voltage and current graph of a scenario with predominantly inductive load, where the current is lagging the voltage.

Continuing with our example of the $10\angle30°$ A rms AC, we can illustrate its sinusoidal AC form by converting the phasor AC rms current value into its sinusoidal form as follows:

$$I(t) = I_{rms} \; Sin(\omega t + \theta°) \; A \; rms, \text{ in the general form,} \tag{3.7}$$

and

$$I(t) = 10 \; Sin(377t + 30°) \; A \; rms, \text{ in this specific case.}$$

Since

$$I_{max} = (\sqrt{2}) \cdot I_{rms} = (1.41) \cdot (10) = 14.1 \; A$$

In peak or maximum value form, the sinusoidal representation would be as follows:

$$I(t) = 14.1 \, \text{Sin}(377t + 30°) \, A$$

where the coefficient 14.1 represents the **maximum value** of the current, in Amp (A), and 30° represents the angle of the current. The 377 coefficient of time "t" represents, ω, the angular frequency of AC in radians per second. This value of 377 rad/s is derived from the standard US AC source frequency of 60 Hz or 60 cycles/s. The AC frequency in Great Britain and some of the former British colonies is 50 Hz. The conversion from 60 Hz to 377 rad/s can be explained mathematically as follows:

$$\omega = 2 \cdot \pi \cdot f$$
$$= 2 \cdot (3.14) \cdot (60) = 377 \, \text{rad/s}$$

(3.8)

Note: The value of ω in the United Kingdom would be 314 1/s, or 314 rad/s. The reader is encouraged to verify this.

EXPONENTIAL FORM

The exponential representation of an AC parameter – whether it is current, voltage, impedance or power – is somewhat similar in form to the polar/phasor form. This is because the exponential form consists of the maximum value and the angle. For instance, the 14.1∠30° A AC example we discussed earlier would be represented in the exponential form as follows:

$$\mathbf{I} = I_m e^{j\theta}$$
$$= 14 \cdot 1 e^{j30} \, A$$

(3.9)

IMPEDANCE ANALYSIS

The concept of impedance was introduced in Chapter 1. In this section, we will cover mathematical and graphical representation and basic computational methods pertaining to impedance. Because impedance is a vector, it can be drawn or depicted as an arrow whose length represents the magnitude of the impedance while its orientation, expressed in degrees, represents the angle of the impedance. Since a vector can be analyzed or split into its horizontal and vertical components, impedance "$\mathbf{Z}$" can be drawn in the vector diagram format as shown in Figure 3.8.

As introduced in Chapter 1, when an AC circuit consists of resistance, capacitance, and inductance, impedance $\mathbf{Z}$ can be stated mathematically in the form of Eqs. 3.10 and 3.11.

The unit for impedance is ohm, or Ω; similar to the unit for resistance, $\mathbf{R}$, capacitive reactance, $\mathbf{X_c}$, and inductive reactance, $\mathbf{X_L}$. Consistency of units between $\mathbf{R}$, $\mathbf{X_L}$, $\mathbf{X_C}$, and $\mathbf{Z}$ is requisite for the following mathematical definitions for $\mathbf{Z}$ to hold true:

$$\mathbf{Z} = \mathbf{R} + \mathbf{Z_l} + \mathbf{Z_c}$$

(3.10)

$$\mathbf{Z} = \mathbf{R} + \mathbf{jX_l} - \mathbf{jX_c}$$

(3.11)

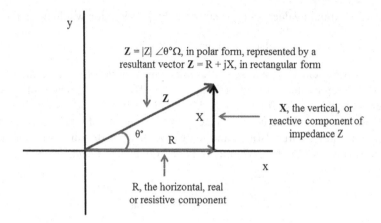

FIGURE 3.8   Graphical representation of impedance **Z** in vector form.

It's obvious from examination of the above two equations that:

$$\mathbf{Z_l} = \mathbf{jX_l} \tag{3.12}$$

$$\mathbf{Z_c} = -\mathbf{jX_c} \tag{3.13}$$

where $\mathbf{Z_l}$ is the impedance contribution by the inductance in the AC circuit and $\mathbf{Z_c}$ is the impedance contribution by the capacitance in the AC circuit. The units for all impedances are ohms, or Ω's.

ENTITY "J" IN AC

The entity "**j**" is in fact an avatar for the mathematical entity "**i**," where "**i**" mathematics denotes the term "iota" and possesses an indefinite or imaginary value of square root of −1. As ostensible, the use of "i" for square root of −1 is avoided in electrical engineering and physics to avoid confusion with use of symbol "i" for current. The values of "j," as applicable in electrical engineering, are as follows:

$$j = \sqrt{-1} = \text{unit vector } 1\angle 90°$$

$$j^2 = -1; \text{ and } 1/j = -j = 1\angle -90°$$

So, when you see a term "2j," it essentially represents a vector or complex entity with a magnitude of "2" and an angle of +90°. In other words, any quantity multiplied by "j" assumes an angle of +90°, while any quantity multiplied by "−j" assumes an angle of − 90°, or 270°. In electrical engineering, the representation of impedance by term "2j" would be construed as a purely inductive impedance of 2 Ω at an angle of +90° or an inductive reactance of 2 Ω. On the other hand, a term "−2j" essentially represents a vector or complex entity with a magnitude of "2" and an angle of −90°. In electrical engineering, the representation of impedance by term "−2j" would be construed as a purely capacitive impedance of 2 Ω at an angle of −90° or a capacitive reactance of 2 Ω. Note that reactance "X" is not equal to impedance "Z" this is affirmed by Eqs. 3.12 and 3.13.

Complete comprehension of AC circuit analysis and associated computations requires basic appreciation of how AC circuit components, such as resistances, capacitances, and inductances, are converted into their respective impedance contributions before they are consolidated into an overall equivalent or combined impedance. In most cases, the derivation of other AC circuit parameters, such as current, voltage, and power, is undertaken after determination of the total or equivalent impedance.

Example 3.2

Determine the equivalent, or total, impedance Z_{Eq} and the source rms current, I, in the AC series circuit below.

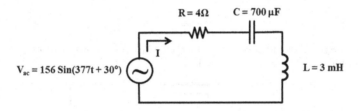

Solution:

According to Eqs. 3.10 and 3.11:

$$Z = R + Z_1 + Z_c \qquad (3.10)$$

or

$$Z = R + jX_1 - jX_c \qquad (3.11)$$

$$\therefore Z_{Eq} = R + \frac{1}{j\omega C} + j\omega \cdot L$$

$$= R + \frac{1}{j(2) \cdot (\pi) \cdot (f) \cdot (C)} + j(2) \cdot (\pi) \cdot (f) \cdot (L)$$

$$= 4 + \frac{1}{j(2) \cdot (3.14) \cdot (60) \cdot (700 \times 10^{-6})} + j \cdot (2) \cdot (3.14) \cdot (60) \cdot (3 \times 10^{-3})$$

$$= 4 + \frac{1}{j(0.2638)} + j \cdot (1.1304)$$

$$= 4 - j(3.79) + j1.1304 = 4 - j2.66 = 4.8\angle - 33.63 \ \Omega$$

Current "**I**" calculation:

$$V(t) = V_{max} \; Sin(\omega t + \theta) = 156 \; Sin(377t + 30°) \text{ - Given}$$

$$\therefore \; |V_P| = |V_{max}| = 156 \text{ V}$$

$$\text{And, } |V_{rms}| = \frac{156}{\sqrt{2}} = 110 \text{ V}$$

$$\therefore V(t), \text{ in rms form, would be} = V_{rms} \; Sin(\omega t + \theta) = 110 \; Sin(377t + 30°)$$

And, V_{rms}, in polar or phasor form would be $= 110 \; \angle 30° V_{rms}$

$$\text{Then, according to Ohm's Law, } I = \frac{V}{Z_{Eq}} = \frac{110 \; \angle 30°}{4.8 \angle - 33.63}$$

Or, $I = 23 \angle \{30° - (-33.63)\}$

$\therefore$ The source rms current, $I = 23 \angle 63.63° \text{ A}$

Example 3.3

Determine the equivalent, or total, impedance $\mathbf{Z_{Eq}}$ and the **rms** value of the source current **I** in the AC parallel circuit below.

$V_{ac} = 156 \; Sin(377t + 30°)$ $R = 4\Omega$ $L = 3 \text{ mH}$

Solution:

Solution strategy: Convert the given inductance value of L = 3 mH into its equivalent inductive reactance X_L and subsequently into its impedance contribution Z_L. Then, combine the resistance R and Z_L values into the equivalent impedance Z_{Eq} by applying the formula for combination of parallel circuit elements.

$$X_L = \omega \cdot L = (377) \cdot (3 \times 10^{-3}) = 1.13 \; \Omega$$

$$\therefore Z_L = j1.13 \; \Omega = 1.13 \angle 90° \; \Omega$$

The parallel combination formula is:

$$\frac{1}{Z_{Eq}} = \frac{1}{R} + \frac{1}{Z_L}$$

or,

$$Z_{Eq} = \frac{(R) \cdot (Z_L)}{R + Z_L} = \frac{(4) \cdot (1.13 \angle 90°)}{4 + j1.13} = \frac{(4.52 \angle 90°)}{4 + j1.13}$$

$$Z_{Eq} = \frac{4.52 \angle 90°}{4.16 \angle 15.77} = 1.09 \angle 74.22° \; \Omega$$

Current "I" calculation:

V(t), in rms form, $= 110\angle 30°$ V_{rms} as determined in the previous example.

Then, according to Ohm's Law, $I = \dfrac{V}{Z_{Eq}} = \dfrac{110\angle 30°}{1.09\angle 74.23}$

∴ The source rms current, $I = 101\angle -44°$ A

TRANSFORMERS

Earlier in this chapter, we introduced the concept of transformers and the historically pivotal role transformers played in the predominant acceptance of AC over DC. In this section, we will explore the technical principles, applications, and computations associated with single-phase transformers. We will touch on some fundamental characteristics of three-phase transformers as well.

Let's begin by exploring the construction of a simple, single-phase AC transformer based on our first introduction of magnetic flux in Chapter 1. We will build upon the toroidal core example we introduced in Chapter 1 by adding an AC voltage source to the conductor coiled onto the left side of the core, as shown in Figure 3.9, and by winding N_s number of turns of a conductor on the right side of the same toroidal core. This converts the toroidal core into a basic transformer with N_p number of turns coiled on the AC source side and N_s turns on the side that is connected to the load.

Primary and Secondary Side of a Transformer: The side or segment of the transformer that is connected to the power source is referred to as the **primary** side. The segment of the transformer that is connected to the load is called the **secondary** side. Note that the toroidal transformer depiction in Figure 3.9 is simplified for voltage transformation principle illustration purpose. In actual, practical, and commercial applications, the primary and secondary windings of a **toroidal core transformer** are concentric, with the core in the center as shown in Figure 3.10.

Even though we used a toroidal core in this basic transformer example, we could have used a more common rectangular or square core which would make this transformer resemble an ordinary transformer as shown in Figure 3.11.

As we examine Figures 3.9 and 3.11, we notice that one of the voltage source (V_{AC}) terminals is grounded. This permits the other terminal to serve as the higher

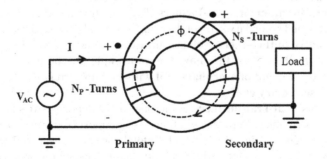

FIGURE 3.9 Induction, Flux and Voltage Transformation

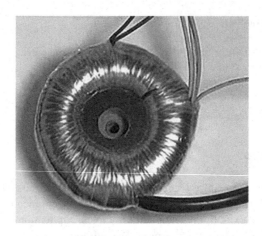

FIGURE 3.10   Toroidal core transformer by Omegatron®.

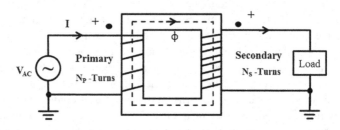

FIGURE 3.11   Ideal transformer with complete primary and secondary circuits.

potential, or "hot," terminal. Hence, the current flows from left to right or in the clockwise direction through the primary circuit of the transformer. Applying the right-hand rule, as explained in Chapter 1, would result in a clockwise magnetic flux ϕ through the core. On the secondary side, through converse application of the right-hand rule, the clockwise core flux ϕ results in the induction of current emerging from the top left side of the secondary circuit. This induced current flows in a clockwise fashion through the load as shown in Figure 3.11. Note that the direction of flow of AC, on the primary or secondary sides of the transformer, signifies the **direction of flow of AC power or AC energy** from the source to the load.

The Dot Convention for Transformers: The dots shown, in Figures 3.9 and 3.11, on the primary and secondary sides of the transformer indicate the **direction of each winding relative to the others**. Voltages measured at the dot end of each winding are in phase. In other words, **they have the same angle**. By convention, the current flowing into the **dot end of a primary coil** will result in current flowing out of the **dot end of a secondary coil**.

Ideal versus Real Transformers: A thorough and comprehensive study of "**real**" transformers is complex. Core and winding losses in most **real** transformers constitute a small percentage – less than 5% – of the total power transformed. Therefore, for simplicity, transformer circuit analyses are often conducted on the premise that they are "**ideal**" and "**lossless**." This means that the power fed into an ideal transformer,

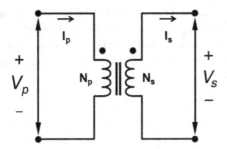

FIGURE 3.12 Ideal transformer model.

on the primary side, is equal to the power put out on the secondary side. In this text, we will limit the scope of our discussion to ideal transformers, with just a brief introduction to the concept of core and winding losses associated with real transformers.

An ideal transformer can be modeled as shown in Figure 3.12.

The two parallel vertical lines, in the middle of the ideal transformer diagram shown in Figure 3.12, represent the assumption that the core of an ideal transformer consists of a ferromagnetic or magnetic material.

Some of the formulas that govern relationships between various basic operational parameters of an ideal transformer are as follows:

$$\text{Turns Ratio} = a = \frac{N_p}{N_s} \tag{3.14}$$

$$V_s = \frac{V_p}{a} = \left(\frac{N_s}{N_p}\right) \cdot V_p \tag{3.15}$$

$$I_s = a \cdot I_p = \left(\frac{N_p}{N_s}\right) \cdot I_p \tag{3.16}$$

$$\left|S_{1\text{-}\phi}\right| = V_s \cdot I_s = V_p \cdot I_p \tag{3.17}$$

where
N_p = Number of winding turns on the primary side of the transformer
N_s = Number of winding turns on the secondary side
V_p = Voltage applied on the primary side of the transformer
V_s = Voltage on the secondary side of the transformer
I_p = Current flowing through the primary side of the transformer
I_s = Current flowing through the secondary side of the transformer
$S_{1\text{-}\phi}$ = Single-phase AC power being transmitted by the transformer

Since turns ratio **a** represents a **ratio** of the primary turns to secondary turns, it is unit-less. Voltages and currents referred to, typically, in transformer circuit analysis and specifications are in the RMS form. Also, when transformer voltages and currents are measured using voltmeters and clamp-on ammeters, respectively, the resulting measurements are in RMS form. This is the reason why many brands of voltmeters and ammeters are labeled as **True RMS Voltmeters** and **True RMS Ammeters**, respectively. In Eq. 3.17, $S_{1-\phi}$ represents the single-phase AC power or, to be more accurate, AC apparent power, **S**. Apparent power is discussed, in greater detail, in Chapter 4. Practical significance of the ideal transformer equations is illustrated through Example 3.4

Example 3.4

The primary of the transformer shown in Figure 3.11 is fed from a 120 VAC source. As apparent from the circuit diagram, the transformer has four (4) turns on the primary and eight (8) turns on the secondary. The current measured on the primary, using a clamp-on ammeter, is 2 A. Determine the following unknown parameters:

a. Turns ratio
b. Secondary current, I_s
c. Secondary voltage, V_s

Solution:

The scenario described in this example can be illustrated pictorially by annotating the circuit diagram shown in Figure 3.11 as follows:

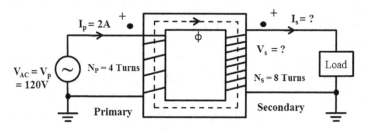

a. According to ideal transformer equation Eq. 3.14,

$$\text{Turns Ratio} = a = \frac{N_p}{N_s}$$

As given, $N_p = 4$ and $N_s = 8$.

$$\therefore \text{Turns Ratio} = \left(\frac{4}{8}\right) = \frac{1}{2} \text{ or } 1:2$$

b. According to ideal transformer equation Eq. 3.16,

$$I_s = a \cdot I_p = \left(\frac{N_p}{N_s}\right) \cdot I_p$$

Since the turns ratio was computed as ½ in part (a), and primary current I_p is given as 2 A in the problem statement,

$$I_s = \left(\frac{1}{2}\right) \cdot I_p = \left(\frac{1}{2}\right) \cdot (2 \text{ A}) = 1 \text{ A}$$

c. According to ideal transformer Eq. 3.15,

$$V_s = \left(\frac{V_p}{a}\right)$$

Since the turns ratio was computed as 1:2 or ½ in part (a) and primary voltage V_p is given as 120 V in the problem statement,

$$V_s = \left(\frac{V_p}{a}\right) = \left(\frac{120 \text{ V}}{1/2}\right) = 240 \text{ V}$$

In Example 3.4, the number of secondary turns, N_s, is twice the number of primary turns, N_p, resulting in the secondary voltage that is twice the magnitude of the primary voltage. This configuration is called a **"step-up"** transformer configuration. In other words, the transformer is being used to boost or step up the primary voltage. Conversely, as shown in Figure 3.13, if the primary windings are greater than the secondary windings, the transformer would constitute a "step-down" transformer. Such configuration is used to reduce incoming utility or source voltage to a desired lower voltage that is compatible with the driven load or equipment. See self-assessment problem 4, at the end of this chapter, for a practical illustration of a step-down transformer scenario and associated analysis.

The ideal transformer model shown in Figure 3.13 can be expanded to include impedance on the primary side of the transformer and the impedance of the load. If those two impedances are introduced into the discussion, as is the case in practical AC electrical systems, the AC transformer circuit would be represented by the circuits shown in Figure 3.14a and b.

Note that the left to right arrows denote the flow of AC power from the source to the load. The explanation of the nomenclature used in Figure 3.14a and b is as follows:

V_{ac} = AC voltage source
V_p = Voltage on the primary side of the transformer
V_s = Voltage on the secondary side of the transformer
a = Turns ratio = N_p/N_s
Z_p = Impedance between the AC voltage source and the primary coil of the transformer.

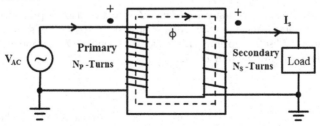

FIGURE 3.13   A step-down transformer configuration.

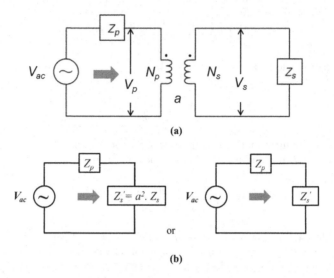

(a)

(b)

FIGURE 3.14 (a) Ideal transformer model with impedance consideration and (b) alternate representations of an ideal transformer equivalent circuit.

$\mathbf{Z_s}$ = Impedance of the load connected to the secondary side of the transformer.

$\mathbf{Z_s}$ = Impedance of the load as seen from the primary side of the transformer = $a^2 \cdot \mathbf{Z_s}$

Impedances $\mathbf{Z_p}$ and $\mathbf{Z_s}$ do not represent the impedances of the transformer itself. In other words, $\mathbf{Z_p}$ represents the combined impedance of all circuit elements connected to the primary side of the transformer, and $\mathbf{Z_s}$ represents the load impedance or the combined impedance of all circuit elements connected to the secondary side of the transformer. Simplification of the transformer circuit involved in the transition from Figure 3.14a and 3.14b can be followed through the basic steps listed below:

1. Combine the impedances of all circuit elements connected between the primary side of the transformer and the AC voltage source. This combined impedance is represented by $\mathbf{Z_p}$.
2. Combine the impedances of all circuit elements connected between the secondary side of the transformer and the load. This combined impedance is represented by $\mathbf{Z_s}$.
3. The combined secondary impedance $\mathbf{Z_s}$ is reflected to the primary side – or appears from the primary side of transformer – as $\mathbf{Z_s'}$. The symbol $\mathbf{Z_s'}$ is read as **"Z" sub "s" prime**. The relationship between $\mathbf{Z_s}$, $\mathbf{Z_s'}$, and the turns ratio **a** is represented by Eq. 3.18:

$$\mathbf{Z_s'} = \mathbf{a}^2 \cdot \mathbf{Z_s} \tag{3.18}$$

4. The original transformer circuit with numerous elements can then be shown in a simplified or condensed form – with only three elements: V_{ac}, $\mathbf{Z_p}$, and $\mathbf{Z_s'}$ – as shown in Figure 3.14b.

Example 3.5

Calculate the equivalent impedance as seen from the vantage point of the AC source V_{ac} in the circuit shown below. The transformer in the circuit is assumed to be ideal. The values of the primary and secondary circuit elements are: $X_{Lp} = 2\ \Omega$, $R_p = 4\ \Omega$, $R_s = 10\ \Omega$, $X_{Ls} = 5\ \Omega$, $N_p = 100$, and $N_s = 200$.

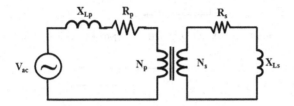

Solution:

Note that, for the sake of simplicity, the primary and secondary inductors are given in the form of their reactances "X_L" instead of actual inductance values in the form of Henries. As explained earlier in this chapter, in the process of reducing the AC circuits to an equivalent impedance form, the individual inductive and capacitive reactances are combined in the form of their respective impedance, or Z, contributions. In other words, X_L's are represented by Z_L's and X_C's are represented by Z_C's.

Employing steps (1) through (4) from the section above results in the following computation and the equivalent circuit:

On the primary side: $Z_{Lp} = jX_{Lp} = j2\ \Omega$
Therefore, $Z_P = R_p + Z_{Lp} = 4 + j2\ \Omega$
On the secondary side: $Z_{Ls} = jX_{Ls} = j5\ \Omega$
Therefore, $Z_s = R_s + Z_{Ls} = 10 + j5\ \Omega$
a = Turns ratio = N_p/N_s = 100/200 = 1/2

$$Z_s' = a^2 \cdot Z_s$$

$$Z_s' = (1/2)^2 \cdot (10 + j5\ \Omega) = 2.5 + j1.25\ \Omega$$

$$Z_{eq} = Z_p + Z_s' = (4 + j2\ \Omega) + (2.5 + j1.25\ \Omega) = 6.5 + j3.25\ \Omega$$

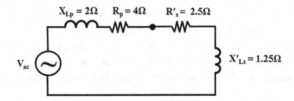

or

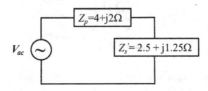

And, combination of the two impedances, as supported by the math above in:

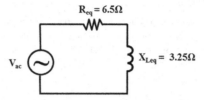

The final equivalent impedance Z_{eq} can be represented in rectangular complex form as:

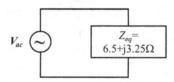

The Z_{eq} derived in the rectangular for above can be stated in polar or phasor form as:

$$Z_{eq} = 6.5 + j3.25 \ \Omega = 7.27\angle 26.6°$$

This conversion from rectangular to phasor form can be accomplished through a scientific calculator, with complex math feature or, as illustrated earlier in this chapter, through application of Pythagorean Theorem and trigonometry.

COMMON SINGLE-PHASE AC TRANSFORMER APPLICATIONS

Three common single-phase transformer configurations are described in this section accompanied by their respective schematics.

Generic step-up or step-down single-phase transformer application: Even though the single-phase transformer schematic shown in Figure 3.15 represents a greater number of turns on the primary side versus the secondary side – implying that it is a step-down transformer – this transformer could be wired or connected in a reverse fashion, resulting in a step-up voltage transformation function.

Step-down voltage transformation – residential power distribution: The single-phase transformer schematic shown in Figure 3.16 represents step-down voltage transformation typically utilized in residential power distribution applications. As shown in the schematic, typically, **7,200 V** – in phase (ϕ) to neutral three-phase Y power supply configuration – are fed to the primary of a single-phase power transformer. The secondary of the residential power distribution transformer is split or

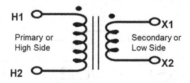

FIGURE 3.15   Generic step-down single-phase transformer configuration.

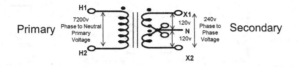

FIGURE 3.16 Step-down, residential power distribution transformer.

"tapped" into two segments, with the center tap connected to a neutral or ground. Such secondary configuration allows electricians to utilize both 120 V, ɸ – neutral, segments for multiple 120 V loads at a particular residence. On the other hand, for the 240 V loads, such as clothes dryers, water heaters, heating, air-conditioning, and ventilation (HVAC) systems, and electric ranges, the electricians connect the line side of the respective load breakers across terminals **X**1 and **X**2, phase to phase.

As shown in Figure 3.16, the 7,200 V presented to the primary of this residential power distribution transformer are stepped down to the secondary voltage of 240 V. Both terminals X1 and X2 are energized or "hot." That is why, if you examine a typical residential breaker panel, you will note that the breakers for loads such as HVAC systems, dryers, water heaters, and ranges are installed in duplex or dual formation. In other words, two breakers are connected per circuit; one on each energized phase. See the breaker interlocking bar – mechanically interconnecting the two breakers together for simultaneous operation – in Figure 3.17. This interlocking or duplexing feature permits simultaneous breaking/operation of both energized phases when the breaker trips or is turned off. The schematic segment of Figure 3.15 shows the electrical interface and location of the breakers in two-phase systems.

Load to source, power isolation transformer: In most industrial and many commercial facilities, equipment such as arc welders, intermittent motor loads, and other types of switching loads tend to feedback electric "spikes" or "noise" to the supply line or supply bus. Such undesirable noise and spikes on the supply line or feeder are subsequently seen by all loads sharing the same supply line. Such switching of electromagnetic circuits also results in generation of electromagnetic radiation. This electromagnetic radiation, sometimes referred to as EMI, or electromagnetic interference, can have an adverse effect on sensitive electronic equipment such as computers, TVs, audio/video stereo systems, VFDs or variable frequency drives, PLCs (Programmable Logic Controllers), PCs (Personal Computers), and other types of electronic equipment. The electrical noise on the utility or power supply lines and EMI often tend to cause trips, faults, and computational errors in sensitive electronic equipment.

There are various methods for attenuating or eliminating the electrical noise on the power supply lines and power distribution systems. Some such approaches include

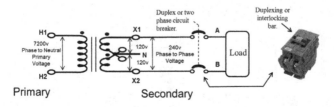

FIGURE 3.17 Step-down, residential power distribution transformer with load.

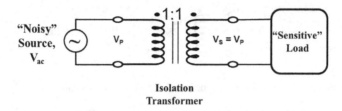

FIGURE 3.18   Load to source, 1:1, power isolation transformer.

the application of capacitors and electrical filters that serve as "sponges" or noise absorbers. Another common method for protecting against electrical noise consists of an isolation transformer. See Figure 3.18 for a schematic of a typical, single-phase isolation transformer. A distinguishing characteristic of an isolation transformer is that its primary to secondary winding turns ration is 1:1. In other words, in isolation transformers, the number of turns on the primary side equals the number of turns on the secondary side. So, the primary and secondary voltages are the same. The net gain is the "cleansing" of voltage or current achieved through conversion of electrical current to magnetic flux ϕ on the primary side and subsequent conversion of the flux to electrical AC in the load or secondary side of the transformer. This flux to current and current to flux conversion is depicted in Figure 3.13.

AUTOTRANSFORMERS

An autotransformer is a type of transformer that requires fewer windings and a smaller core. Therefore, it offers an economical and lighter means for converting voltage from one level to another as compared with its, regular, two winding counterparts. Practical, cost-effective, application of an autotransformer is, typically, limited to voltage ratio of about 3:1.

A typical autotransformer has one winding and four electrical connections, referred to as taps. In autotransformers, a portion of the winding serves as the primary as well as the secondary. See Figure 3.19 where the winding spanning from terminals **X**1 to **X**2 serves as the primary as well as the secondary.

VOLTAGE REGULATION, VOLTAGE REGULATORS, AND BUCK–BOOST TRANSFORMERS

Envision a scenario where you own and manage a small factory that operates on an 8- to 5-day shift. Assume that all power is turned off at the main breaker at the end

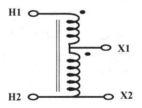

FIGURE 3.19   Autotransformer.

of the day. As the first person to arrive at the facility in the morning, in an effort to assess the power quality of your facility, you measure the voltage at the main breaker before any electrical equipment is turned on. Let's assume that the voltage you measure is 480 V_{RMS}. If you measure the voltage at that main breaker, once again, after turning on all of the routine load, as well as your production machinery, you will most likely read some lower voltage, say 460 V_{RMS}. This reduction in voltage, as load is added to the power distribution system, when quantified as a percentage, is referred to as voltage regulation. As the load on a power distribution system is increased, the current being pulled or demanded, from the source, by the load, increases. Since the ultimate mechanical source of electrical energy has definite or limited capacity, or limited and relatively constant power, P, as the demand for current **I** rises, the voltage sags proportionally. This can be explained on the basis of Eqs. 3.17 and 3.18.

For DC systems, P = V·I, or,

$$I = \frac{P}{V} \text{ or, } V = \frac{P}{I} \tag{3.19}$$

In AC systems, **S**, AC apparent power = **V·I**, or,

$$|I| = \frac{|S|}{|V|} = |V| = \frac{|S|}{|I|} \tag{3.20}$$

As more current is demanded by loads, more current flows through DC supplies and AC power transformers. This results in a voltage drop on the load side of the DC power supply or AC power transformer, as the load approaches the maximum power capacity of the system. For some voltage regulation can be best understood and applied in its mathematical form as stated in Eq. 3.21.

$$\text{Voltage Regulation in } \% = \left[\frac{V_{NL} - V_{FL}}{V_{FL}} \right] \times 100 \tag{3.21}$$

where

V_{NL} = Voltage measured at the source under no-load condition
V_{FL} = Voltage measured at the source under full-load condition

Note that Eq. 3.21 may appear in a slightly different form in some texts, as stated in Eq. 3.22.

$$\text{Voltage Regulation in } \% = \left[\frac{V_{NL} - V_{Rated}}{V_{Rated}} \right] \times 100 \tag{3.22}$$

In Eq. 3.22, V_{Rated} is considered synonymous with the full load voltage, V_{FL}. In other words, the rated voltage is assumed to be the voltage when the system is operating at full load.

As apparent from Eq. 3.21, in an ideal "utopian" scenario, voltage regulation would approach "0." In other words, in an ideal situation, the voltage magnitude at the source would stay constant as the load escalates from no load to rated or full load level. A source that offers a low or negligible regulation is sometimes referred to, by electricians or electrical engineers, as a "**stiff**" source. In a typical power distribution system, a voltage regulation of 1% would be considered as acceptable over input voltage variations of +10% to −20%. Voltage regulation tends to be larger (or worse) in inductive load systems as compared with resistive load systems. Power distribution systems with larger voltage regulation are sometimes referred to as exhibiting "**loose**" voltage regulation.

Example 3.6

Assume that the second (rated or full load) voltage measured in the voltage regulation scenario captured in the first paragraph of this section is 460 V_{RMS}. Determine the voltage regulation.

Solution:

V_{NL} = 480 V, given
V_{FL} = 460 V, given
Apply Eq. 3.9:

$$\text{Voltage Regulation in } \% = \left[\frac{V_{NL} - V_{FL}}{V_{FL}} \right] \times 100$$

$$\text{Voltage Regulation in } \% = \left[\frac{480 - 460}{460} \right] \times 100 = 4.35\%$$

Loose or unacceptable voltage regulation can be remedied through various means, three of such methods are listed below:

1. Application of buck–boost voltage regulating transformers
2. Application of voltage regulating transformers
3. Application of ferroresonant, magnetic saturation-based, transformers

We will elaborate on the more common buck–boost voltage regulating transformer in this text. A schematic diagram of a buck–boost voltage regulating transformer is shown in Figure 3.20. This type of voltage regulator is built around an

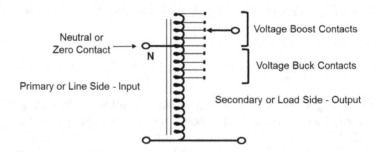

FIGURE 3.20   Buck–boost voltage regulating autotransformers.

autotransformer. It has a movable contact that makes electrical connection with available coil taps.

A buck–boost autotransformer-type voltage regulator can raise voltage (boost) or lower voltage (buck) as necessary. These regulators are, typically, rated ±10%; meaning that they are capable of maintaining the output (or secondary) voltage within ±1% in response to ±10% variation of the input (or primary) voltage.

THREE-PHASE AC

While single-phase AC is based on one energized phase (or conductor) and a ground or neutral, three-phase AC systems consist of three energized phases (or conductors). Some three-phase AC systems consist of three energized conductors and a grounded neutral. Note that electrical engineers and electricians often refer to the energized conductors as "**hot**" conductors. When measuring voltages in single-phase or three-phase systems, the energized or "hot" terminal or conductor is touched with the red (anode) probe of the voltmeter and the black (cathode) probe is connected to the neutral or ground terminal. See Figure 3.21 for a contrast between a single-phase system and a three-phase system. The single-phase circuit, shown in Figure 3.21a, is powered by a 120 V AC source. This single-phase AC source is assumed to have an angle of 0°. Therefore, complete vector or phasor representation of the source voltage would be $120 \angle 0°$ VAC. The load in this single-phase AC circuit is represented by impedance "**Z**."

On the other hand, Figure 3.21b represents a three-phase AC circuit, with a three-phase AC source supplying power to three-phase AC load. Note that the three phases

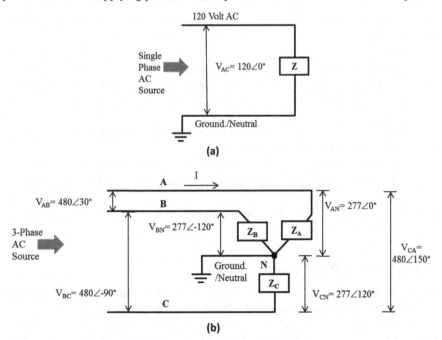

FIGURE 3.21 (a) Single-phase AC system. (b) Three-phase AC system, Y configuration.

in the three-phase AC system are identified as phases A, B, and C. The three-phase load consists of: (1) impedance Z_A, connected between phase A and the neutral, (2) impedance Z_B, connected between phase B and the neutral, and (3) impedance Z_C, connected between phase C and the neutral. In many common three-phase AC load scenarios – such as three-phase motors – the impedance values for each of the three phases are equal. When the three-phases present equal impedance, the three-phase load is said to be balanced. In other words:

$$Z_A = Z_B = Z_C = Z$$

Unequal phase impedances result in an imbalanced load condition, unequal line currents, and, in some cases, unequal line voltages. Under most normal circumstances, the neutral current I_N in Y loads is negligible. However, under unbalanced load conditions, I_N is non-zero. Therefore, neutral line current I_N is often monitored by installing a current transformer (CT) on the neutral leg of the three-phase system. See additional discussion on CTs and their application in the next section. Due to the fact that the line fuses or breakers protecting the three-phase systems are sized typically with the assumption of balanced phase loading, when loads phases become unbalanced, currents in one or more of the three-phases can exceed the fuse or breaker trip threshold, thus resulting in the clearing or "blowing" of the fuse or tripping of the breaker.

THREE-PHASE AC SYSTEMS VERSUS SINGLE-PHASE AC SYSTEMS

The advantages of three-phase AC systems over their single-phase AC counterparts are as follows:

a. Three-phase circuits and power distribution systems are more efficient than single-phase systems. In other words, three-phase power distribution systems can deliver the same magnitude of power with fewer and smaller conductors.
b. Unlike single-phase induction motors, three-phase AC induction motors do not require additional starting windings.
c. Unlike single-phase induction motors, three-phase AC induction motors provide uniform torque. Single-phase motors often tend to deliver pulsating torque.
d. When rectifying AC to DC, three-phase AC rectification yields smoother, relatively ripple-free, DC.

CURRENT TRANSFORMERS

CTs are transformers that are used to measure or sample currents in AC electrical circuits. In measurement applications, a current transformer, abbreviated as "CT," is connected to metering instrumentation to display the magnitude of AC through an AC ammeter or some other form of annunciation, such as a computer-based HMI, Human Machine Interface, system. As shown in Figure 3.22, the conductor whose

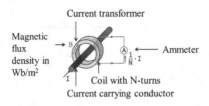

FIGURE 3.22   CT operating principle. By: Beizl, April 15, 2009.

current is to be measured is routed through the CT. When current, **I**, flows through the conductor, it produces magnetic flux – with magnetic flux density **B** – in the cylindrical core of the CT. This magnetic flux, as explained in Chapter 2, initiates the flow of current through the windings of the CT. The current $\mathbf{I_{CT}}$ induced in the windings of the CT is a proportional representation of the current **I** flowing through the conductor being monitored. The proportional relationship between **I** and $\mathbf{I_{CT}}$ can be represented, mathematically, by Eq. 3.23.

$$I_{CT} = \frac{I}{N} \qquad (3.23)$$

When CTs are used to measure phase or line currents in three-phase AC systems, they are installed in a fashion depicted in Figure 3.23. The CT installation photographed and shown in Figure 3.23 pertains to a situation where only the three energized phases need to be metered; a CT on the neutral is, therefore, not needed.

While majority of CT applications pertain to metering or display of AC, in form or another, CTs are often applied for control purposes. When CTs are included in electrical power distribution systems for control purposes, they are terminated at – or connected to – the input/output (I/O) blocks of a control system, such as, a PLC or a Direct Digital Control (DDC) system. In such control applications, the current induced in the CT windings is often scaled or transduced to a low-voltage signal at the I/O blocks. In many control systems, the PLC or DDC's CPU (Central Processing Unit) makes control decisions on the basis of the scaled signal, available at the CT I/O block by continuous monitoring and comparison of the real time current signal

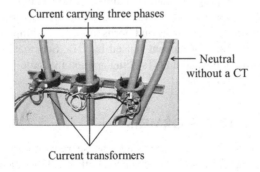

FIGURE 3.23   CTs used as part of metering equipment for three-phase 400 A electricity. By: Ali, December 13, 2004.

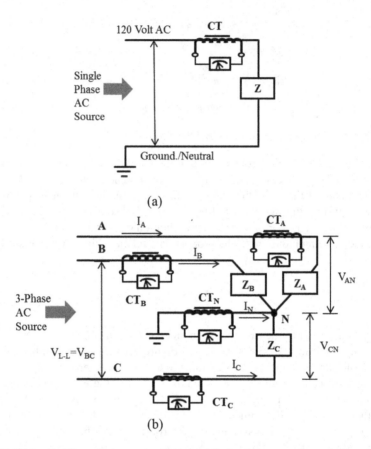

(a)

(b)

FIGURE 3.24 (a) CT applied to a single-phase AC circuit for metering purposes. (b) CTs applied to a three-phase AC circuit for metering purposes.

against "**High**" or "**Low**" set points. These "**High**" or "**Low**" set points are embedded in the program or algorithm of the PLC or DDC where control programs, typically, cycle at the rate of approximately once per 20 ms. The scan rates are a function of the complexity of the program and the number of inputs and outputs included in the architecture of the overall control system. Tripping of breakers, shutting down of equipment, and turning on of alarms and horns are a few examples of control functions executed on the basis of current sensed by CTs. Schematics in Figure 3.24a and b show the distinction between CTs being applied in single-phase AC scenarios versus three-phase AC scenarios.

THREE-PHASE AC TRANSFORMERS

Three-phase transformers, like the basic single-phase transformers, serve the purpose of transforming voltage from one magnitude to another. Like single-phase transformers, three-phase transformers can be used as isolation transformers, where the primary and secondary voltages are the same and the main purpose of the transformer is to insulate and isolate a sensitive load from electrical noise and voltage fluctuations prevalent on the AC supply line. Three-phase transformers' intrinsic and operational parameters, such as impedance, voltage, and current – similar to their single-phase counterparts – are represented by complex numbers and vectors. Upon close examination of Figure 3.21b, it is apparent that in three-phase transformers, the voltages and currents for each of the three phases are separated by 120°. The picture of a typical liquid-cooled three-phase transformer is shown in Figure 3.25a. The anatomy of a three-phase transformer is depicted in Figure 3.25b. The three-phase transformers shown in Figure 3.25 are mid-size and rated approximately 1 MW. These transformers, as many other mid- to large-size transformers, are oil cooled. The oil circulating through the heat dissipating fins in the transformer shown in Figure 3.25a, essentially, carries the heat away from the copper windings and the core to the external fins where heat radiation and dissipation occurs through convection. One of the critical maintenance aspects of such transformers is periodic infrared thermography of these transformers. Infrared thermographic images reveal occlusions in the flow of oil that might inhibit adequate heat dissipation and, subsequently, cause the transformer to overheat and fail. Figure 3.25c shows a thermographic image of a three-phase transformer. In this image, the darker areas represent colder components while the brighter, or lighter, areas represent warmer parts of the transformer. This image represents a transformer that has, more than likely, lost the cooling oil. As a result, the empty heat dissipating fins appear colder (darker) than the rest of the transformer, while the rest of the transformer, where all of the heat is accumulated, appears warm or hot.

Three identical single-phase transformers can be connected to form a three-phase bank. Primary and secondary sides of such a bank of three-phase transformers can be connected in Star (Y)–Delta (Δ), Star (Y)–Star (Y), or Delta (Δ)–Delta (Δ) combinations. Various possible three-phase transformer configurations are shown in Figures 3.26 and 3.28.

When determining the voltages and currents in a three-phase transformer, one must employ the line and phase relationship of star or delta connections along with the ratio of transformation between the coupled windings as explained in the various transformer configurations below.

THREE-PHASE Δ–Δ TRANSFORMER CONFIGURATION

Most transformers, unless otherwise specified, are delivered to customers or installation sites without final terminations or "pre-configuration"; meaning, unless otherwise specified, they are not prewired for Δ–Δ, Δ–Y, Y–Δ, or Y–Y application. One could view an "un-configured" three-phase transformer as a combination of three single-phase transformers. As such, a set of three individual single-phase transformers could be wired or configured as Δ–Δ, Δ–Y, Y–Δ, or Y–Y. If three individual transformers are connected together as depicted in Figure 3.26a, the resulting configuration would be a Δ–Δ three-phase transformer, where the points labeled X1 through X6 and H1 through H3 are terminals, wire termination points, or physical connection points. The schematic version of this Δ–Δ transformer would be represented by Figure 3.26b. The schematic, as usual, conveys the functional information

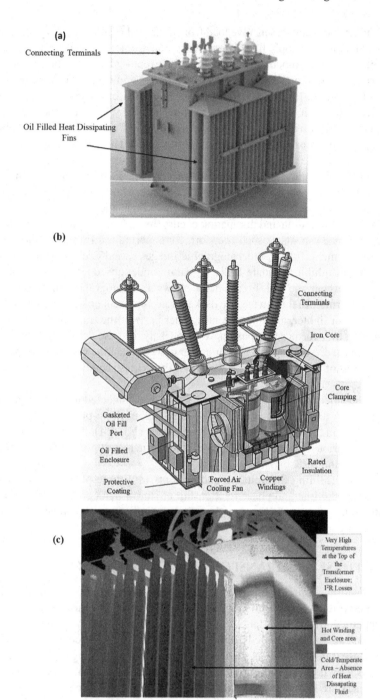

FIGURE 3.25 Three-phase transformer configuration. (a) External view, (b) anatomy of a typical three-phase transformer, and (c) infrared thermographic image of a large three-phase transformer.

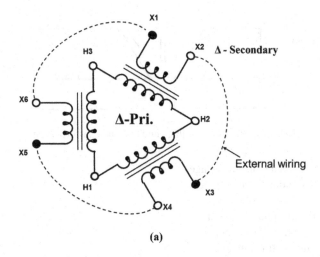

(a)

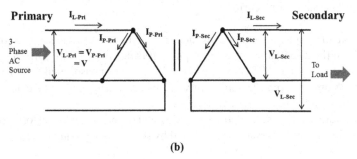

(b)

FIGURE 3.26 Δ–Δ, or Delta–Delta, three-phase transformer configuration. (a) Wiring diagram and (b) schematic.

about the transformer. In other words, the schematic illustrates how the voltage and current get transformed between the primary and secondary of the transformer.

As shown in Figure 3.26 (b), key mathematical relationships between the phase voltages and line voltages, and phase currents and line currents are as follows:

$$V_{P\text{-}Pri} = V_{L\text{-}Pri} = V \tag{3.24}$$

$$V_{P\text{-}Sec} = V_{L\text{-}Sec} \tag{3.25}$$

$$I_{P\text{-}Pri} = \frac{I_{L\text{-}Pri}}{\sqrt{3}} = \frac{I}{\sqrt{3}} \tag{3.26}$$

$$I_{L\text{-}Sec} = \sqrt{3} \cdot I_{P\text{-}Sec} = a \cdot I \tag{3.27}$$

where
$V_{P\text{-}Pri}$ = Primary phase voltage = V
$V_{L\text{-}Pri}$ = Primary line voltage = V

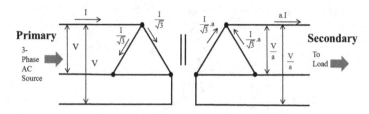

FIGURE 3.27   Δ–Δ, or Delta–Delta, voltage and current transformation.

$V_{P\text{-Sec}}$ = Secondary phase voltage
$V_{L\text{-Sec}}$ = Secondary line voltage
I = Primary line current
$I_{P\text{-Pri}}$ = Primary phase current
$I_{P\text{-Sec}}$ = Secondary phase current
$I_{L\text{-Pri}}$ = Primary line current = I
$I_{L\text{-Sec}}$ = Secondary line current

For the sake of simplicity, and in congruence with the context and scope of this text, we will limit the discussion in this three-phase transformer section to the magnitudes of currents and voltages. Of course, when full vector or phasor analyses of voltages and currents are required, the angles of voltages and currents must be considered in conjunction with the respective magnitudes.

As illustrated in Figure 3.27, the line voltage and line current transformations for Δ–Δ three-phase transformers are governed by the following equations:

$$\frac{V_{L\text{-Sec}}}{V_{L\text{-Pri}}} = \frac{1}{a}$$

or   (3.28)

$$V_{L\text{-Sec}} = \frac{V_{L\text{-Pri}}}{a} = \frac{V}{a}$$

$$\frac{I_{L\text{-Sec}}}{I_{L\text{-Pri}}} = a$$

or   (3.29)

$$I_{L\text{-Sec}} = a \cdot I_{L\text{-Pri}} = a \cdot I$$

THREE-PHASE Δ–Y TRANSFORMER CONFIGURATION

The wiring diagram for a three-phase Δ–Y transformer is depicted in Figure 3.28a. In this diagram, terminals H1 through H3 represent the primary "Δ" connection points and terminals X0 through X3 represent the secondary "Y" connection points. In the wiring diagram, in accordance with electrical convention, the solid, circular, black dot represents an internal connection that determines the Δ or Y configuration of the transformer primary or secondary. The wires connecting the solid black dot

Y - Secondary

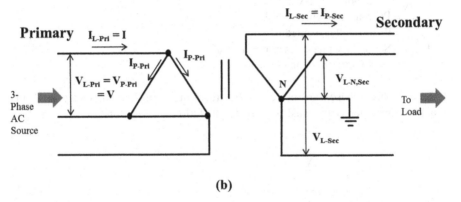

Δ - Primary

(a)

(b)

FIGURE 3.28 (a) Δ–Y, or Delta–Wye, transformer wiring diagram, and (b) Δ–Y, transformer schematic.

terminals are, sometimes, referred to as "jumpers." The hollow, circular, rings represent the terminals where external field connections are terminated. The schematic version of this Δ–Y transformer is shown in Figure 3.28b.

As illustrated in Figure 3.28b, key mathematical relationships between the phase voltages and line voltages, and phase currents and line currents are as follows:

$$V_{P\text{-}Pri} = V_{L\text{-}Pri} \tag{3.30}$$

$$V_{L\text{-}Sec} = V_{P\text{-}Sec} \cdot \sqrt{3} \tag{3.31}$$

$$I_{P\text{-}Pri} = \frac{I_{L\text{-}Pri}}{\sqrt{3}} = \frac{I}{\sqrt{3}} \tag{3.32}$$

$$I_{P\text{-}Sec} = I_{L\text{-}Sec} \tag{3.33}$$

where
  $V_{P\text{-}Pri}$ = Primary phase voltage = **V**
  $V_{L\text{-}Pri}$ = Primary line voltage = $V_{Line\text{-}Delta}$ = **V**

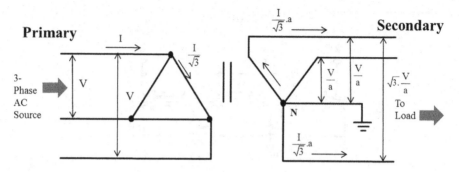

FIGURE 3.29 Δ–Y, or Delta–Star, three-phase voltage and current transformation.

$V_{P\text{-}Sec}$ = Secondary phase voltage
$V_{L\text{-}Sec}$ = Secondary line voltage = $V_{Line\text{-}Y}$
$I_{L\text{-}Pri}$ = Primary line current = I
$I_{P\text{-}Pri}$ = Primary phase current
$I_{P\text{-}Sec}$ = Secondary phase current = $I_{Line\text{-}Y}$
$I_{L\text{-}Sec}$ = Secondary line current = $I_{Line\text{-}Y}$

For a Δ–Y three-phase transformer, as illustrated in Figure 3.29, the voltage and current transformations can be assessed using the following equations:

$$\frac{V_{Line\text{-}Y}}{V_{Line\text{-}Delta}} = \frac{\sqrt{3}}{a}$$

or,   (3.34)

$$V_{Line\text{-}Y} = \frac{\sqrt{3}}{a} \cdot V_{Line\text{-}Delta} = \frac{\sqrt{3}}{a} \cdot V$$

$$\frac{I_{Line\text{-}Y}}{I_{Line\text{-}Delta}} = \frac{a}{\sqrt{3}}$$

or,   (3.35)

$$I_{Line\text{-}Y} = \frac{a}{\sqrt{3}} \cdot I_{Line\text{-}Delta} = \frac{a}{\sqrt{3}} \cdot I$$

THREE-PHASE Y–Δ TRANSFORMER CONFIGURATION

The wiring diagram of a three-phase Y-Δ transformer is shown in Figure 3.30a, where the points labeled H0 through H3 represent the "Y" primary terminals and points X1 through X3 represent the Δ secondary terminals. The schematic version of this Y–Δ transformer is represented by Figure 3.30b.

As illustrated in Figure 3.30b, in a Y–Δ three-phase transformer, the phase and line voltages, and phase and line currents can be calculated using the following equations:

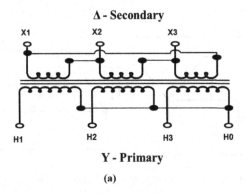

Δ - Secondary

X1   X2   X3

H1   H2   H3   H0

Y - Primary

(a)

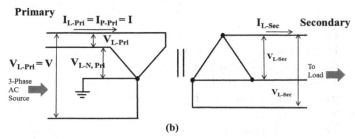

Primary

$I_{L\text{-Pri}} = I_{P\text{-Pri}} = I$

$V_{L\text{-Pri}}$

$V_{L\text{-Pri}} = V$ $V_{L\text{-N, Pri}}$

3-Phase
AC
Source

$I_{L\text{-Sec}}$ **Secondary**

$V_{L\text{-Sec}}$

To
Load

$V_{L\text{-Sec}}$

(b)

FIGURE 3.30 (a) Y–Δ, or Wye–Delta, transformer wiring diagram and (b) Y–Δ, transformer schematic.

$$V_{P\text{-Pri}} = V_{L\text{-N, Pri}} = \frac{V_{L\text{-Pri}}}{\sqrt{3}} = \frac{V}{\sqrt{3}} \qquad (3.36)$$

$$V_{L\text{-Sec}} = V_{P\text{-Sec}} \qquad (3.37)$$

$$I_{P\text{-Pri}} = I_{L\text{-Pri}} \qquad (3.38)$$

$$I_{L\text{-Sec}} = \sqrt{3} \cdot I_{P\text{-Sec}} \qquad (3.39)$$

As illustrated in Figure 3.31, key mathematical formulas for transformations between the primary and secondary voltages and primary and secondary currents are as follows:

$$V_{Line\text{-}Delta} = \frac{V_{Line\text{-}Y}}{a\sqrt{3}}$$

or,   (3.40)

$$V_{Line\text{-}Delta} = \frac{V}{a\sqrt{3}}$$

$$I_{Line\text{-}Delta} = a\sqrt{3} \cdot I_{Line\text{-}Y}$$

or,   (3.41)

$$I_{Line\text{-}Delta} = a\sqrt{3} \cdot I$$

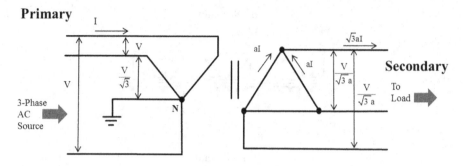

FIGURE 3.31 Y–Δ, or Star–Delta, three-phase voltage and current transformation schematic.

where

$V_{P\text{-}Pri}$ = Primary phase voltage
$V_{L\text{-}Pri}$ = Primary line voltage = V = $V_{Line\text{-}Y}$
$V_{P\text{-}Sec}$ = Secondary, Δ, phase voltage
$V_{L\text{-}Sec}$ = Secondary, Δ, line voltage
$I_{L\text{-}Pri}$ = Primary line current = I = $I_{Line\text{-}Y}$
$I_{P\text{-}Pri}$ = Primary phase current
$I_{P\text{-}Sec}$ = Secondary, Δ, phase current
$I_{L\text{-}Sec}$ = Secondary, Δ, line current

THREE-PHASE Y–Y TRANSFORMER CONFIGURATION

The wiring diagram of three-phase Y–Y transformer is shown in Figure 3.32a, where the points labeled H0 through H3 represent the "Y" primary terminals and points X0 through X3 represent the Y secondary terminals. Note that the neutrals on primary and secondary windings of Y–Y transformers may or may not be grounded. As shown in Figure 3.32a and b, the Y–Y transformer, in this specific example, has grounded neutrals on the primary **and** secondary sides. The decision to ground the neutrals, or to leave them "**floating**," is premised on specific concerns associated with noise, harmonics, and load balancing. Schematic version of this Y–Y transformer is depicted in Figure 3.32b. The term "**floating**," in the electrical domain implies isolation of a circuit or point from a verified low resistance ground. Floating electrical circuits could belong to an AC or DC system; regardless, such points would exhibit voltages above ground, i.e. ±5 VDC, ±10 VDC, ±24 VDC, 14 VAC, 24 VAC, 110 VAC, etc.

For a typical Y–Y three-phase transformer, as depicted in Figure 3.32b, mathematical relationships between the phase and line voltages and phase and line currents are as follows:

$$V_{P\text{-}Pri} = V_{L\text{-}N, Pri} = \frac{V_{L\text{-}Pri}}{\sqrt{3}} = \frac{V}{\sqrt{3}} \qquad (3.42)$$

$$V_{L\text{-}Sec} = \sqrt{3}V_{P\text{-}Sec} = \sqrt{3}V_{L\text{-}N, Sec} \qquad (3.43)$$

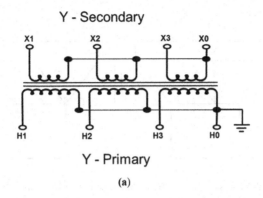

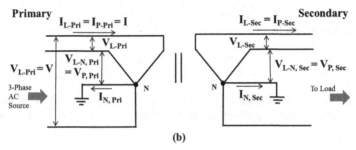

FIGURE 3.32 (a) Y–Y, or Star–Star, three-phase transformer wiring diagram and (b) Y–Y, or Star–Star, three-phase transformer schematic.

$$I_{P\text{-Pri}} = I_{L\text{-Pri}} = I \tag{3.44}$$

$$I_{L\text{-Sec}} = I_{P\text{-Sec}} \tag{3.45}$$

As illustrated in Figure 3.33, key mathematical formulas for transformations between the primary and secondary voltages and primary and secondary currents are as follows:

$$V_{L\text{-sec}} = \frac{V_{L\text{-Pri}}}{a} = \frac{V}{a} \tag{3.46}$$

$$I_{L\text{-sec}} = a \cdot I_{L\text{-Pri}} = a \cdot I \tag{3.47}$$

where
$V = V_{L\text{-Pri}}$ = Primary line voltage
$V_{L\text{-N, Pri}}$ = Primary line to neutral voltage = Primary phase voltage
$V_{L\text{-N, Sec}} = V_{P\text{-Sec}}$ = Secondary phase voltage, or, line to neutral secondary phase voltage
$V_{L\text{-Sec}}$ = Secondary line voltage
I = Primary line current = $I_{L\text{-Pri}}$
$I_{P\text{-Pri}}$ = Primary phase current

Primary   a.I   **Secondary**

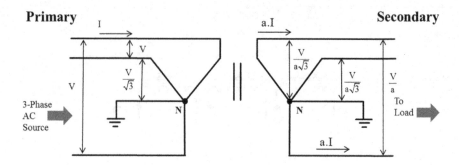

FIGURE 3.33 Y–Y, or Star–Star, voltage and current transformation.

$I_{P\text{-Sec}}$ = Secondary phase current
$I_{L\text{-Sec}}$ = Secondary line current

Example 3.7

Consider the power distribution system shown in the schematic below. Determine the following unknown parameters on the, **Y**, load side of the transformer given that the turns ratio is 2:1:

a. $|I_{L\text{-Sec}}|$ = Magnitude of load or secondary line current
b. $|I_{P\text{-Sec}}|$ = Magnitude of secondary phase current or load phase current
c. $|V_{P\text{-Pri}}|$ = Magnitude of phase voltage on the source or primary side of the transformer
d. $|V_{L\text{-Sec}}|$ = Magnitude of line voltage on the load or secondary side of the transformer
e. $|V_{P\text{-Sec}}|$ = Magnitude of phase voltage on the load or secondary side of the transformer
f. $|V_{L\text{-N, Sec}}|$ = Magnitude of line to neutral voltage on the load or secondary side of the transformer

Primary   I = 10∠30° Amps $I_{L\text{-Sec}}$ **Secondary**

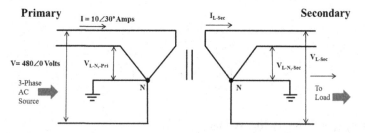

Solution:

a. Using Eq. 3.47

$$I_{P\text{-Pri}} = I_{L\text{-Pri}} = 10\angle 30° \text{ A}$$

$$|I_{L\text{-sec}}| = a \cdot |I_{L\text{-Pri}}| = a \cdot |I| = (2) \cdot |10\angle 30°| = (2) \cdot 10 \text{ A} = 20 \text{ A}$$

b. According to Eq. 3.45 and as computed in part (a):

$$I_{P\text{-}Sec} = I_{L\text{-}Sec} = 20\angle 30° \text{ A}$$

$$\therefore \ |I_{P\text{-}Sec}| = 20 \text{ A}$$

c. According to Eq. 3.42:

$$|V_{P\text{-}Pri}| = |V_{L\text{-}N,\,Pri}| = \frac{|V_{L\text{-}Pri}|}{\sqrt{3}} = \frac{|V|}{\sqrt{3}} = \frac{|480\angle 0°|}{\sqrt{3}} = 277 \text{ V}$$

d. According to Eq. 3.43:

$$|V_{L\text{-}sec}| = \frac{|V_{L\text{-}Pri}|}{a} = \frac{|V|}{a} = \frac{|480\angle 0°|}{2} = \frac{480}{2} = 240 \text{ V}$$

e. As computed in part (d), $|V_{L\text{-}Sec}| = 240$ V. Then, using Eq. 3.43:

$$V_{L\text{-}Sec} = \sqrt{3}V_{P\text{-}Sec} = \sqrt{3}V_{L\text{-}N,\,Sec}$$

And, by rearranging Eq. 3.43

$$V_{P\text{-}Sec} = V_{L\text{-}N,\,Sec} = \frac{V_{L\text{-}Sec}}{\sqrt{3}}$$

$$\therefore \ |V_{P\text{-}Sec}| = |V_{L\text{-}N,\,Sec}| = \frac{|V_{L\text{-}Sec}|}{\sqrt{3}}$$

and,

$$|V_{P\text{-}Sec}| = \frac{240}{\sqrt{3}} = 139 \text{ V}$$

f. According to the rearranged version of Eq. 3.29, in part (e):

$$V_{P\text{-}Sec} = V_{L\text{-}N,\,Sec}$$

and

$$|V_{P\text{-}Sec}| = |V_{L\text{-}N,\,Sec}| = \frac{240}{\sqrt{3}} = 139 \text{ V}$$

SELF-ASSESSMENT PROBLEMS AND QUESTIONS – CHAPTER 3

1. A plating tank with an effective resistance of 100 Ω is connected to the output of a full-wave rectifier. The AC supply voltage is 340 V_{peak}. Determine the amount of time, in hours, it would take to perform 0.075 faradays worth of electroplating.

2. Determine the source current I_{rms} in the AC circuit below.

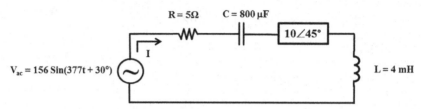

3. Calculate the impedance Z_{EQ} as seen by the AC voltage source in the circuit below:

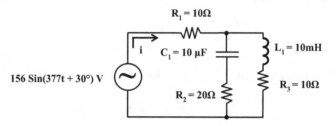

4. A single-phase 1 kVA resistive load, designed to operate at 240 VAC, has to be powered by a 480 VAC source. A transformer is applied as shown in the diagram below. Answer the following questions associated with this scenario:

   a. Would the transformer be connected in a "**step-up**" configuration or a "**step-down**" configuration?
   b. When installing the transformer, what turns ratio, **a**, should it be connected for?
   c. What would be the secondary current, I_s, when the load is operating at full capacity?
   d. What would be the primary current at full load?

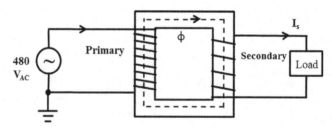

5. Calculate the equivalent impedance as seen from the vantage point of the AC source V_{ac} in the circuit shown below. The transformer in the circuit is assumed to be ideal. The values of the primary and secondary circuit elements are: $X_{lp} = 1\,\Omega$, $R_p = 4\,\Omega$, $R_s = 10\,\Omega$, $X_{Ls} = 5\,\Omega$, $X_{Cs} = 10\,\Omega$, $N_p = 100$, and $N_s = 200$.

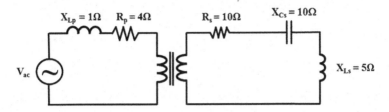

6. The no load voltage at the main switch yard of a manufacturing facility is 13,400 V_{AC}. The voltage regulation of the main switch yard is 4%. What is the rated full load voltage that is most likely to be measured on the load side of the main switch yard?

7. Consider the power distribution system shown in the schematic below. Determine the following unknown parameters on the, **Y**, load side of the transformer given that the turns ratio is 2:1:

a. $|I_{L-Sec}|$ = Magnitude of load or secondary line current
b. $|I_{P-Sec}|$ = Magnitude of secondary phase current or load phase current
c. $|V_{P-Pri}|$ = Magnitude of phase voltage on the source or primary side of the transformer
d. $|V_{L-Sec}|$ = Magnitude of line voltage on the load or secondary side of the transformer
e. $|V_{P-Sec}|$ = Magnitude of phase voltage on the load or secondary side of the transformer
f. $|V_{L-N, Sec}|$ = Magnitude of line to neutral voltage on the load or secondary side of the transformer

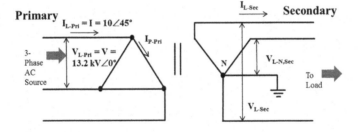

4 DC and AC Power, Efficiency, and Energy Flow

INTRODUCTION

In this chapter, we begin with an introduction to concepts of power, energy, and efficiency. Direct current (DC) power and efficiency discussion sets the stage for the more complex concepts of single-phase alternating current (AC) power and AC energy. And, discussion of single-phase AC serves as a suitable segue for consideration of the more complex subject of three-phase AC power. Significance of efficiency and energy will be explained within the context of DC and AC powers. Once the basic understanding of AC and DC powers is clear, we will be better poised to examine the flow of power and energy from the point of generation to the point of consumption. The flow of power and energy from some common sources such as steam and water to electricity is described, with illustrations in the form of example problems and self-assessment problems. The flow of power from utility to pumps is discussed in the "Wire to Water" section. Conversely, the flow of power and energy from water to wire and steam to wire is illustrated through hydroelectric and steam turbines, respectively.

POWER

Power can be defined in multiple ways. Power can be defined as the rate of performance of work, rate of generation of energy, rate of consumption of energy, and even as rate of application of torque. This is congruent with the fact that work, energy, and torque are equivalent with respect to the magnitude. All three can be quantified or measured – in the Metric or SI system – as N-m (Newton-meters) or J (Joules). In the US unit realm, work, energy, and torque can be measured in ft-lbf, in-lbf, BTU (British Thermal Unit), Hp-hr (horsepower-hour), etc. Since power is a derivative of energy with respect to time, its units in the SI or Metric realm are N-m/s (Newton-meters/second) or J/s (Joules/second). And, the units for power in the US domain are ft-lbf/s (foot-pound force/second), in-lbf/s (in-pound force/second), BTU/s (British Thermal Units/second), and hp (horsepower).

The power and energy relationships stated above gain practical significance when stated mathematically as follows:

$$E = W = |\tau| \tag{4.1}$$

where E represents energy, W represents work, and $|\tau|$ denotes torque.

Two common forms of energy in the mechanical and civil realm are as follows:

$$Potential\ Energy = mgh \tag{4.2}$$

$$Kinetic\ Energy = \frac{1}{2}mv^2 \tag{4.3}$$

$$Work = F \times d \tag{4.4}$$

$$|\tau| = F \times l \tag{4.5}$$

where
  m = Mass of an object, in kg (SI unit system)
  g = Acceleration due to gravity = 9.8 m/s² (SI unit system)
  h = Elevation of the object, in meters
  v = Velocity in m/s
  F = Force in N (Newtons)
  d = Distance, in meters, over which the object is pushed or pulled by force, F
  l = Moment arm, or perpendicular distance, in meters, from the fulcrum, at
      which the force is applied

In order to illustrate that potential energy, kinetic energy, work, and torque can ultimately be measured in joules and kWh (unit for electrical energy), let's expand Eqs. 4.2–4.5 into N-m and Joules:

$$Potential\ Energy = mgh = (kg)\left(\frac{m}{s^2}\right)(m) = N\text{-}m = J$$

$$Kinetic\ Energy = \frac{1}{2}mv^2 = \frac{1}{2}(kg)\left(\frac{m}{s}\right)^2 = \frac{1}{2}(kg)\left(\frac{m}{s^2}\right)(m)$$

$$= \frac{1}{2}N\text{-}m = \frac{1}{2}J$$

$$Work = F \times d = N\text{-}m = J$$

$$\tau = F \times l = N\text{-}m = J$$

Since the four forms of energy correspond and correlate to the basic unit of energy, Joule, the basic unit of electrical power can be derived as follows:

$$Power = \frac{Energy}{Time} = \left(\frac{J}{s}\right) = Watt;$$

and

$$1,000\ Watts = 1\ kW, or\ 1\ kilo\text{-}Watt$$

Conversely,

$$\text{Energy} = (\text{Power})\cdot(\text{Time}) = (\text{W})\cdot(\text{s}) = \left(\frac{\text{J}}{\text{s}}\right)(\text{s}) = \text{J}$$

Or,

$$1\,\text{kWh of Energy} = (1,000\ \text{W})\cdot(1\,\text{h})$$

$$= (1,000)\left(\frac{\text{J}}{\text{s}}\right)(3,600\ \text{s}) = 3,600\ \text{kJ}$$

Now, in order to demonstrate the interchangeability between electrical energy and hydrocarbon fuel (i.e. fuel oil, natural gas, propane, gasoline, etc.), let's extrapolate the conversion of 1 kWh into the Btu's and dekatherms:

Since $1,055\ \text{J} = 1\ \text{Btu}$,

$$1\,\text{kWh of Energy} = 3,600\ \text{kJ} = 3,600,000\ \text{J} = \left(\frac{3,600,000\ \text{J}}{1,055\ \text{J/Btu}}\right)$$

$$= 3,412\ \text{Btu}$$

And since $1,000,000\ \text{Btu} = 1\ \text{MMBtu} = 1\ \text{DT}$,

$$1\,\text{kWh of Energy} = 3,600\ \text{kJ} = 3,412\ \text{Btu} = 0.003412\ \text{MMBtu}$$

$$= 0.003412\ \text{DT(Dekatherm)}$$

Following the energy and power unit conversion methods illustrated above, one can perform operating energy cost comparison between alternative pieces of equipment.

Example 4.1

Which of the following two water heaters would cost the least to operate, on annual cost basis, under the given assumptions?

A. **Electric Water Heater:**
   Estimated annual energy consumption: **5,000 kWh**
   Efficiency: **100%**
   Cost Rate: **$0.10/kWh**
B. **Natural Gas Water Heater:**
   Estimated annual energy consumption: **Same as the electric water heater**
   Efficiency: **100%**
   Cost Rate: **$10.87/DT**

Solution:

Since the cost rate for electrical energy and the annual electrical energy consumption for the electric water heater are given:

Total annual cost for operating the **electric water heater**

$$= (5,000 \text{ kWh}) \cdot (\$0.10/\text{kWh}) = \$500$$

The annual energy consumption by the *gas water heater*, assumed to be the same as the electric water heater = **5,000 kWh**

Then, the annual energy consumption by the gas water heater in DT or MMBtu would be = **(5,000 kWh)·(3,412 Btu/kWh)(1 DT/1,000,000 Btu) = 17.06 DT**.

Since the natural gas cost rate is given as **\$10.87/DT**, the annual operating cost for the gas water heater would be = **(17.06 DT)·(\$10.87/DT) = \$185.44**.

Answer: The gas water heater would cost substantially less to operate than the electric water heater.

ELECTRIC MOTOR HORSEPOWER REQUIRED TO MOVE/CONVEY MASS[1]

Conveyance of material through powered conveyors, aside from the selection and design of other pieces of equipment, requires sizing and specification of conveyor motors. Two simple formulas that can be used to assess the brake horsepower required to move *"loose"* (i.e. aggregate bulk materials like sand, flour, grain, sugar, etc.) or discrete *"unit"* mass (i.e. solid objects like steel parts, fabricated mechanical components, wooden objects, etc.) are listed below:

$$P = F \cdot \Delta v \qquad (4.6)$$

$$F = \dot{m} \cdot \Delta v \qquad (4.7)$$

where

P = Power in J/s or Watts = Brake horsepower required at the motor shaft.

F = Linear force required to move the conveyor (or conveyor belt), loaded with the aggregate or unit mass to be moved

Δv = Change in the velocity of aggregate or unit mass being conveyed, in the direction of the applied force and in the direction the material is being moved.

$\dot{m}$ = Mass flow rate.

Application of this method for assessing the size of a conveyor motor is illustrated through Example 4.2.

---

[1] *Thermodynamics Made Simple for Energy Engineers*, by S. Bobby Rauf.

Example 4.2

Sand drops at the rate of 20,000 kg/min onto a conveyor belt moving with a velocity of 2.0 m/s.

   a. What force is required to keep the belt moving?
   b. What is the minimum motor size that should be specified for this application?

Solution:

a.
$$F = \dot{m} \cdot \Delta v = (20,000 \text{ kg/min}) \cdot (2.0 \text{ m/s} - 0 \text{ m/s}) \cdot (1/60 \text{ min/s})$$
$$= 666.67 \text{ kg} \cdot \text{m/s}^2 \text{ or, } F = 666.67 \text{ N}$$

b.
$$\text{Power} = P = F \cdot v = (666.67 \text{ N}) \cdot (2 \text{ m/s})$$
$$= 1,333.3 \text{ N-m/s} = 1,333.3 \text{ J/s} = 1,333.3 \text{ W}$$

Since 746 W = 1 hp, minimum motor horsepower required:

$$= 1,333.3 / 746 = 1.79 \text{ hp}$$

∴ Specify a 2 hp (standard size) motor

DC POWER

Power in DC realm is equivalent to power in the mechanical realm, for most practical purposes. DC electrical power is also referred to as the "real" power. DC power is called real because, unlike AC power, it can be transformed entirely into work or other forms of energy, i.e. heat energy, potential energy, and kinetic energy, etc. Such transformation can be assessed through power and energy relationships discussed earlier in this chapter. Equation 4.8 represents one way to correlate DC, real, power to mechanical work performed.

$$W = \text{Energy} = P \cdot t \qquad (4.8)$$

where
   W = Mechanical work performed under DC electrical power, measured in Joules (or N-m).
   P = DC electrical power, in Watts (or J/s).
   t = DC power application duration, in seconds (s).

DC power can be defined mathematically as stipulated in Eq. 4.9.

$$P = V \cdot I \qquad (4.9)$$

where

P = DC electrical power, in Watts (or J/s)
V = DC voltage measured in volts (V)
I = DC current measured in Amp (A)

We can quantify mechanical work in terms of voltage, current, and time by substituting Eq. 4.9 into Eq. 4.8:

$$W = \text{Energy} = V \cdot I \cdot t \qquad\qquad (4.10)$$

In the SI or Metric unit system, DC power or "real" power is traditionally measured in Watts, kW, MW, GW, TW (10^{12} W), where k = 1,000, M = 1,000,000, G = 1 billion, and T = 1 trillion.

Some of the more common power conversion factors that are used to convert between SI system and US system of units are listed below:

1.055 kJ/s = 1.055 kW = 1 BTU/s
1 hp = 746 W = 746 J/s = 746 N-m/s = 0.746 kW = 550 ft-lbf/s

In the SI or Metric unit system, DC energy or "real" energy is traditionally measured in Wh, kWh, MWh, GWh, and TWh (10^{12} Wh).

Some mainstream conversion factors that can be used to convert electrical energy units within the SI realm or between the SI and US realms are referenced below:

1000 kW × 1 h = 1 MWh
1 BTU = 1,055 J = 1.055 kJ
1 BTU = 778 ft-lbf
1 hp × 1 hour = 1 hp-hour

Example 4.3

An automobile is parked with parking lights left on for 1 hour. The automobile battery is rated 12 V_{DC} and the lamps are incandescent. If all of the parking lamps are drawing a combined current of 4 A, what is the total energy consumed by the parking lights, in the form of heat and light?

Solution:

Apply Eq. 4.10:

$$\text{Energy} = V \cdot I \cdot t = (12 \text{ V}) \cdot (4 \text{ A}) \cdot (1 \text{ h}) = (48 \text{ W}) \cdot (3,600 \text{ s})$$

$$= \left(48 \frac{J}{s}\right)(3,600 \text{ s}) = 172,800 \text{ J}$$

SINGLE-PHASE AC POWER

AC, single and three-phase, was introduced in Chapter 3. In Chapter 3, we also discussed the differences between DC, single-phase AC, and three-phase AC. In this chapter, in addition to the DC power discussion in the foregoing section, we will focus mostly on single-phase and three-phase power. While single-phase AC power is, primarily, power or rate of performance of work, it differs from DC power, real power, or mechanical power in the fact that it is not devoted *entirely* to performance of work or conversion into other forms of energy. In most cases, a percentage of AC power is "sequestered" in the form of electromagnetic energy and power. This percentage of AC power is not available for performance of mechanical work or conversion to other forms of energy; instead, it is dedicated to charging and discharging inductances and capacitances in the AC electrical system. Yet, the utility or the power company must generate and supply the *entire* AC power to sustain *all* customers regardless of the size of inductances and capacitances in their systems. That is where the concept of power factor gains its significance, as explained in this chapter and then in more detail in Chapter 5.

The term "single-phase AC" implies that the AC power source consists of one energized or higher electrical potential conductor or terminal. The other conductor, wire, or terminal serves as a neutral and is typically connected to building or power distribution system ground. Mechanical engineers, with thermodynamic background, could view the energized line or terminal as the "heat source" and could apply the analogy of "heat sink" to the neutral or ground. This analogy is premised on the fact that, for heat engines to perform work through steam turbines, the superheated steam must have a heat source and a heat sink to traverse between.

As introduced briefly in earlier chapters, total AC power is called apparent power and is denoted by $\mathbf{S}$. Note that the apparent power symbol $\mathbf{S}$ is bold faced. The bold font signifies the fact that apparent power is a *vector* and can be "completely" represented by a *complex number* that consists of a *magnitude* and an *angle*, where these two entities are stated in combination using sinusoidal, polar/phasor, rectangular, or exponential forms.

Fundamental mathematical definition of single-phase AC power is represented in terms of Eqs. 4.11 and 4.12. The magnitude and angle of single-phase AC apparent power can be determined through Eqs. 4.13 and 4.14, respectively. The vector values of single-phase AC voltage and current can be calculated using Eqs. 4.15 and 4.16, respectively. The magnitudes of single-phase AC voltage and current can be calculated using Eqs. 4.17 and 4.18, respectively. Values of P, Q, and |S| can also be calculated in terms of $\mathbf{V_{RMS}}$, $\mathbf{R}$, $\mathbf{X}$, and $\mathbf{Z}$ as shown in Eqs. 4.19–4.21.

$$\vec{S} = \vec{V} \cdot \vec{I}^* \tag{4.11}$$

$$\vec{S} = P + jQ \tag{4.12}$$

$$|S| = |V| \cdot |I| \tag{4.13}$$

$$\text{Angle of S} = \angle\theta_S = \text{Tan}^{-1}\left(\frac{Q}{P}\right) \tag{4.14}$$

$$\vec{V} = \frac{\vec{S}}{\vec{I}^*} \tag{4.15}$$

$$\vec{I}^* = \frac{\vec{S}}{\vec{V}} \tag{4.16}$$

$$|V| = \frac{|S|}{|I|} \tag{4.17}$$

$$|I| = \frac{|S|}{|V|} \tag{4.18}$$

$$P = \frac{V_{RMS}^{2}}{R} \tag{4.19}$$

$$Q = \frac{V_{RMS}^{2}}{X} \tag{4.20}$$

$$|S| = \frac{|V_{RMS}|^{2}}{|Z|} \tag{4.21}$$

where
$\vec{S}$ = Complex, vector, representation of single phase AC power, measured in VA, or Volt-Amperes, kVA or MVA.
$\vec{V}$ = Complex, vector, representation of single phase AC voltage. measured in volts.
$\vec{I}$ = Complex, vector, representation of single phase AC current. measured in amperes.
$|V|$ = Magnitude or absolute value of single phase AC voltage, in volts.
$|V_{RMS}|$ = Magnitude of the RMS value of single phase AC voltage.
$|I|$ = Magnitude or absolute value of single phase AC current, in amps.
$|S|$ = Magnitude or absolute value of single phase AC apparent power, measured in VA, or Volt-Amperes, kVA or MVA.
θ = Angle of apparent power S.
P = Real power, or real component of apparent power S, measured in Watts, kW, or MW.
Q = Reactive power, or imaginary component of apparent power S, measured in VARs, or, Volt-Amperes Reactive, kVAR, or MVAR. See the carbonated beverage analogy in Figure 4.1 for another perspective on the relationship between apparent power, real power, and reactive power.
j = Unit vector, with a magnitude of "1" or unity, portending an angle of 90° with respect to the x-axis. As explained in Chapter 3, unit vector j can be written in polar or phasor form as $1\angle 90°$.
X = Reactance, in Ω.
R = Resistance in Ω.
$|Z|$ = Magnitude or absolute value of AC impedance in Ω.

Total AC Power, **S**, is analogous to a "Full" glass of your favorite carbonated beverage.

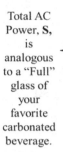

Reactive Power, **Q**, is analogous to the "foam" in your favorite carbonated beverage – it doesn't perform work!

Real Power, **P**, is analogous to the "liquid" in your favorite carbonated beverage.

FIGURE 4.1 Your favorite carbonated beverage analogy for AC power!

Since apparent power **S** is represented as a vector with horizontal component "P" and vertical component "Q" in Eq. 4.12, an alternate method for calculating the magnitude of apparent power **S** (vector) can derived on the Pythagorean Theorem as stipulated by Eq. 4.22. The angle of apparent power vector **S** could still be computed using Eq. 4.14.

$$|S| = \text{Magnitude of AC Apparent Power} = \sqrt{P^2 + Q^2} \qquad (4.22)$$

Power Factor: A concept that is inherently important in most analyses and consideration of AC power is "Power Factor." While power factor is discussed, in detail, in Chapter 5, as a prelude to that detailed discussion and to complete the discussion of single-phase AC power, mathematical definition of power factor is stated below in the form of Eqs. 4.23 and 4.24.

$$PF = \text{Power Factor} = \frac{|P|}{|S|} \qquad (4.23)$$

$$PF = \text{Power Factor} = \cos(\theta_V - \theta_I) \qquad (4.24)$$

where
  $|P|$ = Magnitude of the real component of AC apparent power
  $|S|$ = Magnitude of the oveall AC apparent power
  θ_v = Angle of AC voltage
  θ_I = Angle of AC current

Example 4.4

A 156Sin377t sinusoidal voltage is connected across a load consisting of a parallel combination of a 10 Ω resistor and a 5 Ω inductive reactance.

  a. Determine the real power dissipated by the resistor.
  b. Determine the reactive power sequestered in a 5 Ω parallel reactance.
  c. Calculate the total apparent power delivered to this parallel R and X circuit by the AC voltage source.

Solution:

The circuit diagram for this scenario would be as depicted below:

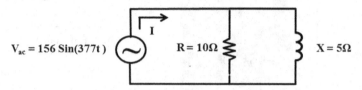

$V_{ac} = 156 \text{ Sin}(377t)$ $R = 10\Omega$ $X = 5\Omega$

  a. We can apply Eq. 4.19 to determine the power dissipated or consumed in
     the 10 Ω resistor. However, we must first derive the V_{RMS} from the given
     AC voltage of 156Sin377t. This is due to the fact that, by convention,
     coefficient 156 stated in the give AC voltage function of 156Sin400t is the
     peak or maximum voltage, V_m.
       As discussed in Chapter 3 and stipulated by Eq. 3.3:

$$V_m = \sqrt{2}V_{RMS} \text{ or } V_{RMS} = \frac{V_m}{\sqrt{2}}$$

$$\therefore V_{RMS} = \frac{156}{\sqrt{2}} = 110.3 \text{ V}$$

Then, according to Eq. 4.19:

$$P = \frac{V_{RMS}^2}{R} = \frac{(110.3)^2}{10} = 1,217 \text{ W}$$

  b. Apply Eq. 4.20 to determine the reactive power sequestered in the 5 Ω
     parallel reactance.

$$Q = \frac{V_{RMS}^2}{X} = \frac{(110.3)^2}{5} = \frac{12,166}{5} = 2,433 \text{ VAR}$$

  c. Apply Eq. 4.12 to calculate the total apparent power **S** delivered to this
     parallel R and X circuit by the AC voltage source.

$$\bar{S} = P + jQ = 1,217 + j2,433 = 2,720\angle63.4° \text{ }\Omega$$

Note: We can also apply Eq. 4.22 to verify the magnitude of the total apparent
power **S** delivered to this parallel R and X circuit:

$$|S| = \text{Magnitude of AC Apparent Power} = \sqrt{P^2 + Q^2}$$

$$= \sqrt{1,217^2 + 2,433^2} = 2,720 \text{ VA}$$

Note that jQ reactive power entity is entered into the apparent power calculation
as +jQ because of the fact that inductance in the given AC circuit results in positive impedance contribution or "+jX."

Ancillary: The reader is encouraged to verify the apparent power of 2,720 VA by applying Eq. 4.21. **Hint:** The Z in this case must be computed through parallel combination of R and Z_L as shown below:

$$|Z| = \left| \frac{R \cdot Z_L}{R + Z_L} \right| \qquad (4.25)$$

Example 4.5

The AC circuit shown below depicts a simplified, single-phase, one-line diagram of a "special-purpose" power-generating station. Assume that there is no voltage drop between the generator and the power distribution system. The line current is measured to be 1,000 A rms. Calculate the following if the power factor is known to be 0.9:

    a. Magnitude of the apparent power presented to the power distribution system.

    b. Magnitude of the real power presented to the power distribution system.

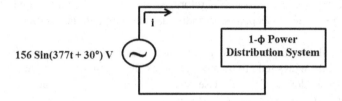

Solution:

    a. Magnitude of the apparent power presented to the power distribution system:

$$|S|_{1\text{-}\phi} = \sqrt{(P)^2 + (Q)^2} = |V \cdot I^*|$$

Since we are interested in the magnitude of the single-phase AC power:

$$|S|_{1\text{-}\phi} = |V \cdot I^*| = |V| \cdot |I|$$

where V and I are RMS values. Note that, by convention, the 156 V in the given AC voltage function is V_P or the peak value.

$$V_{rms} = \frac{V_p}{\sqrt{2}} = \frac{156}{\sqrt{2}} = 110 \text{ V}$$

$$|S|_{1\text{-}\phi} = |V \cdot I^*| = |V| \cdot |I|$$

$$= (110 \text{ V}) \cdot (1,000 \text{ A}) = 110,000 \text{ VA}$$

$$= 110 \text{ kVA}$$

b. Magnitude of the real power presented to the power distribution system can be determined by rearranging and applying 4.23 as follows:

$$PF = \text{Power Factor} = \frac{|P|}{|S|} \tag{4.26}$$

$$|P| = |S| \cdot PF = (110 \text{ kVA}) \cdot (0.9) = 99.34 \text{ kW}$$

THREE-PHASE AC POWER

Having explored the concepts, principles, analytical techniques, equations, and conventions associated with single-phase AC, in the last section, we are now adequately prepared to explore basic yet practical concepts, principles, analytical techniques, equations, and conventions associated with three-phase AC. Because conveyance and transformation of electrical energy into mechanical and other forms of energy is more effective and efficient in three-phase configuration than single-phase configuration, larger loads and large transfers of electrical power and energy are mostly conducted in three-phase form. Therefore, we find most large motors and other types of loads, with a few exceptions, in industrial and commercial facilities to be three phase.

Analytical techniques, mathematical formulas, schematics, and other types of three-phase drawings tend to be more complex than their single-phase counterparts. Due to the complexity of three-phase AC, we will limit our discussion of three-phase AC power, in this section, to the basic level.

Efficient, effective, and sustainable application of three-phase AC requires that the load on each of the three-phases be as equal as possible. When the loads on each of the three phases are equal, three-phase AC system is said to be balanced. On the other hand, when the currents and impedances on the three-phases are NOT equal, three-phase AC system is said to be unbalanced.

Basic three-phase power formulas are listed below:

$$\vec{S}_{3\text{-}\phi} = \text{Three phase apparent power} = \sqrt{3}\,\vec{V}_{\text{L-L}}\,\vec{I}_{\text{L}}^{\;*} \tag{4.27}$$

$$\left|S_{3\text{-}\phi}\right| = \sqrt{3}\left|V_{\text{L-L}}\right| \cdot \left|I_{\text{L}}^{\;*}\right| \tag{4.28}$$

$$\vec{S}_{3\text{-}\phi} = 3 \cdot \vec{S}_{1\text{-}\phi} \tag{4.29}$$

$$\left|\vec{S}_{3\text{-}\phi}\right| = 3\left|\vec{S}_{1\text{-}\phi}\right| = 3\left(\frac{V_{\text{L-L}} \cdot I_{\text{L}}}{\sqrt{3}}\right) = \sqrt{3}V_{\text{L-L}} \cdot I_{\text{L}}, \text{ for three phase Y and } \Delta \text{ circuits.} \tag{4.30}$$

where

$\vec{S}_{3\text{-}\phi}$ = Complex, vector, representation of three phase AC power, measured in VA, or Volt-Amperes, kVA or MVA.

$\left|\vec{S}_{3\text{-}\phi}\right|$ = Magnitude of three phase AC power, measured in VA, or Volt-Amperes, kVA or MVA.

$\bar{S}_{1-\phi}$ = Complex, vector, representation of one of the three phases of AC power, measured in VA, or Volt-Amperes, kVA or MVA.

$\left|\bar{S}_{1-\phi}\right|$ = Magnitude of one of the three phases of AC power, measured in VA, or Volt-Amperes, kVA or MVA.

$\bar{V}_{L-L}$ = Complex, vector, representation of line to line, or phase to phase, RMS voltage, measured in volts.

$\left|\bar{V}_{L-L}\right|$ = Magnitude of the Line to Line, or Phase to Phase, RMS voltage, measured in volts.

$\bar{I}_{L}$ = Complex, vector, representation of line RMS current, with magnitude measured in amperes, typically, using a clamp-on ammeter or a CT, Current Transformer.

$\bar{I}_{L}^{*}$ = Conjugate of vector, representation of line RMS current. Sign of the angle of the conjugate current is opposite of the sign of the original current.

$\left|\bar{I}_{L}^{*}\right| = \left|\bar{I}_{L}\right|$ = Magnitude of the RMS value of the line current.

Example 4.6

The AC circuit shown below depicts a three-phase, one-line schematic of a hydro-electric power generating station, modeled after the Three Gorges Dam, China. Assume that there is no voltage drop between the generator and the primary side of the transmission system transformer. The line current is measured to be **22,453 A rms**. Calculate the following if the power factor is known to be **0.9**:

    a. Magnitude of the apparent power presented to the transmission lines.
    b. Magnitude of the real power presented to the transmission lines.

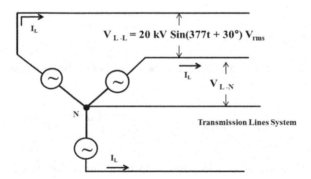

Solution:

    a. Magnitude of the apparent power presented to the transmission lines:
       Note that the AC voltage function is specified in RMS form and not in "peak" or "maximum" form. Therefore, we don't need to derive the RMS voltage. The line current is given in RMS form as well. ·
       According to Eq. 4.30:

$$\left|\bar{S}_{3-\phi}\right| = 3\left|\bar{S}_{1-\phi}\right| = 3\left(\frac{V_{L-L} \cdot I_{L}}{\sqrt{3}}\right) = \sqrt{3}V_{L-L} \cdot I_{L}, \text{ for three phase Y and } \Delta \text{ circuits.}$$

Therefore,

$$\left|\vec{S}_{3\text{-}\phi}\right| = \sqrt{3}V_{L\text{-}L} \cdot I_L = \sqrt{3}(20,000 \text{ V}) \cdot (22,450 \text{ A}) = 777,690,813 \text{ VA}$$

$$= 777,690 \text{ kVA} = 777.69 \text{ MVA}$$

b. Magnitude of the real power presented to the transmission lines can be determined by rearranging and using Eq. 4.23:

$$PF = \text{Power Factor} = \frac{|P|}{|S|}$$

or

$$|P| = PF \cdot |S| = (0.9) \cdot (777.69 \text{ MVA}) = 700 \text{ MW}$$

EFFICIENCY

Efficiency is defined, generally, as the ratio of output to input. The output and input, in general, could be in the form of power, energy, torque, or work. The concept of efficiency, when applied in the electrical engineering domain, typically, involves power or energy.

In electrical engineering, when *power* is the subject of analysis, efficiency is defined as follows:

$$\text{Efficiency in percent} = \eta = \frac{\text{Output Power}}{\text{Input Power}} \times 100 \qquad (4.31)$$

where
η (Eta) is a universal symbol for efficiency

In electrical engineering, when energy is the subject of analysis, efficiency is defined as follows:

$$\text{Efficiency in percent} = \eta = \frac{\text{Output Energy}}{\text{Input Energy}} \times 100 \qquad (4.32)$$

Although work is not used as commonly in the computation of efficiency, in the electromechanical realm, where applicable, the overall system efficiency calculation, based on work, could be stated as follows:

$$\text{Efficiency (\%)} = \eta = \frac{\text{Work Performed by the Electromechanical System}}{\text{Input Energy}} \times 100 \quad (4.33)$$

As obvious from the definitions of efficiency above, since energy cannot be created, efficiency cannot exceed "1" or 100%. The decimal result for efficiency is often converted to, and stated as, a percentage value.

In the following section, we will explore the relationship between power and efficiency in steam, mechanical, and electrical systems and develop better understanding of the flow of power in steam-type electrical power generating systems.

POWER CONVERSION FROM STEAM TO ELECTRICAL FORM – STEAM TO WIRE

The power delivered by steam to the turbine blades, call it P_{Steam}, in a simplified – no heat loss, no kinetic head loss, no potential head loss, and zero frictional head loss – scenario can be represented by the mathematical relationship stated in the form of Eq. 4.34. In the context of flow of energy from steam to electricity, functional relationship between electrical power, $P_{Electrical}$, generator efficiency $\eta_{Generator}$, steam turbine efficiency $\eta_{Turbine}$, and P_{Steam} can be expressed in the form of Eq. 4.35.

$$P_{Steam} = (h_i - h_f) \cdot \dot{m} \qquad (4.34)[2]$$

$$P_{Electrical} = (P_{Steam}) \cdot (\eta_{Turbine}) \cdot (\eta_{Generator}) \qquad (4.35)[3]$$

See the power flow diagram depicted in Figure 4.2. Even though this diagram refers to the flow of power, in conformance with Eq. 4.32, it applies just as well to flow of energy. This diagram shows the harnessing of power and energy contained in superheated steam – with high enthalpy or heat content – and their conversion to mechanical brake horsepower in the steam turbine. The brake horsepower thus imparted onto the turbine shaft is conveyed to the electrical power generator, typically, through direct coupling of the turbine and generator shafts. The generator, subsequently, converts the brake horsepower and mechanical energy into electrical power (W, kW, MW, etc.) and energy (Wh, KWh, MWh, etc.).

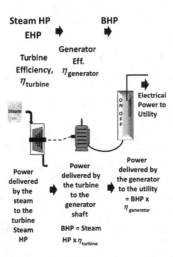

FIGURE 4.2 Steam to wire power and energy flow – steam power generation system.

---

[2] *Thermodynamics Made Simple for Energy Engineers*, by S. Bobby Rauf.
[3] *Thermodynamics Made Simple for Energy Engineers*, by S. Bobby Rauf.

However, as annotated in Figure 4.2, the transformation of power and energy from steam to electricity is de-rated first in the turbine and later in the generator by respective efficiencies.

The flow of power and energy from – steam to electricity – as depicted in Figure 4.2 is also referred to as "steam to wire" flow of power and energy. Of course, the electrical power galvanized by the generator is presented to the power distribution grid via necessary switchgear and transformers.

POWER CONVERSION FROM WATER TO ELECTRICAL FORM – WATER TO WIRE

In hydroelectric power plants, power and energy is transferred from water to electrical form via hydraulic turbines. This flow of power and energy – referred to as **"Water to Wire"** flow of power – is illustrated in Figure 4.3. The power carried by the water to the turbine blades is called **"Water Horsepower,"** or **WHP**; sometimes, also referred to as *hydraulic horsepower* or *fluid horsepower*. As explained in the next section, when energy or power flows from electrical power source to the fluid, through a hydraulic pump, the **WHP** is considered equivalent to – or referred to as – P_P, the *pump horsepower*.

The transformation of power from hydraulic form to electrical can be computed through equations similar to the ones listed below:

$$WHP = \frac{(h_A)(\gamma)(\dot{V})}{550} \tag{4.36}$$

$$WHP = \frac{(\Delta P)(\dot{V})}{550} \tag{4.37}$$

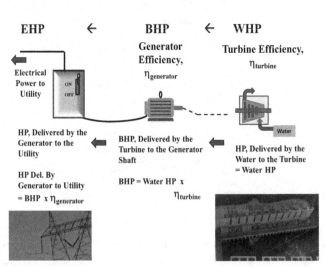

FIGURE 4.3   Water to wire power and energy flow.

where

WHP = Water horsepower, imparted the water onto the turbine, in hp

h_A = Net head added to the water, in ft, by virtue of the height of water in the reservoir

γ = Specific weight or weight density = 62.4 lbf/ft³

ΔP = Differential pressure, across the turbine, in pounds per ft²

$\dot{V}$ = Volumetric flow rate of water flowing through the turbine, measured in ft³/s.

In the context of flow of energy from water to electricity, functional relationship between electrical power, $P_{Electrical}$, generator efficiency $\eta_{Generator}$, turbine efficiency $\eta_{Turbine}$, and **WHP** can be expressed in the form of Eq. 4.38.

$$P_{Electrical} = (\mathbf{WHP}) \cdot (\eta_{Turbine}) \cdot (\eta_{Generator}) \qquad (4.38)$$

See the power flow diagram depicted in Figure 4.3. Even though this diagram refers to the flow of *power*, in conformance with Eq. 4.38, it applies just as well to flow of energy. This diagram shows the harnessing of power and energy contained in water – including potential head, kinetic head, pressure heads, and their conversion to mechanical brake horsepower in the turbine. The brake horsepower thus imparted onto the turbine shaft is conveyed to the electrical power generator, typically, through direct coupling of the turbine and generator shafts. The generator, subsequently, converts the brake horsepower and mechanical energy into electrical power (W, kW, MW, etc.) and energy (Wh, KWh, MWh, etc.).

However, as annotated in Figure 4.3 and supported by Eq. 4.38, the transformation of power and energy from water to electricity is first depreciated in the turbine and later in the generator by respective efficiency values of the turbine and generator.

The flow of power and energy from water to electricity – as depicted in Figure 4.3 – is also referred to as *"water to wire"* flow of power and energy. Electrical power generated by the generator is routed to the power grid through the necessary switchgear and transformers.

POWER CONVERSION IN HYDRAULIC PUMP SYSTEMS – WIRE TO WATER TRANSFORMATION OF POWER AND ENERGY

Hydraulic pumps, driven by electric motors, represent the exact converse of the hydroelectric power generating systems. The flow of power in electric pump applications follows the path depicted in Figure 4.4. Equations 4.36 and 4.37 are still, mathematically, pertinent. However, as annotated in the form of Eqs. 4.39 and 4.40, the WHP equations are interpreted in a context that is opposite of Eqs. 4.36 and 4.37.

WHP = Fluid horse power delivered by the pump to the water

$$= \frac{(h_A)(\gamma)(\dot{V})}{550} \qquad (4.39)$$

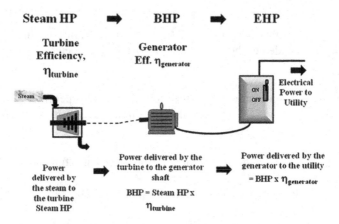

FIGURE 4.4   Wire to water power flow – hydraulic pump system.

WHP = Fluid horse power delivered by the pump to the water

$$= \frac{(\Delta P)(\dot{V})}{550} \qquad\qquad (4.40)$$

$$\mathbf{WHP} = (\mathbf{P_{Electrical}})(\eta_{Motor})(\eta_{Pump}) \qquad\qquad (4.41)$$

Discussion on Eqs. 4.39–4.41, continued:

where

WHP = P_P = Water horsepower imparted by the pump and motor system onto the water, measured in hp. Also referred to as "pump horsepower" or "fluid horsepower"

h_A = Net head added by the pump and motor system onto the water, measured in ft

γ = Specific weight or weight density = 62.4 lbf/ft^3

ΔP = Differential pressure, across the pump, measured in pounds per ft^2

$\dot{V}$ = Volumetric flow rate of water through the motor and pump system, measured in ft^3/sec.

$P_{Electrical}$ = Electrical power drawn by the motor, from the power source or utility

η_{Motor} = Motor efficiency

η_{Pump} = Pump efficiency

As evident from Figure 4.4, Eqs. 4.39–4.41 in electric motor and hydraulic pump systems, power, and energy are transferred from an electrical power source to a hydraulic system. Because in this scenario the energy stems from the power grid – transported by power transmission wires – and terminates into the fluid, it is referred to as ***wire to water*** flow of energy and power.

In electric motor and pump systems – as annotated in Figure 4.4 and supported by Eq. 4.41 – during the transformation of power and energy from electricity to water, energy is first depreciated in the motor and later in the pump by respective efficiency values of the motor and the pump.

Some of the efficiency, energy flow, and power flow principles and methods described above are illustrated through Examples 4.7 and 4.8 and self-assessment problems at the end of the chapter.

Example 4.7

As an energy engineer, you are charged with the task to estimate the heat content or enthalpy, h_i, of the superheated steam that must be fed to a steam turbine in order to supply **10 MW** (MegaWatt) of electrical power to the electrical grid. Assume that there is no heat loss in the turbine system and that difference between the enthalpies on the entrance and exit ends of the turbine is converted completely into work, minus the inefficiency of the turbine. All of the data available and pertinent to this project is listed below:

- Electrical power generator efficiency: **90%**
- Steam turbine efficiency: **70%**
- Mass flow rate for steam, ṁ: **25 kg/s (55 lbm/s)**
- Estimated exit enthalpy, h_f, of the steam: **2,875 kJ/kg (1,239 BTU/lbm)**

Solution:

Strategy: In order to determine the estimated enthalpy, h_i, of the incoming steam, we need to start with the stated output **(10 MW)** of the generator and work our way upstream to derive the energy delivered to the vanes of the turbine. The assumption that there is no heat loss in the turbine system and that the difference between the enthalpies on the entrance and exit ends of the turbine is converted completely into work, minus the inefficiency of the turbine, implies that the energy delivered by the steam is equal to the net energy delivered to the turbine vanes. Also, note that net energy delivered to the turbine vanes is reduced or de-rated according to the given efficiency of the turbine.

SOLUTION IN SI/METRIC UNITS

Since, 1J/s = 1W and 1 kJ/s = 1kW,

$$\text{Power output of the generator} = 10 \text{ MW} = 10,000 \text{ kW}$$

$$= 10,000 \text{ kJ/s}$$

Brake horsepower delivered by the turbine to the generator, through the turbine shaft, is determined as follows:

$$BHP = \frac{\textbf{Generator Output}}{\textbf{Generator Efficiency}}$$

$$= \frac{10,000 \text{ kJ/s}}{0.9} = 1.11 \times 10^4 \text{ kJ/s or } 11,111 \text{ kJ/s}$$

Power delivered by the steam to the turbine vanes is determined as follows:

$$P_{steam} = \frac{BHP}{Turbine\ Efficiency}$$

$$= \frac{1.11 \times 10^4\ kJ/s}{0.7} = 1.5873 \times 10^4\ kJ/s\ or\ 15,873\ kJ/s$$

Of course, we could obtain the same result, in one step, by rearranging and applying Eq. 4.35 as follows:

$$P_{steam} = \frac{P_{Electrical}}{(\eta_{Turbine}) \cdot (\eta_{Generator})} = \frac{10,000\ kJ/s}{(0.7) \cdot (0.9)} = 15,873\ kJ/s$$

Since the difference in the turbine entrance and exit enthalpies, in this scenario, is equal to the energy delivered to the turbine vanes:

$$P_{Steam} = (h_i - h_f) \cdot \dot{m} \tag{4.34}$$

$$P_{steam} = 15,873\ kJ/s = (h_i - 2,875\ kJ/kg) \cdot 25\ kg/s$$

$$h_i = \frac{15,873\ kJ/s}{25\ kg/s} + 2,875\ kJ/kg$$

$$h_i = 3,509\ kJ/kg$$

SOLUTION IN US/IMPERIAL UNITS

Power output of the generator = 10 MW = 10,000 kW = 10,000 kJ/s
Since 1.055 kJ = 1.0 BTU,

$$Power\ output\ of\ the\ generator = (10,000\ kJ/s) \cdot (1 / 1.055\ kJ/BTU)$$

$$= 9,479\ BTU/s$$

Brake horsepower delivered by the turbine to the generator:

$$BHP = \frac{Generator\ Output}{Generator\ Efficiency} = \frac{9,479\ BTU/s}{0.9} = 10,532\ BTU/s$$

Power delivered by the steam to the turbine vanes is determined as follows:

$$P_{steam} = \frac{BHP}{Turbine\ Efficiency} = \frac{10,532\ BTU/s}{0.7} = 15,046\ BTU/s$$

Since the difference in the turbine entrance and exit enthalpies, once again, is equal to the energy delivered to the turbine vanes:

$$P_{Steam} = (h_i - h_f) \cdot \dot{m} \tag{4.34}$$

$$15,046\ BTU/s = (h_i - 1,239\ BTU/lbm) \cdot (55\ lbm/s)$$

$$h_i = \frac{15,046\ BTU/s}{55\ lbm/s} + 1,239\ BTU/lbm$$

$$= 1,512\ BTU/lbm$$

Verification: The accuracy of the initial enthalpy, h_i, value of **1,512 BTU/lbm** can be verified by simple unit conversion to the metric form as follows:

Since 1.055 kJ = 1.0 Btu, or, there are 0.9479 Btu/kJ

$$\frac{1,512 \text{ Btu/lbm}}{0.9479 \text{ Btu/kJ}} \times 2.2046 \text{ lb/kg} \Rightarrow 3,517 \text{ kJ/kg}$$

$$\approx 3,509 \text{ kJ/kg (as computed in SI units at the outset)}$$

Example 4.8

Pressures on the intake and exit ends of a turbine are measured to be 50 and 10 psia, respectively. The volumetric flow rate for the fluid (water) is 120 ft³/s. See the diagram below. The turbine is driving an electric generator with nameplate efficiency of 90%. Calculate the following:

a. Water horsepower delivered by the water to the turbine.
b. Brake horsepower delivered by the turbine to the shaft driving the generator. The efficiency of the turbine is 65%.
c. Maximum power generated, in KWs, by this hydroelectric power generating system.

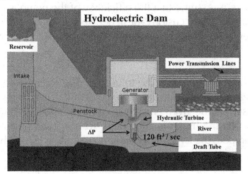

Solution:

Given:

P_{in} = 50 psia
P_{out} = 10 psia
ΔP = 40 psia
ΔP Conv. to psf = 40 lb/in² × 144 in²/ft² = 5,760 psf
$\dot{V}$ = Volumetric Flow Rate = 120 cu-ft/s
Turbine efficiency = 65%
Generator efficiency = 90%

a. According to Eq. 4.32, WHP delivered by the water to the turbine can be stated as follows:

$$WHP = P_p = \frac{(\Delta P) \cdot (\dot{V})}{550}, \text{ provided the differential pressure,}$$

ΔP, is expressed in psf, or pounds per square foot

$$\therefore P_p = \frac{(5,760 \text{ lbf/ft}^2) \cdot (120 \text{ ft}^3/s)}{550} = 1,257 \text{ hp}$$

b. Brake horsepower delivered by the turbine to the generator – via the shaft – at a turbine efficiency, η_t, of 65%:

$$BHP = WHP \times \eta_t$$

$$= 1,257 \text{ hp} \times 0.65$$

$$= 817 \text{ hp}$$

c. Maximum power generated, in kWs, by this hydroelectric power generating system, with the generator efficiency, η_g, of 90%:

Electrical Power Generated(**EHP**), in kW $= BHP \times \eta_g \times 0.746 \text{ kW/hp}$ [4]

$$= 817 \text{ hp} \times 0.9 \times 0.746 \text{ kW/hp}$$

$$= 549 \text{ kW}$$

Example 4.9

A pump is to be installed to supply a maximum of 200 ft³/s at a gage pressure (ΔP) of 8,000 psf. Determine the size of the motor required and the size (kVA) of the transformer for this system. Assume that the efficiency of the pump is 85% and that the nameplate efficiency of the "custom"-designed special motor is 92%. Power factor of the transformer is 0.8. The frictional and other hydraulic losses are negligible.

Solution:

Solution strategy in this case would be to use Eq. 4.32 to compute the WHP. Then, the amounts of real power "**P**" drawn by the motor and delivered by the motor would be computed based on the given efficiencies of the pump and the motor. Once the real power drawn by the motor is computed, we can use the power factor of the transformer to compute the kVA, or apparent power rating "**S**," of the transformer.

Given:

$\Delta P = 8,000$ psf
$V = $ Volumetric flow rate $= 200$ ft³/s
Pump efficiency $= 85\%$
Motor efficiency $= 92\%$
Power factor of the transformer $= 0.8$

According to Eq. 4.32, WHP delivered by the pump to the water can be stated as follows:

$$WHP = P_p = \frac{(\Delta P) \cdot (\dot{V})}{550}, \text{ provided the differential pressure,}$$

ΔP, is expressed in psf, or pounds per square foot

$$\therefore P_p = \frac{(8,000 \text{ lbf/ft}^2) \cdot (200 \text{ ft}^3/s)}{550} = 2,909 \text{ hp}$$

---

[4] Electric Horse Power, abbreviated as EHP, is loosely used to refer to generated electrical power measured in Horse Power, hp or kW.

According to the wire to water (hydraulic pump) power flow diagram in Figure 4.4:

$$\text{BHP delivered by the motor to the pump must be} = \frac{\text{WHP}}{\eta_p} = \frac{2,909}{0.85}\ \text{hp}$$

$$= 3,422\ \text{hp}\{\text{Custom, non-standard, size}\}$$

∴ the size of this custom motor would be 3,422 hp.

Once again, according to the wire to water power flow diagram in Figure 4.4:

$$\text{EHP drawn by the motor, from the transformer} = \frac{\text{BHP}}{\eta_{\text{motor}}} = \frac{3,422}{0.92}\ \text{hp} = 3,720\ \text{hp}$$

or,

EHP, in kW = (3,720 hp) · (0.746 kW/hp) = 2,775 kW

Then, according to Eq. 4.23, the apparent power, **S** (kVA) rating of the feeder transformer would be as follows:

$$\text{PF} = \text{Power Factor} = \frac{|P|}{|S|}$$

or,

$$|S| = \frac{|P|}{\text{PF}} = \frac{2,775\ \text{kW}}{0.8} = 3,469\ \text{kVA}$$

SELF-ASSESSMENT PROBLEMS AND QUESTIONS – CHAPTER 4

1. Consider a hydroelectric reservoir where water is flowing through the turbine at the rate of 1,100 ft³/s. See the diagram below. The turbine exit point is 700 ft lower than the elevation of the reservoir water surface. The turbine efficiency is 90% and the total frictional head loss through the penstock shaft and turbine system is 52 ft.

   a. Calculate the power output of the turbine in MWs.
   b. If the efficiency of the electric power generator is 92%, what would the electric power output be for this hydroelectric power generating system?

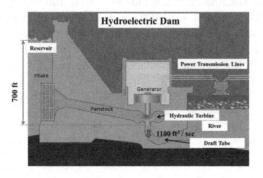

2. Which of the following two water heaters would cost the least to operate, on annual cost basis, under the given assumptions?
  A. Electric water heater:
     Estimated annual energy required to heat the water: **9,000 kWh**
     Efficiency: **95%**
     Cost rate: **$0.10/kWh**
  B. Natural gas water heater:
     Estimated annual energy required to heat the water: Same as the electric water heater
     Efficiency: **98%**
     Cost rate: **$10.87/DT**

3. A computer manufacturing company is testing a prototype for the amount of heat it dissipates as wasted energy over a 10-hour period of operation. The computer is powered by a 24 V DC power supply and is designed to draw 3 A of current. Determine the total energy dissipated in Btu.

4. In response to a significant near-miss incident and midair fire on a new commercial jet aircraft, a governmental agency is performing forensic analysis on the type of lithium–ion aircraft battery suspected to be the root cause. Estimate the amount of *current* involved in the suspected fault on the basis of the following forensic data:
  • Total energy released in the catastrophic failure of the battery: **866 kJ**
  • Estimated duration of fault: **2 seconds**
  • Rated voltage of the battery: **3.7 V_{DC}**

5. A **156Sin377t** sinusoidal voltage is connected across a load consisting of a parallel combination of a **20 Ω** resistor and a 10 Ω capacitive reactance.
  a. Determine the real power dissipated by the resistor.
  b. Determine the reactive power stored in a **10 Ω** parallel capacitive reactance.
  c. Calculate the total apparent power delivered to this parallel R and X circuit by the AC voltage source.

6. A **156Sin400t** sinusoidal voltage is connected across an unknown resistive load. If the power dissipated in the resistor is 1,000 W, what is the resistance of the resistive load?

7. The AC circuit shown below depicts a three-phase, one-line, schematic of a hydroelectric power generating station. Assume that there is no voltage drop between the generator and the primary side of the transmission system transformer. The line current is indicated by an EMS system to be **10 kA**, RMS. Calculate the following if the power factor is known to be **0.95**:
  a. Magnitude of the apparent power presented to the transmission lines.
  b. Magnitude of the real power presented to the transmission lines.
  c. The RMS line to neutral voltage at the source.

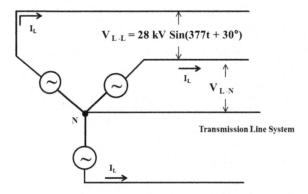

8. A pump is to be installed on the ground floor of a commercial building to supply 200 ft³/s of water up to an elevation of 100 ft. Determine the minimum size of the motor for this application. Assume that the efficiency of the pump is 80%. The weight density of water $\gamma = 62.4$ lbf/ft³.

5 Power Factor and Its Significance in Practical Applications

INTRODUCTION

In this chapter, we will explore the physical aspects of power factor in greater detail, and we will illustrate the difference between lagging power factor and leading power factor. We will explain and demonstrate potential economic benefits that can be derived from improvement of power factor in AC systems. We will explain the consequence of low power factor on the energy productivity,[1] energy cost, and the life of electrical equipment. We will conclude the power factor discussion with the introduction of conventional alternatives for power factor improvement.

Power factor stems from the fact that AC lags behind the AC voltage in AC circuits or systems that are predominantly inductive. Conversely, AC leads the AC voltage in AC circuits or systems that are predominantly capacitive. If an AC circuit is purely resistive, or if the inductive reactance is completely offset by the capacitive reactance in the circuit, the voltage and current would be completely coincident, resulting in the perfect "utopian" power factor as shown in Figure 5.1.

In AC electrical systems where inductive reactances X_L and X_C are not equal, AC and AC voltage will not be coincident. The current will lag behind the voltage

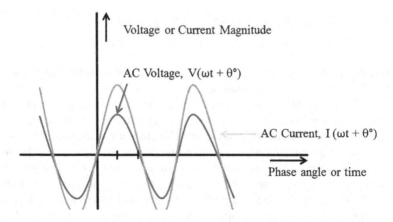

FIGURE 5.1 Coincident AC voltage and current and perfect power factor.

---

[1] *Finance and Accounting for Energy Engineers*, by S. Bobby Rauf.

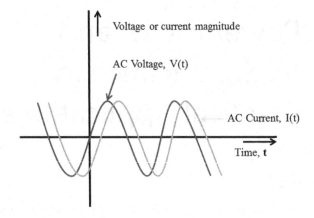

FIGURE 5.2   Lagging power factor graph as a function of time.

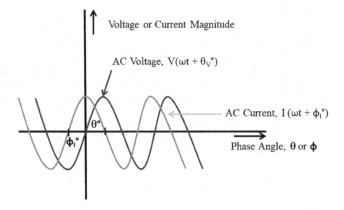

FIGURE 5.3   Lagging power factor graph as a function of phase angle.

with respect to time and phase angle, resulting in a power factor that is less than "1" or 100% when X_L is greater than X_C. See Figures 5.2 and 5.3. The graph shown in Figure 5.2 shows AC voltage and current plotted as a function of time, while Figure 5.3 depicts AC voltage and current plotted as a function of phase angle. Upon closer examination, one can discern the fact that, in both of these graphs, the current function (or graph) **follows** the voltage function in the rise and fall below the horizontal (abscissa) axis. This would constitute a lagging power factor scenario, explained in greater detail later in this chapter. At this point, note that the "**leading**" and "**lagging**" qualifier associated with the term power factor is always assessed from current's vantage point. In other words, it is always the AC that is lagging behind, leading or in synch with the AC voltage.

One way to remember that the current lags behind voltage in predominantly inductive circuits and that the current leads the voltage in predominantly capacitive circuits is through a well-known mnemonic.

ELI the ICE Man

- The L in **ELI** means **inductance**. The **E** (voltage) comes first, and then the **I** (current) lags behind. Inductive reactance produces a lagging power factor.
- The C in **ICE** means **capacitance**. The **I** (current) comes first (leads), and then the **E** (voltage) comes later. Capacitive reactance produces a leading power factor.

POWER FACTOR

Power factor of an AC system may be defined "qualitatively" as the capacity of an AC system to covert transmitted or delivered apparent power to actual work or other forms of energy. The mathematical or "quantitative" definition of power factor – as introduced in Chapter 4 – is represented by Eqs. 5.1 and 5.2.

$$PF = \text{Power Factor} = \frac{|P|}{|S|} \tag{5.1}$$

$$PF = \text{Power Factor} = \text{Cos}(\theta_V - \theta_I) \tag{5.2}$$

where
$|P|$ = Magnitude of the real component of AC apparent power
$|S|$ = Magnitude of the oveall AC apparent power
θ_v = Angle of AC voltage
θ_I = Angle of AC, also denoted as ϕ_I

LAGGING POWER FACTOR

Lagging power factor is substantially more common than a leading power factor on the "consumer side of the fence." Since loads in most industrial, commercial, and institutional facilities consist of inductive equipment such as motors, transformers, and solenoids, the inductive reactance, X_L, in such facilities exceeds any capacitive reactance, X_C, that might be present. When AC voltage is applied to such predominantly inductive systems, the resulting AC lags behind the voltage due to the reluctance in electromagnetic systems. Understanding of this phenomenon can also be facilitated by recalling one of the definitions of inductance. Inductance – which is an innate characteristic of coils, transformers, and the like – **resists the change** in the flow of current. Hence, the current lags behind the voltage in predominantly inductive electrical systems. If this lag were measured in terms of phase angle, the voltage angle θ_V would be greater than the current angle θ_I, the angular difference would be positive, and would be quantified as $(\theta_V - \theta_I)$. The cosine of this angular difference would represent the lagging power factor of the AC system and would be a stated in the form of Eq. 5.2. Alternatively, power factor or the "efficiency" of conversion of AC apparent power, **S**, to real power **P** (rate of performance of work, or delivery of torque and energy) could be represented by the ratio of the magnitudes of the real power **P** and apparent power **S** as stipulated by Eq. 5.1.

Note: Since cosine of any angle that lies in the first or fourth quadrants of the Cartesian coordinates is positive, power factor would be a positive number – declining with increasing angular difference, positive or negative – whether or not the current lags behind the voltage.

Example 5.1

In an AC system, the voltage angle $\theta_V = 65°$ and the current angle $\phi_I = 20°$.

    a. Calculate and define the power factor.
    b. Calculate and define the power factor if the voltage and current angles
       are reversed.

Solution:

    a. $\text{Cos}(65° - 20°) = \text{Cos}(45°) = 0.707$; or 70.7% lagging
    b. $\text{Cos}(20° - 65°) = \text{Cos}(-45°) = \text{Cos}(+45°) = 0.707$; or 70.7% leading

Lagging Power Factor – Impedance Perspective

A basic AC circuit with a lagging power factor, as depicted in Figure 5.4, would consist of resistance and inductance, represented by $\mathbf{R}$ and $\mathbf{X_L}$, respectively. As shown in Figure 5.4, the AC voltage source, $\mathbf{V_{ac}}$, drives AC $\mathbf{I}$ through the total resistance $\mathbf{R}$ and the total inductance $\mathbf{L}$ in the AC system. Customarily, $\mathbf{R}$ would represent not only the resistive loads in the overall system but also the inherent resistance of the power distribution system, i.e. resistance of the conductors and other current-carrying devices. The total inductance in the system, $\mathbf{L}$, could be an aggregate of the inductances of all inductors in the system, i.e. motors, solenoids, transformers, etc. The total inductance, $\mathbf{L}$, could be represented by its inductive reactance $\mathbf{X_L}$ as shown in Figure 5.4 or by its impedance contribution $\mathbf{Z_L}$ as annotated in Figure 5.6a. While, in practice, there may be some capacitance and capacitive reactance in the system, for the sake of simplicity, it is assumed that this AC system is predominantly inductive, $\mathbf{X_L} \gg \mathbf{X_C}$ or $\mathbf{X_C}$ is negligible.

One salient effect of the dominant inductive reactance $\mathbf{X_L}$ in this circuit is that it causes the current to lag behind the voltage. The current lags behind the voltage in terms of **time** as well as the **angle** as depicted graphically in Figure 5.5a and b. Note that the current dives down into the negative magnitude **after** the voltage does.

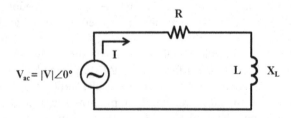

FIGURE 5.4 AC circuit with a lagging power factor.

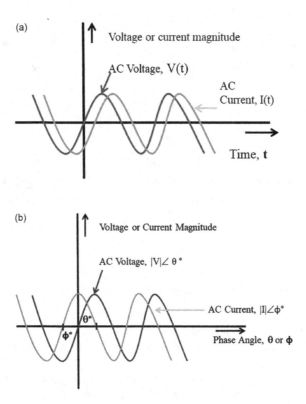

FIGURE 5.5 AC circuit with a lagging power factor; current lagging the voltage in phase angle and time domains. (a) Current lagging the voltage in time domain. (b) Current lagging the voltage in phase angle domain. Note: the angle of voltage vector is greater than the angle of the current vector.

While Figure 5.4 illustrates the physical – circuit element – aspect of a predominantly inductive AC electrical system and Figure 5.5 highlights the resulting discordance between the voltage and current waveforms, Figure 5.6 helps us visualize the formation of a, less than perfect, non-unity power factor in terms of vectors.

The combination of the resistance and the inductance in the circuit to form the total impedance of the circuit is depicted in the form of vectors $\mathbf{R}$ and $\mathbf{Z_L}$ and $\mathbf{Z}$, respectively, in Figure 5.6a. Note that the impedance contribution of inductance in the circuit is represented by the vector $\mathbf{Z_L}$ pointing straight up at an angle of $90°$. As discussed earlier in this text, impedance contribution $\mathbf{Z_L}$ by the inductance $\mathbf{L}$, or inductive reactance $\mathbf{X_L}$ in an AC system, is denoted in complex (rectangular) number form as $j\mathbf{X_L}$, where unit vector "$\mathbf{j}$" is represented in polar/phasor form as $1\angle90°$ and is responsible for assigning the $90°$ angle to $\mathbf{Z_L}$.

As always, resistance $\mathbf{R}$ possesses an angle of zero; therefore, it is represented by a horizontal vector along the x-axis as shown in Figure 5.6a.

Vector addition of vectors $\mathbf{Z_L}$ and $\mathbf{R}$ results in the resultant vector $\mathbf{Z}$, where $\mathbf{Z}$ represents the total or equivalent impedance of the predominantly inductive AC circuit. As ostensible from Figure 5.5a, magnitude of $\mathbf{Z}$, denoted as $|\mathbf{Z}|$, is a geometric

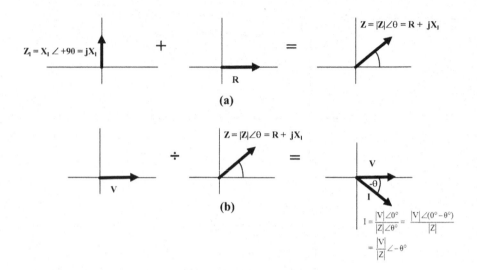

FIGURE 5.6 (a and b) Graphical analysis of lagging power factor.

hypotenuse of the triangle formed by component vectors $\mathbf{R}$ and $\mathbf{Z_L}$. Therefore, as annotated in Figure 5.5a, $\mathbf{Z}$ can be represented in polar form as $|\mathbf{Z}|\angle\theta°$ or in rectangular form as $\mathbf{R} + \mathbf{jX_L}$.

One of the methods for assessment of power factor requires the knowledge of AC voltage and AC in polar or phasor form. The voltage is assumed to be at an angle of $\mathbf{0°}$ and as shown in Figure 5.4 represented in vector/polar form as $|\mathbf{V}|\angle\mathbf{0°}$. The current is not known. The current can be determined by applying Ohm's law as follows:

$$I = \frac{|V|\angle 0°}{|Z|\angle \theta°} = \frac{|V|\angle(0° - \theta°)}{|Z|}$$

$$= \frac{|V|}{|Z|}\angle -\theta°$$

Then, by definition and in accordance with Eq. 5.2,

$$PF = \text{Power Factor} = \text{Cos}(\theta_V - \theta_I)$$

Since in this lagging power factor case,

$$\theta_V = 0°$$

and

$$\phi_I = \theta_I = -\theta°$$

$$PF = \text{Power Factor} = \text{Cos}(\theta_V - \theta_I) = \text{Cos}\{0 - (-\theta°)\}$$

$$= \text{Cos}(\theta°)$$

or

Power Factor = Cosine of the angle of the impdance, Z.

$$PF = \text{Cos}(\theta_Z) \tag{5.3}$$

Example 5.2

A $110\angle0°$ voltage is applied to load consisting of a $10\ \Omega$ resistance and $10\ \Omega$ inductive reactance in series. Calculate the power factor as seen by the source.
   Ancillary: Verify the result using Eq. 5.1.

Solution:

The circuit diagram for this system would appear as follows:

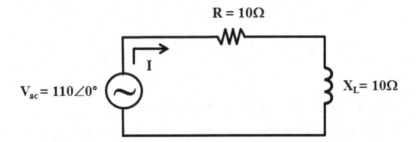

Equation 5.3, derived above, can be applied to determine the power factor if the angle of the total impedance were known. The total impedance can be assessed, in polar/phasor form, using a scientific calculator or through the angle computation formula covered in the preceding chapters.

$$|\theta_Z| = \text{Tan}^{-1}\left(\frac{R}{|X_L|}\right) = \text{Tan}^{-1}\left(\frac{10\ \Omega}{10\ \Omega}\right) = \text{Tan}^{-1}(1) = 45°$$

$$\therefore \text{Power Factor} = PF = \text{Cos}(\theta_Z) = \text{Cos}(45°) = 0.707, \text{ or } 70.7\% \text{ lagging.}$$

Ancillary: In order to verify the power factor computed above, as obvious from Eq. 5.1, we need to find the real power component **P** and the total apparent power **S** dissipated in the given circuit.
   Apparent power S can be calculated using Eq. 4.11 introduced in Chapter 4.

$$\vec{S} = \vec{V} \cdot \vec{I}^* \tag{4.11}$$

However, we must calculate the phasor value of the AC first.

$$I = \frac{110\angle 0°}{|Z|\angle \theta_Z°}$$

$$|Z| = \sqrt{R^2 + X_L{}^2} = \sqrt{10^2 + 10^2} = \sqrt{200} = 14.14$$

$$\therefore \ I = \frac{110\angle 0°}{14.14\angle 45°} = 7.78\angle -45°$$

Note: The negative angle of the current, with the voltage angle being 0°, signifies the fact that the current is **lagging** behind the voltage, and therefore, the power factor will be a **lagging** power factor.

$$S = V \cdot I^* = (110\angle 0°) \cdot (7.78\angle 45°) = 856\angle 45°$$

Note: "*" stands for conjugation of the original current angle

Then, $P = |S| \cdot \text{Cos}(45°) = |856| \cdot \text{Cos}(45°) = 605$ W

and,

$$Q = |S| \cdot \text{Sin}(45°) = |856| \cdot \text{Sin}(45°) = 605 \text{ W {Redundant}}$$

$$\therefore \ PF = \frac{P}{|S|} = \frac{605}{856} = 0.707 \text{ or } 70.7\%. \text{ lagging.}$$

LEADING POWER FACTOR AND IMPEDANCE

A basic AC circuit with a leading power factor, as depicted in Figure 5.7, would consist of resistance and capacitance, represented by **R** and **X$_C$**, respectively. As shown in Figure 5.7, the AC voltage source, **V$_{ac}$**, drives AC **I** through the total resistance **R** and the total capacitance **C** in the AC system. The resistance, **R**, would represent not only the resistive loads in the overall system but also the inherent resistance of the power distribution system, i.e. resistance of the conductors and other current-carrying devices. The total capacitance in the system, **C**, could be an aggregate of the capacitances in the system, transmission line capacitance, etc. Leading power factor can also be caused by machines (i.e. certain power generators) that might be operating at a leading power factor, computer switched mode power supplies, etc. The total capacitance, **C**, is represented by its capacitive reactance, **X$_C$**, as shown in Figure 5.7 or by its impedance

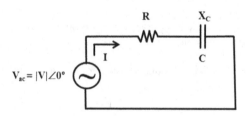

FIGURE 5.7   AC circuit with a leading power factor.

contribution, $\mathbf{Z_C}$, as annotated in Figure 5.9. While, in practice, there may be some inductance and inductive reactance in the system, for the sake of simplicity, it is assumed that this AC system is predominantly capacitive, $\mathbf{X_C} \gg \mathbf{X_L}$, or $\mathbf{X_L}$ is negligible.

One distinct effect of the dominant capacitive reactance $\mathbf{X_C}$ in this circuit is that it causes the current to lead the voltage. The current leads the voltage in terms of **time** as well as the **angle**, as depicted graphically in Figure 5.8a and b.

While Figure 5.7 illustrates the physical – circuit element – aspect of a predominantly capacitive AC electrical system and Figure 5.8 highlights the resulting discordance between the voltage and current waveforms, Figure 5.9 depicts the formation of a, less than perfect, non-unity power factor in terms of vectors.

The combination of the resistance and the capacitance in the circuit to form the total impedance of the circuit is depicted in the form of vectors $\mathbf{R}$ and $\mathbf{Z_C}$ and $\mathbf{Z}$, respectively, in Figure 5.9a. Note that the impedance contribution of capacitance in the circuit is represented by the vector $\mathbf{Z_C}$ pointing directly downward at an angle of $-90°$. As discussed earlier in this text, impedance contribution $\mathbf{Z_C}$ by the capacitance

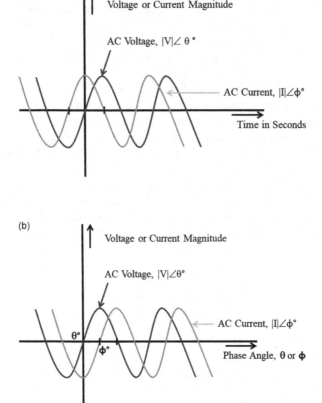

FIGURE 5.8 AC circuit with a leading power factor; current leading the voltage in phase angle and time domains. (a) Current leading the voltage in time domain. (b) Current leading the voltage in phase angle domain. Note: current angle is greater than the voltage angle.

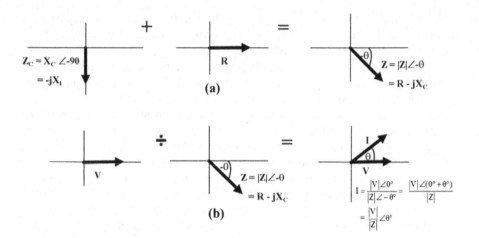

FIGURE 5.9   (a and b) Graphical analysis of leading power factor.

C or capacitive reactance X_C in an AC system is denoted in complex (rectangular) number form as $-jX_C$, where unit vector "$-j$" is represented in polar/phasor form as $1\angle-90°$ and is responsible for assigning the $-90°$ angle to Z_C.

Resistance **R** portends an angle of zero; therefore, it is represented by a horizontal vector along the x-axis as shown in Figure 5.9a.

Vector addition of vectors Z_C and **R** results in the resultant vector **Z**, where **Z** represents the total or equivalent impedance of the predominantly capacitive AC circuit. As obvious from Figure 5.9a, magnitude of **Z**, denoted as |**Z**|, is a geometric hypotenuse of the triangle formed by component vectors **R** and Z_C. Therefore, as annotated in Figure 5.9a, **Z** can be represented in polar form as $|Z|\angle-\theta°$, or in rectangular form as $R - jX_C$.

One approach for assessment of power factor requires the knowledge of AC voltage and AC in polar or phasor form. The voltage is assumed to be at an angle of $0°$ and as shown in Figure 5.7, represented in vector/polar form as $|V|\angle0°$. The current is not known. The current – as with the lagging power factor scenario – can be determined by applying Ohm's law as follows:

$$I = \frac{|V|\angle0°}{|Z|\angle-\theta°} = \frac{|V|\angle(0°+\theta°)}{|Z|}$$

$$= \frac{|V|}{|Z|}\angle\theta°$$

Then, by definition and in accordance with Eq. 5.2,

$$PF = \text{Power Factor} = Cos(\theta_V - \theta_I)$$

In this leading power factor case,

$$\theta_V = 0°$$

and

$$\phi_1 = +\theta°, \text{ or just } \theta°$$

$$PF = \text{Power Factor} = \text{Cos}(\theta_V - \theta_1) = \text{Cos}(0 - \theta°)$$

$$= \text{Cos}(-\theta°) = \text{Cos}(\theta°)$$

or

Power Factor = Cosine of the angle of the impdance, Z.

(5.3)

PF = $\text{Cos}(\theta_Z)$ which is the same as in the lagging power factor scenario

Example 5.3

A $110\angle0°$ voltage is applied to load consisting of a 10 Ω resistance and 10 Ω capacitive reactance, in series. Calculate the power factor as seen by the source.
Ancillary: Verify the result using Eq. 5.1.

Solution:

The circuit diagram for this system would appear as follows:

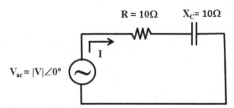

Equation 5.3, derived above, can be applied to determine the power factor if the angle of the total impedance were known. The total impedance can be assessed, in polar/phasor form, using a scientific calculator or through the angle computation formula covered in the preceding chapters.

$$|\theta_Z| = \text{Tan}^{-1}\left(\frac{R}{|X_L|}\right) = \text{Tan}^{-1}\left(\frac{10\,\Omega}{10\,\Omega}\right) = \text{Tan}^{-1}(1) = 45°$$

As shown in Figure 5.9b, **Z** in the leading power factor scenario lies in the fourth quadrant; therefore, $\theta_Z = -45°$.

∴ Power Factor = PF = $\text{Cos}(\theta_Z) = \text{Cos}(-45°) = 0.707$, or 70.7% leading.

Note: Because power factor computation is based on "Cosine" of impedance angle, power factor remained the same as we transitioned from 10 Ω inductive reactance system in Example 5.2 to a 10 Ω capacitive reactance system in Example 5.3; it simply transformed from being lagging to leading.
   Ancillary: In order to verify the power factor computed above, as obvious from Eq. 5.1, we need to find the real power component **P** and the total apparent power **S** dissipated in the given circuit.

Apparent power **S** can be calculated using Eq. 4.1 introduced in Chapter 4.

$$\vec{S} = \vec{V} \cdot \vec{I}^* \qquad (4.11)$$

However, as in Example 5.2, we must calculate the phasor value of the AC first.

$$I = \frac{110\angle 0°}{|Z|\angle - \theta_z°}$$

$$|Z| = \sqrt{R^2 + X_L^2} = \sqrt{10^2 + 10^2} = \sqrt{200} = 14.14$$

$$\therefore I = \frac{110\angle 0°}{14.14\angle - 45°} = 7.78\angle + 45° \text{ or } 7.78\angle 45°$$

Note: The positive angle of the current, with the voltage angle being 0°, signifies the fact that the current is **leading** the voltage, and therefore the power factor will be a **leading** power factor.

$$S = V \cdot I^* = (110\angle 0°) \cdot (7.78\angle - 45°) = 856\angle - 45°$$

Note: * stands for conjugation of the original current angle

Then, $P = |S| \cdot \text{Cos}(-45°) = |856| \cdot \text{Cos}(-45°) = 605 \text{ W}$

and,

$$Q = |S| \cdot \text{Sin}(-45°) = |856| \cdot \text{Sin}(-45°) = -605 \text{ W \{Redundant\}}$$

or,

$$S = P + jQ = 605 + j(-605) = 605 - j605 \text{ VA}$$

$$\therefore PF = \frac{P}{|S|} = \frac{605}{856} = 0.707 \text{ or } 70.7\%. \text{ leading.}$$

Example 5.4

An $X_C = 10 \ \Omega$ capacitive reactance worth of capacitance is added in series with the series resistor and inductive reactance AC circuit from Example 5.2.

   a. Draw the circuit diagram for the new circuit.
   b. Determine the net impedance of the new circuit.
   c. Calculate the new power factor.

Solution:

   a. **Ancillary Note:** Even though in this example the net capacitance is in **series** with the resistive load, in many cases, the net capacitance is in **shunt** or parallel with the overall resistive load. Calculations associated with the net capacitance in shunt with the overall resistive load are illustrated in the power factor correction section later in this chapter.

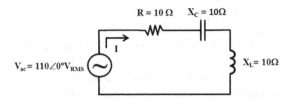

b. Since all elements in the modified circuit are connected in series, the total or net impedance in this AC circuit would be as follows:

$$Z = R + jX_L - jX_C = 10\,\Omega + j10\,\Omega - j10\,\Omega = 10\,\Omega,$$

or,

$$Z = 10\,\Omega\angle 0°\ \Omega$$

c. Since, in this case, the inductive reactance is completely offset by the capacitive reactance, the net or total impedance is composed solely of pure resistance of 10 Ω. As discussed earlier in this chapter and Chapter 4, power factor in circuits that are purely resistive or have "0" net reactance is "1" or 100%.

This can be proven by applying Eq. 5.3 derived earlier and recognizing the fact that θ_Z, the angle of total impedance Z, as derived in part (b) is $0°$:

Power Factor = Cosine of the angle of the impdance, Z

or

$$PF = Cos(\theta_Z) = Cos(0) = 1$$

As obvious by now, in essence, Example 5.4 shows us how capacitive reactance can be added to improve or correct power factor in a predominantly inductive AC circuit. This serves as an appropriate segue into the next section that pertains to the power factor correction.

POWER FACTOR CORRECTION

There are multiple methods for correcting power factor. Some of the more common approaches for correcting power factor are as follows:

a. Addition of power factor correcting capacitors
b. Addition of power factor correcting inductors
c. Frequency manipulation
d. Utilization of synchronous motors or generators

The power factor correction method chosen depends on the composition or design of the AC system. In general, if the power factor is less than "1," or 100%, due to the predominance of inductive reactance, X_L, capacitance or capacitive reactance, X_C, must be added in series or shunt to raise the power factor closer to 100%. In fact,

Example 5.4 – where $10\,\Omega$ of $\mathbf{X_L}$ is offset by the addition of $10\,\Omega$ of $\mathbf{X_C}$ – is a classic illustration of this approach. Frequency manipulation method is applicable when, in addition to resistance, both inductive and capacitive reactances are present. As described later in this chapter, when the power factor is less than 100% due to the dominance of capacitance in an AC electrical system, inductive reactance may be introduced into the circuit to offset the capacitive reactance and improve the system's power factor.

ADDITION OF CAPACITANCE OR CAPACITIVE REACTANCE

Since power factor correcting capacitors, essentially, impact the AC circuits or systems through addition of capacitive reactance $\mathbf{X_C}$, or $\mathbf{Z_C} = -\,j\mathbf{X_C}$, their addition to a predominantly inductive circuit yields different results depending upon whether they are connected in series with the load or in parallel (or shunt).

Power factor correcting capacitors can also be viewed as devices used to reduce current in power systems. Capacitors do so by providing a leading current to loads that require lagging current. In other words, capacitors provide "off-setting" current for inductive loads.

Another side benefit in the use of a capacitor is in the fact that they can reduce voltage drop, especially in motors starting situations and other high-impulse loads. This can be attributed to the fact that capacitors, as discussed earlier in this text, don't permit instantaneous drop or rise in voltage. It is partly due to this rationale that, sometimes, the power factor correcting capacitors are referred to as "voltage boosting" capacitors.

There are multiple approaches for assessment of capacitance required for specific improvement of power factor. One method relies on the knowledge of existing and desired reactive powers, $\mathbf{Q_1}$ and $\mathbf{Q_2}$, respectively, the operating voltage and the frequency. The formula for determination of the required per phase capacitance is as follows:

$$C = \frac{(Q_1 - Q_2)}{2\pi f V^2} \qquad (5.4)$$

where

C = Capacitance, in Farads, required to reduce the reactive power from Q_1 to Q_2 per phase

Q_1 = Initial, higher reactive power, in VARs, per phase

Q_2 = Improved, lower reactive power, in VARs, per phase

V = Voltage, in volts

f = Frequency, in Hz

This formula is premised on the following mathematical definitions of reactive power Q:

$$|Q| = \frac{V^2}{X_C} = V^2 2\pi f C \qquad (5.5)$$

Or for inductive reactance:

$$|Q| = \frac{V^2}{X_L} = \frac{V^2}{2\pi f L} \qquad (5.6)$$

An electrical engineer may use Eq. 5.4 to determine the capacitance required to improve the power factor, by a specific amount, in lieu of the apparent power (S) reduction approach. The apparent power-based power factor calculation method is illustrated, later, through Example 5.7.

Example 5.5

The HMI (Human Machine Interface) monitor of an EMS system, monitoring a chilled water pump motor, is showing that the motor is drawing a reactive power, Q_1, of 50 kVARs. Funding has been allocated as a part of a DSM (Demand Side Management) energy conservation project to improve the power factor of the motor branch circuit such that the reactive power is reduced to Q_2 of 10 kVARs. The branch circuit is operating at 240 V_{RMS}. The system frequency is 60 Hz. Calculate the amount of capacitance that must be added, per phase, in shunt, phase to phase, to achieve the power factor improvement project objective.

Solution:

Apply Eq. 5.4:

$$C = \frac{(Q_1 - Q_2)}{2\pi f V^2} = \frac{(50\text{ kVAR} - 10\text{ kVAR})}{2\pi(60\text{ Hz})(240\text{ V})^2}$$

$$= \frac{(50,000\text{ VAR} - 10,000\text{ VAR})}{2\pi(60\text{ Hz})(240\text{ V})^2} = 0.001843\text{ F} = 1.84\text{ mF}$$

Example 5.6

Power Factor Improvement and Cost Savings: An air compressor station is consuming 2,000 kW at a power factor of 0.7. The utility company charges a \$5.00/kVA per month as penalty for poor power factor. What would be the annual, pre-tax, savings if capacitors could be installed and power factor improved to 0.9?

Solution:

Since the objective is to assess the cost savings on the basis of apparent power S (kVA) reduction, we must determine the apparent power S_1 being drawn by the air compressor motor at the existing power factor of 0.7 (70%) and the apparent power S_2 at the desired power factor of 0.9 (90%). We can compute S_1 and S_2 by rearranging and applying Eq. 5.1.

$$PF = \text{Power Factor} = \frac{|P|}{|S|} \tag{5.1}$$

Rearrange Eq. 5.1:

$$\text{Magnitude of Apparent Power } |S| = \frac{|P|}{PF}$$

Therefore,

$$\text{Magnitude of existing apparent power } |S_1| = \frac{|P|}{PF_1} = \frac{2,000\,\text{kW}}{0.7}$$

$$= 2,857\,\text{kVA}$$

$$\text{Magnitude of desired apparent power } |S_2| = \frac{|P|}{PF_2} = \frac{2,000\,\text{kW}}{0.9}$$

$$= 2,222\,\text{kVA}$$

Note that real power drawn, which is a function of the actual work performed or torque generated, remains unchanged, at 2,000 kW.

Annual savings attributed to power factor improvement

$$= (2,857\,\text{kVA} - 2,222\,\text{kVA}) \cdot \left(\frac{\$5}{\text{kVA} - \text{Month}}\right) \cdot (12\,\text{Months/Year})$$

$$= \$38,095$$

AUTOMATIC SWITCHING PF CORRECTING CAPACITOR SYSTEM

Imagine a plant engineering project that entails installation of a large, 500 hp, air compressor. The 500 hp rating of the air compressor is premised on the size of the motor. In other words, the 500 hp rating, in most cases, would be real power (**P**) delivered by the motor to the compressor. However, as we discussed earlier, motors add **inductive reactance** to electrical system. So, the branch circuit the motor pertains to, in this facility, would have a non-unity, or less than 100% power factor, say 80%. Sizeable loads, such as large air compressors, therefore, degrade the aggregate power factor of the entire electrical system. On new installations, such as the one being considered here, degradation of power factor can be prevented by including automatic switching power factor correcting capacitor banks such as the one shown in Figure 5.10. Automatic switching power factor correcting capacitor banks constitute a more optimal and effective approach for automatic introduction and removal of capacitance to maintain the system power factor at an optimal level.

The controls in an automated power factor correcting system connect or disconnect capacitors as the inductive load appears and disappears from the overall electrical circuit. This ensures that the power factor correcting capacitors – and the associated capacitive reactance – drop out of the circuit when the compressor motor load (and associated inductive reactance) switches off. **Important point**: If the power factor capacitors were to stay connected to the compressor motor branch circuit when the motor is off, the circuit would become predominantly capacitive, as X_C would exceed X_L, resulting in a low leading power factor. In other words, if the capacitor bank is not disconnected when the compressor-based inductive reactance, X_L, drops out, it would result in a **capacitive reactance "overdose."**

As annotated in Figure 5.10, there are six key functional components in the pictured automatic power factor correcting system:

FIGURE 5.10 Automated power factor correcting capacitor bank.

1. Control Contactor or Relay: For connecting and disconnecting of the capacitors.
2. Terminal Strip or Connection Points: For interface to the motor circuit, sensors, and other field connections.
3. Slow Blow Fuses: For overload protection.
4. Inrush Limiting Contactors: For limiting the motor inrush current.
5. Capacitors (single-phase or three-phase units), delta connection: Power factor correcting capacitors (X_C).
6. Control Transformer: For obtaining control voltage, i.e. 110 VAC, from the existing power source, operating at power circuit voltage, i.e. 480 VAC.

General Information on Power Factor and the Application of Power Factor Correcting Capacitors

What is a suitable power factor goal for industrial, commercial, and institutional facilities? It would be difficult to present a specific and perfect answer to this question. While one could find enough data to support the claim that the overall average power factor in the United States is approximately 82%, the best answer varies from situation to situation and industry to industry. For example, on one end of the spectrum, one might observe power factors approaching 90%–97% in the institutional and commercial structures, while on the low end of the power factor spectrum it is not uncommon to come across power factors to the tune of 50% or lower in certain foundry, plastics, and chemical industries, where the melting processes employ electrical inductive heating.

What is the financial return and payback associated with power factor correction projects? Some utilities base demand charges on kVA demanded by the customer facility. In such cases, an approximate 10% demand (kVA) reduction can be realized if a power factor correction capacitor bank is installed to improve the power factor from 82% to 90%. Such an investment has a typical payback of, approximately, 2 years. Note, however, that the payback periods, on power factor correction-based energy projects – or for that matter, on any energy project – will be *shorter* for consumers located in higher energy cost regions, i.e. the Northeastern United States, the Western states, and Hawaii. Conversely, the payback periods will be longer in lower energy cost regions, i.e. the Southeast, the Dakota's, etc. See *Finance and Accounting for Energy Engineers*, by S. Bobby Rauf.

Application, installation, operation, and maintenance of power factor correcting capacitors: Application of power factor correcting capacitors should be preceded by proper analysis and design. Improper application of power factor correcting capacitors can damage the electrical system. Improper application of power factor correcting capacitors can cause fuses to blow and can amplify system harmonics. After installation, capacitor banks must be included in the facility's Preventive Maintenance Program. Loose connections and inherent defects, when ignored, can lead to unanticipated failures, faults, downtimes, and arc flash hazards.[2]

Higher power factor means lower apparent power, S (in kVA), demanded and lower "unproductive" current in the electrical system. This "unproductive" current is wasted, as heat, as it flows through conductors, protective devices, power monitoring devices, transformers, and loads. As we know, in general, operation of equipment above normal temperatures results in shortening of the life of equipment. Case in point and analogy: overheated machines, such as automobile engines, tend to cease and fail when overloaded or operated at abnormally high temperatures. Therefore, improving the power factor not only results in improved energy efficiency but also extends service life and reduces maintenance cost of equipment.

If inductive reactance (X_L, in Ω) and reactive power (Q, in VARs) are encountered through a distribution transformer, application of PF capacitor is recommended at the distribution bus feeding the transformer. When the inductive reactance (X_L, in Ω) and reactive power (Q, in VARs) are contributed by large motors, the greatest PF corrective effect and energy loss reduction are realized when the capacitors are installed at each motor load (on line side) or between the motor starters and the motors. However, the initial cost for a number of small banks of capacitors would tend to be higher than the initial cost of a single, large, bank of capacitors, catering to several motor loads. In such cases, the higher initial cost must be weighed against gains in energy cost and PF correction benefits. Systems that experience large load swings, throughout the day or month, are good candidates for automatic switching correction systems. As explained earlier, these systems, automatically, correct the PF as load variations occur.

ADDITION OF INDUCTANCE OR INDUCTIVE REACTANCE TO CORRECT FOR LEADING POWER FACTOR

If the power factor is less than "**1**" due to the predominance of capacitance or capacitive reactance, X_C, inductance or inductive reactance, X_L, must be added in series or shunt to raise the power factor closer to 100%.

In some AC electrical systems, such as the power transmission lines, the total impedance as seen from the utility side is indeed predominantly capacitive. In such cases, *inductance* must be added to improve the power factor.

---

[2] NFPA® 70E, 2018.

Addition of power factor correcting inductive reactance, and analytical methods to correct power factor in predominantly capacitive electrical systems, is illustrated in self-assessment problems 2 and 3.

FREQUENCY MANIPULATION APPROACH FOR CORRECTION OF POWER FACTOR

While pursuit of a perfect *unity* power factor in low power factor situations is not always feasible or practical, for simplicity, we will limit our discussion associated with frequency-based power factor correction to scenarios where power factor is, indeed, raised to 1 or 100%. As stated earlier, frequency manipulation method pertains to AC electrical systems where both the inductive and capacitive reactances are present, in addition to resistance.

Consider the AC circuit shown in Figure 5.11 with an existing power factor that is less than 1.

In order for this circuit's power factor to be unity or 100%, the circuit must be purely resistive. In other words, $Z_{Eq} = R$ and Z_C must cancel or offset Z_L. So, when inductive reactance X_L and capacitive reactance X_C are fixed, the electrical frequencies "f" and "ω" may be changed such that the magnitudes of Z_C and Z_L become equal and cancel each other. The frequency at which $|Z_C|$ and $|Z_L|$ equate is called *resonant frequency*, f_0 or ω_0. An electrical circuit operating at resonance frequency is called a *resonant circuit*. The derivation and mathematical proof for resonance frequency is as follows:

In order for Z_L and Z_C to cancel each other when added together in a

series combination: $Z_L = -Z_C$

Since $Z_L = j2\pi fL$ and $Z_C = -jX_C = -j\dfrac{1}{2\pi fC}$

$$j2\pi fL = -\left(-j\dfrac{1}{2\pi fC}\right)$$

or

$$X_L = X_C \tag{5.7}$$

$$2\pi fL = \dfrac{1}{2\pi fC} \text{ and, } f^2 = \dfrac{1}{(2\pi)^2 LC}$$

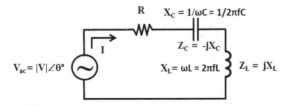

FIGURE 5.11   An RLC (Resister, Inductor, and Capacitor) circuit.

and

$$f = f_0 = \frac{1}{2\pi\sqrt{LC}}, \text{ or, since } 2\pi f = \omega \qquad (5.8)$$

$$\omega = \omega_0 = \frac{1}{\sqrt{LC}} \qquad (5.9)$$

where
  L = Inductance in henry, or H
  C = Capacitance in Farads, or F
  ω = Angular frequency, in general, in rad/s
  ω_0 = Resonance angular frequency, in rad/s
  f = Electrical frequency, in general, in cycles/s, hertz, or Hz; 60 Hz in the
    United States
  f_0 = Resonance electrical frequency, in hertz
  Z_C = Impedance contribution by capacitance in an AC circuit, in Ω
  Z_L = Impedance contribution by inductance in an AC circuit, in Ω
  X_C = Reactance due to capacitance in an AC circuit, in Ω
  X_L = Reactance due to inductance in an AC circuit, in Ω

As obvious, the frequency method for correcting power factor is feasible only in
circumstances where the electrical frequency can be varied through the use of electromechanical systems such as electrical power generators or via electronic variable
frequency drives.

Example 5.7

The output of a variable frequency drive, as shown in the circuit below, is 157Sinωt.
The VFD output is currently set at 60 Hz. This drive is connected to a resistive
load, capacitive reactance, and an inductive reactance.

  a. What should be the new frequency setting to attain a power factor of 1
    or 100%?
  b. What is the resonance frequency for this AC system?
  c. Could a power factor other than unity be attained through variation of
    frequency?

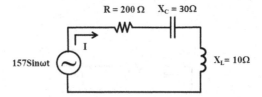

Solution:

  a. As explained earlier in this section, the power factor of an AC cir-
    cuit, consisting of inductive and capacitive reactance, peaks to the

maximum value of unity, or 100%, at resonance frequency, f_0. However, in this case, because the inductance and capacitance are represented as inductive and capacitive reactances, respectively, we cannot apply **Eq. 5.8**, directly, to compute f_0. We must convert reactances X_C and X_B to corresponding capacitance, **C**, and inductance, **L**, values.

$$\text{Since } X_L = 2\pi fL, L = \frac{X_L}{2\pi f}, \text{ or, } L = \frac{10\,\Omega}{2\pi(60\text{ Hz})} = 0.02654\text{ H}$$

and,

$$X_C = \frac{1}{2\pi fC}, C = \frac{1}{2\pi fX_C}, \text{ or, } C = \frac{1}{2\pi(60\text{ Hz})(30\,\Omega)} = 88.46 \times 10^{-6}\text{ F}$$

$$f = f_0 = \frac{1}{2\pi\sqrt{LC}} = \frac{1}{2(3.14)\sqrt{(0.02654).(88.46\times10^{-6})}} \qquad \text{Eq. 5.8}$$

$$\therefore\ f = f_0 = 104\text{ Hz}$$

b. As explained earlier, resonance frequency *is* the AC frequency at which the total capacitive reactance in an AC circuit cancels or offsets the total inductive reactance. Therefore, resonance frequency, f_0, for the given AC circuit is the frequency calculated in part (a). In other words, $f_0 = 104$ **Hz**.

c. The power factor does indeed vary with frequency with all other entities held constant. For instance, with all parameter held constant, if the frequency is lowered to 30 Hz, the power factor drops down to, approximately, 96%.

Ancillary: The reader is encouraged to prove this. Also see self-assessment problem 4.

UTILIZATION OF SYNCHRONOUS MOTORS OR GENERATOR FOR POWER FACTOR CORRECTION

This approach involves the operation of synchronous motors or generators such that AC leads the voltage. When synchronous motors are operated in this manner, they act like generators. So, in essence, when synchronous motors are operated in leading current mode, the excess demand for apparent power "**S**," due to predominance of inductive reactance, can be satisfied by the leading current supplied by the synchronous motor. This excess demand for apparent power "**S**" would, otherwise, be sourced from the utility company at the cost of lower power factor as seen by the utility. With all that said, note that application of synchronous motors for the exclusive purpose of improving power factor is not advisable and not economically suitable. In other words, the cost of power factor correcting capacitors is far less than synchronous motors.

SELF-ASSESSMENT PROBLEMS AND QUESTIONS – CHAPTER 5

1. Determine the power factor of the circuit shown below, as seen by the AC source.

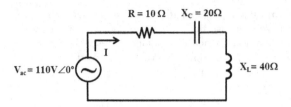

2. Assume that the circuit depicted below represents one phase of a special power transmission line. Determine the power factor of the circuit shown below, as seen by the AC source.

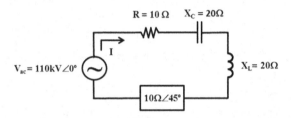

3. If the power factor in problem 2 is less than 1.0, how much capacitance or inductance must be added in series to raise the power factor to unity?

4. The output of a variable frequency drive, as shown in the circuit below, is $157 Sin\omega t$. The VFD output is currently set at 50 Hz. This drive is connected to a resistive load, capacitive reactance, and an inductive reactance.

   a. What should be the new frequency setting to attain a power factor of 1, or 100%.

   b. What is the existing power factor at 50 Hz?

   c. What would be the power factor if all circuit elements remain unchanged and the VFD frequency is lowered to 30 Hz?

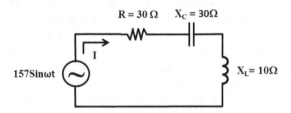

5. The HMI monitor of an automated HVAC system, monitoring an air washer supply fan motor, is indicating a reactive power, Q_1, of 60 kVARs. This system is located in the United Kingdom, where the AC frequency is 50 Hz. Determine the amount of capacitance that must be added to improve the power factor of the motor branch circuit such that the reactive power is reduced to Q_2 of 20 kVARs. The branch circuit is operating at 240 V_{RMS}.

6. **Power Factor Improvement and Cost Savings:** In conjunction with the local utility company DSM program, a manufacturing plant is being offered $2 per kVA for improvement in power factor from 0.75 to 0.85. The plant is operating at its contract level of 30 MW. Determine estimated annual pre-tax revenue if the plant accepts the offer.

7. The output of a variable frequency drive, as shown in the circuit below, is 157Sinωt. The VFD output is currently set at 60 Hz.This drive is connected to a resistive load, capacitive reactance, inductive reactance, and a "black box" load, Z_B, of 10Ω∠45°. What should be the new frequency setting to attain a power factor of 1 or 100%?

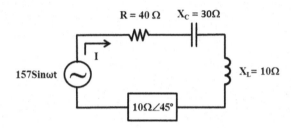

6 Power Quality and Power Management

INTRODUCTION

In this chapter, we will discuss important power quality and power management concepts and methods. Introduction to the concepts of demand and peak demand should poise us well for investigation of the concept of load factor, its role in power quality considerations, and electrical energy cost reduction through peak shaving. Later in this chapter, we will explain the concept of service factor as it applies to electrical equipment. This chapter concludes with discussion and analysis of large industrial power bill computation. The bill calculation examples are designed to help the reader develop skills and acumen to analyze and understand important components of large power bills, in their efforts to identify cost reduction opportunities. As with other electrical concepts presented in this text, we will substantiate our discussion with analogies, mathematical and analytical models, as applicable.

DEMAND

In the electrical power distribution and energy realms, generally, the term "demand" implies electrical power demanded by electrical loads. The term "power" could be construed as **apparent power**, **S**, measured in VA, kVA, MVA, etc., or it could be interpreted as real power, **P**, measured in W, kW, MW, etc. Many electrical power utilities tend to apply the term **demand** in the context of **real power**, **P**, demanded from the grid. However, some electrical power utilities use the term demand to signify **apparent power** demanded from the grid, measured in kVA. When demand is known to represent real power in kWs or MWs, average demand is computed in accordance with Eq. 6.1.

$$\text{Average Demand, in kW} = \frac{\text{Energy (kWh) consumed during the billing month}}{\text{Total number of hours in the month}} \quad (6.1)$$

PEAK DEMAND

The term "peak demand" has two common or mainstream interpretations. The first interpretation is associated, mostly, with electrical power load profiling, such as the real power, P (kW), load profile or graph included among the various power monitoring screens available through EMS, Energy Monitoring Systems,[1] or BMS, Building Management Systems. The graph in Figure 6.1 is a screen capture of load profile

---

[1] *Finance and Accounting for Energy Engineers* by S. Bobby Rauf.

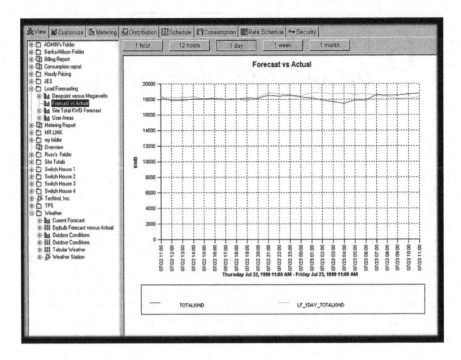

FIGURE 6.1   Actual versus forecasted power demand (kW).

screen in an EMS system. The graph depicts a plot of real power load (in kWs) plotted as a function of time. The darker line spanning from the left to right in the graph represents the **actual power demand (in kW),** recorded, by the hour, over a 24-hour period. The lighter line in the graph represents **demand forecasted** on the basis of the actual load profile measured and recorded by the EMS system, over the long term. The forecasted demand plays an essential role in energy conservation-related peak shaving programs.[2] As obvious, demand forecast premised on actual, measured, load profile – for a specific facility – would tend to be more reliable than general load profiles based on theoretical models. This constitutes an important justification for EMS.

The second interpretation of the term peak demand is associated more directly with the computation of electrical power bills for large industrial and commercial consumers of electricity. The peak demand, in the billing context, is computed by most power companies on the basis of demand intervals.

Some utilities (power companies) base their peak demand computation on the basis of 30-minute intervals, while others base the demand computation on the basis of 15-minute intervals. We will limit our consideration of the peak demand interval to 30-minute intervals.

Under one of the more common large power consumer contract rate schedules, referred to as the OPT, Time of the Day, or Time of Use (TOU) rate schedule, the power company captures and records the energy consumed by the consumer, in kWh,

---

[2] *Finance and Accounting for Energy Engineers* by S. Bobby Rauf.

over half-hour (30-minute) intervals. Each energy data point thus captured is converted to the corresponding demand value, in kW, as stated under Eq. 6.2.

$$\text{Demand} = \text{Power in kW} = \frac{\text{Energy in kWh}}{\text{Time in Hours}} = \frac{\text{Energy in kWh}}{\frac{1}{2}\text{Hour}} \qquad (6.2)$$

This could amount to, approximately, 1,440 data points for a 30-day billing month, with each data point representing the demand (kW) for a specific 30-minute interval during the billing month. The utility provider (power company) selects the highest demand data point during the declared "on-peak" periods and labels it **Peak Demand** for the month. See additional information on peak demand under the bill calculation section.

LOAD FACTOR

The most common and general definition of load factor is that it is the ratio of **average power** to **peak demand**. This definition of load factor can be translated into Eq. 6.3.

$$\text{Load Factor} = \frac{\text{Average Demand for the Month, in kW or MW}}{\text{Peak Demand for the Billing Month, in kW or MW}} \qquad (6.3)$$

Equation 6.3 can be modified to define load factor (LF) in terms of **Energy Usage (in kWh)** and **Demand (in kW)**.

$$\text{Load Factor} = \frac{(\text{kWh Used in the Billing Period})/(\text{Hours in Billing Period})}{\text{Peak Demand in kWs}} \qquad (6.4)$$

The result of Eqs. 6.3 and 6.4 can be multiplied with 100% to represent the calculated load factor in %, similar to representation of power factor in %. A closer look at Eqs. 6.3 and 6.4 highlights the following facts:

1. High peak demands (large denominator) result in low load factors. Low load factors result in higher energy cost due to high peak demand and, sometimes, due to low load factor triggered penalties. In essence, low load factor signifies greater opportunity for energy cost reduction through peak shaving or peak management.
2. Higher load factors signify better power management, better power quality, higher existing energy productivity, and lower energy cost. In other words, a high load factor would mean that there is a lower likelihood of finding peak shaving-related "low-hanging fruit." For instance, a commercial or industrial facility that operates 24/7 is likely to have a high load factor, approaching 100%, and fewer peak-shaving opportunities.
3. High load factor also implies relatively leveled and controlled demand (kWs), short peaks with low amplitude, accomplished in some cases through "peak shaving" and targeted peak avoidance.
4. Note that, in some quarters, load factor is quantified as a product of the decimal value acquired through Eq. 6.3 or 6.4, above, and 24 hours. The

resulting product would be a number representing load factor expressed in "hours" instead of a decimal value or %. So, for instance, in a facility where the load factor is calculated to be 70%, the load factor can also be expressed as: 0.7 × 24 hours, or 16.8 hours.

A load factor of unity or "1," 100%, or 24 hours indicates that demand peaks are negligible resulting in steady and relatively leveled power (kW) demand. Since most facilities don't operate 24 hours a day, and since the demand (kW) varies during a 24-hour period and the billing month, load factor – similar to power factor – in most practical applications is usually less than one "1." In some large industrial plants, operating 24/7, load factor is vitiated by sporadic operation of certain large loads such as large air compressors and large chillers.

Low load factor is a good indicator of the cost savings potential in shifting some electric loads to off-peak hours, thus reducing on-peak demand. In a utopian situation, if the load factor of a facility is already almost unity, further demand reduction can be accomplished through replacement of existing **low-efficiency** equipment with **higher-efficiency** equipment.

Example 6.1

The EMS at an automotive plastic component manufacturing plant that operates around the clock, 365 days a year, is displaying following electrical power data:
  Billing Days in the Current Month: 30
  On-Peak Energy Consumption: 4,320,000 kWh
  Off-Peak Energy Consumption: 17,280,000 kWh
  Highest 30-minute Energy Recorded: 17,500 kWh
  Calculate the following assuming that this facility is on OPT, TOU, contract with 30 minutes demand interval:

  a. Average demand.
  b. Peak demand.
  c. The load factor for the current month.

Solution:

a. Average demand can be calculated by applying Eq. 6.1 as follows:

$$\text{Average Demand, in kW} = \frac{\text{Energy (kWh or MWh) consumed during the billing Month}}{\text{Total number of hours in the billing month}}$$

$$= \frac{\text{On Peak Energy} + \text{Off Peak Energy Consumption}}{\text{Total number of hours in the billing month}}$$

$$= \frac{4{,}320{,}000 \text{ kWh} + 17{,}280{,}000 \text{ kWh}}{(24 \text{ hours/Day}) \cdot (30 \text{ Days/Month})}$$

$$= 30{,}000 \text{ kW or } 30 \text{ MW}$$

b. Peak demand can be calculated by applying Eq. 6.2 to the 30-minute interval during which the highest energy consumption is recorded:

$$\text{Demand} = \text{Power in kW} = \frac{\text{Energy in kWh}}{\text{Time in hours}} = \frac{\text{Energy in kWh}}{\frac{1}{2}\text{hour}}$$

$$= \frac{17{,}500 \text{ kWh}}{0.5 \text{ h}} = 35{,}000 \text{ kW or } 35 \text{ MW}$$

c. Load factor can be calculated by applying Eq. 6.3 as follows:

$$\text{Load Factor} = \frac{\text{Average Demand for the Month, in kW or MW}}{\text{Peak Demand for the Billing Month, in kW or MW}}$$

Using the Average Demand, calculated in part (a) as 30 MW, and the

Peak Demand calculated in part (b) as 35 MW:

$$\text{Load Factor} = \frac{30 \text{ MW}}{35 \text{ MW}} = 0.8571 \text{ or } 85.71\%$$

SERVICE FACTOR

Service factor of electrical equipment such as motors, transformers, and switch-gear can be defined as the ratio of load the equipment can sustain continuously and the load rating of that equipment. Another way to view service factor is that it is a ratio of "safe" operating load to standard (nameplate) load. Service factor is typically expressed in decimal. Occasionally, the decimal value of load factor is presented in percentage form. The formula for service fact is stated as Eq. 6.5 below.

$$\text{Service Factor} = \frac{\text{Safe or Continuous Load, in kW, kVA or hp}}{\text{Nameplate rating of equipment, in kW, kVA or hp}} \tag{6.5}$$

Example 6.2

A 20 hp motor has been tested by a motor manufacturer to safely and continuously sustain a load of 25 hp. What service factor should the manufacturer include on the nameplate of the motor?

Solution:

$$\text{Service Factor} = \frac{\text{Safe or Continuous Load, in kW, kVA or hp}}{\text{Nameplate rating of equipment, in kW, kVA or hp}}$$

$$= \frac{\text{Safe Operating Load}}{\text{Full Load Rating of the Motor}} = \frac{25 \text{ hp}}{20 \text{ hp}} = 1.25$$

COMPUTATION OF LARGE INDUSTRIAL OR COMMERCIAL ELECTRICAL POWER BILLS

Electrical bill calculation for large electrical power consumers is different and more complex than the assessment of monthly residential bills. As supported by the electric bill shown in Figure 6.2, residential electrical bill of an average size American home in the Southeast can be calculated using Eq. 6.6.

$$\text{Baseline Charge} = (\text{Present Reading} - \text{Previous Reading}) \cdot (\text{Rate in \$/kWh}) \qquad (6.6)$$

$$\text{Total Bill} = \text{Baseline Charge} + \text{Special Riders} + \text{Taxes} \qquad (6.7)$$

Example 6.3

If the "present reading" in the residential bill depicted in Figure 6.2 were 85,552, what would be the baseline bill for the month? The energy cost rate for this property remains unchanged, at ¢11.9/kWh. (**Hint:** The baseline charge does not include riders or taxes.)

Solution:

According to Eq. 6.6,

$$\text{Baseline Charge} = (85,552 - 84,552) \cdot (\$0.119/\text{kWh})$$

$$= (1,000) \cdot (\$0.119/\text{kW}) = \$119$$

Service From: FEB 26 to MAR 26 (28 Days)				Your next scheduled meter reading will occur between APR 26 and MAY 01		
PREVIOUS BILL AMOUNT		PAYMENTS (-)		NEW CHARGES (+)	ADJUSTMENTS (+ OR -)	AMOUNT DUE (=)
$51.94		$51.94		$47.35	$0.00	$47.35
METER NUMBER	METER READINGS: PREVIOUS PRESENT	MULTI-PLIER		TOTAL USAGE	RATE SCHEDULE DESCRIPTION	AMOUNT
431400	84552 84938	1		386 KWH	RS - Residential Service Renewable Energy Rider Sales Tax	45.75 .22 1.38
					Amount Due To be drafted on or after	47.35 Apr 06, 2013

Electricity Usage	This Month	Last Year	Our records indicate your telephone number is ???-???-???? . If this is incorrect, please follow the instructions on the back of the bill.
Total KWH	386	N/A	
Days	28	N/A	A late payment charge of 1.0 % will be added to any past due utility balance not paid within 25 days of the bill date.
AVG KWH per Day	14	N/A	
AVG Cost per Day	$1.63	N/A	

FIGURE 6.2   Residential electrical power bill.

Case Study 6.1: EMS-Based Electrical Power Bill – Large Industrial Power Consumer[3]

This case study is modeled on a scenario with large industrial customer in the southeastern region of the United States. The rates in $/kW or $/kWh and the fixed charges are assumed to represent that time frame, during non-summer months. This consumer is assumed to have a 45 MW (megawatt) contract with the power company. This facility is a 24/7, 365-day, operation and a significant portion of its load can be shifted from on-peak periods to off-peak periods, during its daily operations. This overall service contract of 45 MW is split into the following two components:

A. A 30 MW uninterruptible power service on OPT, TOU schedule.
B. A 15 MW, HP, hourly pricing schedule. This schedule applies to incremental, interruptible load and was offered to this customer as a part of DSM (Demand Side Management) program.

The monthly electrical bill computation for this facility is performed automatically by an existing EMS. A screen print of one such bill is shown in Figure 6.3.

The calculation methods associated with the various line items in this bill are explained through a specific bill calculation scenario described and analyzed in Case Study 6.2.

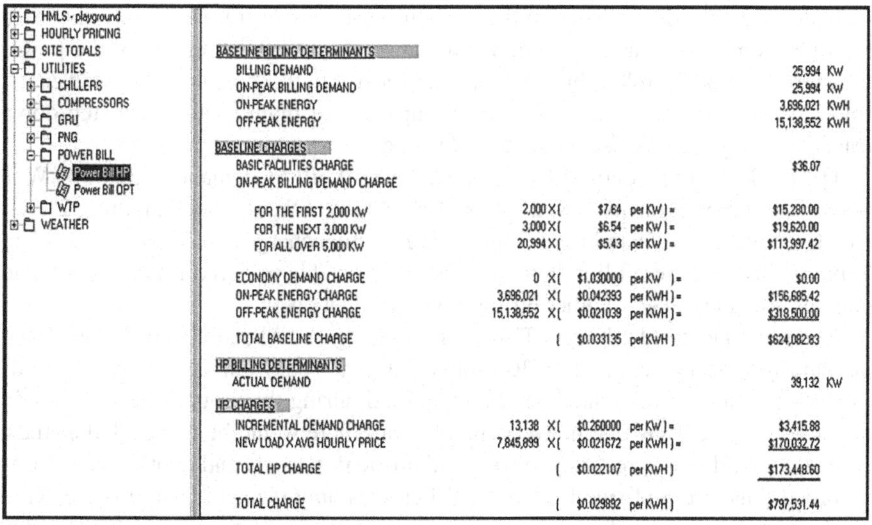

BASELINE BILLING DETERMINANTS			
BILLING DEMAND			25,994 KW
ON-PEAK BILLING DEMAND			25,994 KW
ON-PEAK ENERGY			3,696,021 KWH
OFF-PEAK ENERGY			15,138,552 KWH
BASELINE CHARGES			
BASIC FACILITIES CHARGE			$36.07
ON-PEAK BILLING DEMAND CHARGE			
FOR THE FIRST 2,000 KW	2,000 X(	$7.64 per KW) =	$15,280.00
FOR THE NEXT 3,000 KW	3,000 X(	$6.54 per KW) =	$19,620.00
FOR ALL OVER 5,000 KW	20,994 X(	$5.43 per KW) =	$113,997.42
ECONOMY DEMAND CHARGE	0 X(	$1.030000 per KW) =	$0.00
ON-PEAK ENERGY CHARGE	3,696,021 X(	$0.042393 per KWH) =	$156,685.42
OFF-PEAK ENERGY CHARGE	15,138,552 X(	$0.021039 per KWH) =	$318,500.00
TOTAL BASELINE CHARGE	(	$0.033135 per KWH)	$624,082.83
HP BILLING DETERMINANTS			
ACTUAL DEMAND			39,132 KW
HP CHARGES			
INCREMENTAL DEMAND CHARGE	13,138 X(	$0.260000 per KW) =	$3,415.88
NEW LOAD X AVG HOURLY PRICE	7,845,899 X(	$0.021672 per KWH) =	$170,032.72
TOTAL HP CHARGE	(	$0.022107 per KWH)	$173,448.60
TOTAL CHARGE	(	$0.029892 per KWH)	$797,531.44

Sidebar tree:
- HMLS - playground
- HOURLY PRICING
- SITE TOTALS
- UTILITIES
  - CHILLERS
  - COMPRESSORS
  - GRU
  - PNG
  - POWER BILL
    - Power Bill HP
    - Power Bill OPT
  - WTP
- WEATHER

FIGURE 6.3 Industrial electrical power bill – EMS based.

[3] *Finance and Accounting for Energy Engineers*, by S. Bobby Rauf.

Case Study 6.2 – Electrical Power Baseline Parameters and Bill Calculation – Large Industrial Power Consumer[4]

In an effort to illustrate the interpretation and arithmetic behind the various components of an OPT-based electrical power bill, let us consider data from a specific billing period, recorded and processed in Table 6.1. This table shows all of the pertinent measured data, derived billing parameters, standard charges, tiered demand charge rates, energy charge rates, and computed line item charges for the billing month. In the following sections, we will highlight the arithmetic behind the derived or extended line item charges.

Baseline Billing Determinants: The on-peak billing demand and the billing demand are typically the same. This segment of the bill represents derived and measured data. The measured portion is in the form of the energy consumption (kWh) measured in 15- or 30-minute intervals, which are then divided by 0.25 or 0.5 hours, respectively. This yields demand (kW) for each of the intervals for the entire billing month. The utility company then selects the highest of these demands – registered during the on-peak periods – as the peak demand or billing demand for the month. The other billing determinants consist of the total energy usage during on-peak and off periods. The consumed energy data is measured and recorded for the billing month, in kWh.

Basic Facilities Charge: This charge could be considered to represent administrative cost associated with the generation and processing of the bill. This charge stays, relatively, constant over time.

Extra Facilities Charge: This charge is a means for the utility to recoup its cost embedded in providing "extra facilities" to the customer. In this case study, the extra facilities consisted of a set of redundant transmission lines installed by the power company to enhance reliability of the power service to this customer. In some cases, this type of charge is associated with the upgrade of main switch yard step-down transformers, regulators, separate metering, etc.

On-Peak Billing Demand Charge: The overall billing demand of 26,000 kW is tiered into three segments: first 2,000 kW, next 3,000 kW, and the remaining 21,000 kW. Each of these demand tiers is applied respective charges, in $/kW, as shown in Table 6.1 to yield tiered line item demand charges. These tiered line item demand charges are later added to other billing line item charges.

Economy Demand Charge: This charge is triggered when the highest integrated demand (or the highest 15- or 30-minute interval peak) during off-peak periods exceeds the highest integrated demand recorded during the on-peak periods. So, in essence, one might say that in every monthly bill there would be two peak demands, a peak demand that is recorded during the **on-peak** periods and another one that is recorded during the **off-peak** periods. When economy demand charge is triggered on the premise described above, the difference between the two demands is labeled as the economy demand. The economy demand figure thus derived is multiplied with the stated $1.03 rate multiplier to compute the economy demand charge for

---

[4] *Finance and Accounting for Energy Engineers* by S. Bobby Rauf.

TABLE 6.1
Large Industrial Electric Bill Calculation

Description	Demand and Energy Parameters	Rates	Line Item Charge
Baseline billing determinants			
Billing demand	26,000 kW		
On-peak billing demand	26,000 kW		
On-peak energy usage	4,334,573 kWh		
Off-peak energy usage	14,500,000 kWh		
Baseline charges			
Basic facilities charge		$36.07	$36.07
Extra facilities charge		$13,000	$13,000
On-peak billing demand charge			
For the first 2,000 kW	2,000 ×	$7.64/kW =	$15,280.00
For the next 3,000 kW	3,000 ×	$6.54/kW =	$19,620.00
For demand over 5,000 kW	21,000 ×	$5.43/kW =	$114,030
Economy demand charge	0 ×	$1.03/kW =	$0
On-peak energy usage	4,334,573 ×	$0.042393/kWh =	$183,755.55
Off-peak energy usage	14,500,000 ×	$0.021039/kWh =	$305,065.50
Total baseline charge		$0.033135/kWh	**$650,787.12**
HP billing determinants			
Actual demand	40,000 kW		
HP charges			
Incremental demand charge	14,000 kW ×	$0.26/kW =	$3,640.00
New load × average hourly price	8,805,800 ×	$0.021672/kWh =	$190,839.30
HP subtotal			**$194,479.3**
Total bill for the month			**$845,266.42**

the month. Since, in this case study, the on-peak demand is assumed to be greater than the demands set during the off-peak periods, economy demand is **not** triggered; hence, the economy demand charge is zero.

Energy Charge: This charge accounts for the actual (measured) energy consumed during the billing month. This charge comprises of two components as shown in Table 9.3: the on-peak energy charge and the off-peak energy charge. As apparent from examination of the table, the on-peak rate ($0.042/kWh) is almost twice the magnitude of the off-peak rate ($0.021/kWh).

Total Baseline Charge: The total baseline charge is simply the sum of all line items calculated or identified to that point, which is $650,787.12 in this case.

HP, Hourly Pricing, Billing Determinants: This portion of the bill represents the demand and energy charges associated with the load that fall under the HP contract. The HP demand was measured and billed at 14,000 kW in this case. The energy consumed under the HP contract was 8,805,800 kWh.

HP Charges: Special HP rates are applied to the recorded demand and energy under HP contract, yielding respective line item charges as shown on Table 6.1.

Total Bill for the Month: The last line of the bill represents the sum of total baseline charge and the HP charge amounting to the total bill of **$845,266.42** for the month.

SELF-ASSESSMENT PROBLEMS AND QUESTIONS – CHAPTER 6

1. The BMS at a truck assembly plant that operates 365 days a year is displaying following electrical power and energy consumption data:

    Billing Days in the Current Month: 30

    On-Peak Energy Consumption: 2,880,000 kWh

    Off-Peak Energy Consumption: 11,520,000 kWh

    Three highest 30-minute energy usages for the billing month are (i) 12,500 kWh, (ii) 12,300 kWh, and (iii) 12,290 kWh.

    Assuming this facility is on OPT, TOU, contract with 30-minute demand interval, determine the following:

    a. Average demand.

    b. Peak demand.

    c. The load factor for the current month.

    d. Average annual demand.

2. A 200 kVA transformer has been tested by the manufacturer to safely and continuously sustain a load of 230 kVA. What service factor should the manufacturer include on the nameplate of this transformer?

3. A 5 hp single-phase AC motor, rated at a service factor of 1.10, is being tested at maximum safe load, powered by 230 V_{AC} source. Determine the amount of current drawn by this motor from the power source if the motor efficiency is 90% and the power factor is 0.85.

4. Consecutive electrical power meter readings at a home in Hawaii are listed below. Determine the total electrical power bill for the month of this residence if the flat $/kWh cost rate is ¢ 21/kWh. The renewable energy rider is $15, and the energy sales tax rate is 4%.

    Previous reading: 45,000

    Current or present reading: 46,000

5. If the peak demand in Case Study 6.2 is reduced by 10% through implementation of peak-shaving measures, what would be the baseline cost for the demand portion of the bill?

7 Electrical Machines – Motors and Generators

INTRODUCTION

Electromechanical rotating machines can be generators or motors. Rotating machines are called **motors** when they **consume** electrical energy or convert electrical energy into mechanical energy, work, or torque. Rotating machines are referred to as **generators** when they **produce** electrical energy from mechanical energy, work, or torque. In practical applications, while direct current (DC) machines are almost always single-phase, alternating current (AC) machines can be single-phase or three-phase. In this chapter, we will explore fundamental operating principles and concepts associated with DC and AC motors and generators. The electromagnetic principles behind the operation of generators and motors will be illustrated through simplified electrical diagrams. Basic principles and equations governing important and practical functions and operational parameters of motors and generators will be introduced. Common calculations involving electric motors will be illustrated. Concept of induction motor slip is explained and associated calculations are covered.

DC GENERATOR

A DC generator, also referred to as a **dynamo**, is an electromagnetic device designed to convert mechanical energy or mechanical power – namely, brake horsepower – to electrical energy or electrical power. The electrical energy and power developed in DC dynamos consist of DC and DC voltage. A DC generator is, fundamentally, an AC generator. The feature that differentiates a DC generator's function and output from an AC generator is called a "commutator." Common commutator consists of two rings as shown in Figure 7.1. As shown in Figure 7.1, the current, **I**, and associated voltage are generated in the dynamo by virtue of the motion, movement, or rotation of the coil windings within a magnetic field setup by permanent magnets or, in some cases, by "field windings." Of course, the rotation of the winding coils is caused by force, torque, work-producing system, or energy source such as steam turbines, gas turbines, hydroelectric turbines, and wind turbines. The end result is conversion of mechanical work or energy into electrical energy or electrical power.

As shown in Figure 7.1, and as stipulated by Eqs. 7.1–7.3, key variables, the interaction of which, results in the production of voltage across the windings are as follows:

- Magnetic flux density **B**
- Number of turns, **N,** the coil consists of
- The area of cross-section **A**

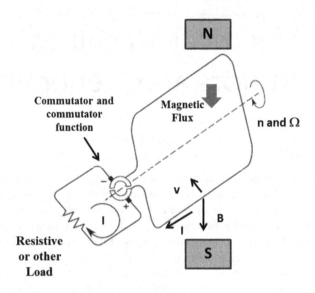

FIGURE 7.1   A DC generator or dynamo.

- The electrical frequency, ω, at which the armature is rotated by external work producing force
- Rotational speed of the armature, Ω,
- The number of poles, **p**.

$$V(t) = V_m \operatorname{Sin}\omega t = \omega\, NAB \operatorname{Sin}\omega t \qquad (7.1)$$

Voltage generated by the dynamo, or DC generator, can be expressed in terms of the rotational speed of the armature, Ω (in rad/s), by applying the ω to Ω conversion formula expressed as Eq. 7.2.

$$\omega = \frac{p}{2}\Omega \qquad (7.2)$$

$$V(t) = \frac{p}{2}\Omega\, NAB \operatorname{Sin}\left(\frac{p}{2}\Omega\right)t \qquad (7.3)$$

Since Ω (in rad/s) is related to the RPM of the DC generator as stated in Eq. 7.4, Eq. 7.3 can be expanded in the form of Eq. 7.5.

$$\Omega = \frac{2\pi n}{60} \qquad (7.4)$$

$$V(t) = \pi n \frac{p}{60} NAB \operatorname{Sin}\left(\frac{p}{2}\Omega\right)t \qquad (7.5)$$

Since V(t) represents the sinusoidal form of the voltage generated by the dynamo, to derive the magnitude of DC, work producing, effective, or RMS portion of this voltage, we can equate the coefficient of the sinusoidal term in Eq. 7.5 to V_{peak}, V_{max} or, simply, V_p or V_m and apply Eq. 1.3. This yields Eq. 7.6 for computation of the DC voltage produced by a dynamo or DC generator.

$$V_{DC} = V_{RMS} = V_{eff} = \frac{V_p}{\sqrt{2}} = \left(\frac{\pi}{2}\right) \cdot \left(\frac{n}{30}\right) \cdot \frac{pNAB}{\left(\sqrt{2}\right)} = \frac{\pi np NAB}{(60) \cdot \left(\sqrt{2}\right)} \qquad (7.6)$$

where

$$V_m = V_P = \left(\frac{\pi}{2}\right) \cdot \left(\frac{n}{30}\right) \cdot pNAB = \frac{\pi np NAB}{60} \qquad (7.7)$$

and

n = Rotational speed of the dynamo, in rpm.
p = Number of poles in the design and construction of the armature. For instance, **one** coil or set of winding with N turns consist of two (2) poles.
N = Number of turns constituting and armature coil.
A = Cross-sectional area portended by the coil in m².
B = Magnet field intensity, in Tesla, or T.

Graphical depiction of the DC voltage computed through Eq. 7.7 is shown in Figure 7.2. If the AC supply voltage in this instance were 120 V, the fully rectified DC voltage would be approximately 108 V. The reader is encouraged to prove that utilizing the equations introduced in earlier chapters.

The relationship between the rotational speed of the generator, n_s, number of poles, **p**, the electrical frequency, **f**, and the angular speed, ω, corresponding to the electrical frequency is given by Eqs. 7.8 7.9.

$$n_s = \text{Synchronous speed} = \text{Rotaional speed} = \frac{120f}{p} \qquad (7.8)$$

$$\omega = 2\pi f \qquad (7.9)$$

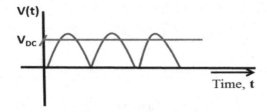

FIGURE 7.2 DC voltage output of a DC generator or dynamo.

Example 7.1

The rotor of a single-phase alternator is rotating at an actual mechanical rotational speed of 2,400 rpm. The alternator consists of four pole construction. The effective diameter of the coil is 0.13 m and the length of the coil loop is 0.22 m. The coil consists of 20 turns. The magnetic flux density has been measured to be 1.15 T. Calculate the RMS, effective, or V_{DC} produced by this generator.

Solution:

The RMS, effective, or DC voltage produced through an alternator or generator can be computed by applying Eq. 7.6:

$$V_{DC} = V_{RMS} = V_{eff} = \frac{V_P}{\sqrt{2}} = \frac{\pi n p\, NAB}{(60)\cdot\left(\sqrt{2}\right)}$$

Given:
  n = 2,400 rpm
  p = 4
  N = 20
  B = 1.15 T
  A = (Eff. diameter of the coil conductor) × (Eff. length of the coil)
    = (0.22 m) × (0.13 m) = 0.0286 m²

$$V_{DC} = V_{RMS} = \frac{\pi n p\, NAB}{(60)\cdot\left(\sqrt{2}\right)}$$

$$= \frac{(3.14)(2,400 \text{ rpm})(4)(20)(0.0286 \text{ m}^2)(1.15 \text{ T})}{(60)\cdot\left(\sqrt{2}\right)} = 233.7 \text{ V}$$

Example 7.2

A two-pole alternator/generator is producing electrical power at an electrical frequency of 60 Hz. (a) Determine the angular speed corresponding to the generated electrical frequency. (b) Determine the rotational (synchronous) speed of the armature/rotor. (c) Determine the angular velocity of the armature/rotor, assuming slip = 0.

Solution:

  a. Angular speed, ω, corresponding to the generated electrical frequency, f, can be calculated using Eq. 7.9:

$$\omega = 2\pi f = (2)\cdot(3.14)\cdot(60 \text{ Hz}) = 377 \text{ rad/s}$$

  b. The rotational or synchronous speed of the armature/rotor is given by Eq. 7.8:

$$n_s = \frac{120 f}{p} = \frac{(120)(60 \text{ Hz})}{2} = 3,600 \text{ rpm}$$

c. Angular velocity of the armature/rotor is simply the rotational speed, in rpm, converted into rad/s. Since there are 2π radians per revolution:

$$\omega_s = 3{,}600 \text{ rev./min} = \left(\frac{3{,}600 \text{ rev.}}{\text{min}}\right) \cdot \left(\frac{2\pi \text{ rad.}}{\text{rev.}(60 \text{ s/min})}\right) = 377 \text{ rad/s}$$

DC MOTOR

A DC motor can be perceived as a DC generator or dynamo operating in reverse. As in the case of a dynamo, a magnet – serving as a stator – provides the magnetic field (**B** and **H**) that interacts with the DC flowing through the rotor. The rotating torque in motors due to the interaction of the flow of current through the coil and the magnetic field and this phenomenon is governed by Lorentz's law. The DC flowing through the rotor windings is supplied from an external DC voltage or current source, via the commutator rings that are stationary. Basic construction of a DC motor is illustrated in Figure 7.3.

In other words, a DC motor is a mechanically commutated electric motor powered from DC. The current in the rotor is switched by the commutator. The relative angle between the stator and rotor magnetic flux is maintained near 90°, which generates the maximum torque.

In DC motors, different connections of the field and armature winding provide different inherent speed/torque regulation characteristics. Insofar as the control of the **speed of a DC motor** is concerned, it can be controlled by changing either the **voltage applied** to the armature or by changing the **field current**. Since voltage is related to current by Ohm's law, $V = I \cdot R$, speed control can be accomplished through introduction of variable resistance in the armature circuit or field circuit. However, modern DC motors are often controlled by power electronics systems called DC drives.

Historically, the introduction of DC motors to run machinery eliminated the need for steam or internal combustion engines. Case in point, the application of DC

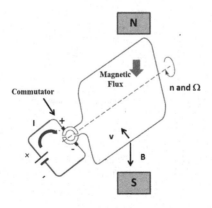

FIGURE 7.3 DC motor.

motors as the motive power in locomotives. An advantage of DC motors is that they can be operated **directly** from DC power supply, with variable voltage, without the need for an inverter. This positions DC motors well for electric vehicles' application. Despite the dominance of AC induction motors in myriad applications, DC motors continue to play an important role in providing motive power in small toys and disk drives. Overall, DC motors continue to provide an alternative to inverter-driven AC motors where motor speed variation is needed.

Some important properties and differences between different types of DC motors are listed below:

As somewhat obvious from the circuit diagrams of different types of DC motors depicted in Figure 7.4:

- A "**Separately Excited**" DC motor is one in which the field winding has a separate DC voltage source for energization of the field windings.
- A "**Series Excited**" DC motor is one in which the field winding is in series with the armature windings. So, the field current and the armature current are the same. It's ostensible then that the independent armature and field current-based control is not available in series wound DC motors. As the load is increased, the speed drops.
- A "**Shunt Excited**" motor is one in which the field winding and the armature windings are in parallel, which implies that the field and the armature

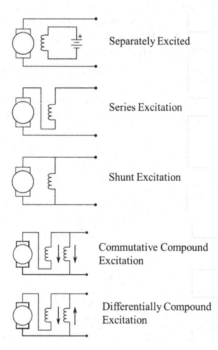

FIGURE 7.4   DC motor types.

winding voltage cannot be changed independently. With shunt-wound field windings, the speed stays constant as more load is added.

- A "**Commutative Compound Excited**" DC motor, AKA, Long Shunt Compound Wound DC Motor, is one in which both types of field windings are present: a **shunt field** winding and a **series field** winding. The currents through shunt and the series field windings are flowing in the same direction. With shunt and series field windings, commutative compound motors exhibit the speed regulation attribute of a typical shunt field winding DC motor as well as the high-staring torque characteristic of the series excited DC motor. The commutative compound DC motor is more common than the differentially compound DC motor, discussed next.
- A "**Differentially Compound Excited**" DC motor, AKA, Short Shunt Compound Wound DC Motor is one in which, like the commutative compound motor, both types of field windings are present: a shunt field winding and a series field winding. However, in the differentially compound excited motor, the currents and fluxes, through shunt and the series field windings, are flowing in the opposite direction. This type of compound DC motor is **not** commonly used due to the fact that the opposing fluxes in the shunt and series field windings result in loss of speed control under overload, or high armature current, conditions.

Circuit Models for Series and Shunt DC Motors: The electrical circuit models, utilized to analyze DC motors performance, are depicted in Figure 7.5 along with

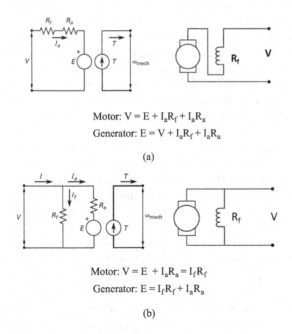

Motor: $V = E + I_a R_f + I_a R_a$

Generator: $E = V + I_a R_f + I_a R_a$

(a)

Motor: $V = E + I_a R_a = I_f R_f$

Generator: $E = I_f R_f + I_a R_a$

(b)

FIGURE 7.5 Series and shunt DC field excitation circuits. (a) Series DC field excitation machine. (b) Shunt DC field excitation machine.

associated equations – based on the application of Kirchoff's voltage law under each respective scenario – for motor and generator operation of the DC series and shunt excited machines.

where
  E = EMF
  I_a = Armature current, positive for a motor and negative for a generator
  I_f = Field current
  T = Mechanical torque that results in motor brake horsepower.
  R_a = Armature resistance
  R_f = Field winding resistance
  ω_m = Mechanical frequency in rad/s

Note that the brush resistance is also considered to be in series but is often included in the armature resistance specification and the magnetic flux varies with the armature current.

Brushed and Brushless DC Motors: At the most basic level, there are brushed DC motors and brushless DC motors. So, what are "brushes?" Brushes, unlike in the layperson context, are semicircular or circular electrically conductive rings, in contact with the circumference of the motor shaft, designed to pass DC voltage and current from an external DC source to the rotating shaft of the rotor to facilitate the set-up of magnetic flux in the rotor windings and core. Brushes, because of their mechanical wear, have limited life, and therein lies the advantage of brushless motors. In brushless DC motors, permanent magnets are used to generate magnetic flux around the rotor, and that magnetic flux, in turn, interacts with the magnetic field/flux produced in the stationary windings of the stator or armature. In essence, a brushless DC motor is essentially flipped inside out, eliminating the need for brushes to flip the electromagnetic field. A brush-type DC motor, typically small, is fed from a regular constant DC power supply or a battery. Multiple coils or poles in a regular DC motor help the motion or rotation to be smooth so that when current through the coil is perpendicular, the tangential speed is parallel to the magnetic lines of flux.

Brushes eventually wear out, sometimes causing dangerous sparking, limiting the lifespan of a brushed motor. Brushless DC motors are quiet, lighter, and more durable. Because, in most DC brushless motor applications, microprocessor-based control systems are used to control the electrical current through the armature of the brushless DC motors, one can achieve much more precise motion control with brushless DC motors.

Operation of almost all motors involves the generation of back electromotive force (EMF). In accordance with laws of electromagnetics, back EMF opposes the applied voltage and, therefore, tries to reduce the armature current. The back EMF is directly proportional to the speed of the motor. Therefore, back EMF is very small at the start of the motor and then gradually increases as the motor speeds up. Because at the start of the DC motor the back EMF is very small, the armature current can become very high. Therefore, some sort of load or active mechanism must be included to reduce the voltage at the start.

AC ALTERNATOR

The basic construction and premise of operation is similar to a dynamo with the exception of the fact that the commutator is **unnecessary** and therefore absent. Another salient difference between a dynamo and an AC alternator is that the roles and the properties of the stator and rotor are reversed. In an AC alternator, as shown in Figure 7.6a, the magnetic field is produced by the **rotating rotor** and the **stator serves as an armature**. The key reason for the armature – the segment of the generator where the generated current and EMF (voltage) are harnessed – to serve as a stationary "exoskeleton" is that large induced currents require robust insulation of the armature windings. In addition, with large currents, larger magnetic forces and torques are in play, which makes it important to secure or anchor the windings in a rugged structure. As with a DC generator, the power in AC alternators is fundamentally produced in **sinusoidal form**. Since the output of an AC alternator does not need to be rectified to DC form, commutator function is not needed. Construction of a basic AC alternator or generator is shown in Figure 7.6a. The output of a typical AC alternator is depicted in Figure 7.6b.

As observable in Figure 7.6a, the magnetized rotor is being rotated counterclockwise by external means. As the magnetized rotor rotates, the magnetic field emanating from the north pole and terminating into the south pole cuts through the three-phase coils in the armature. Movement of the magnetic field through the coils initiates current flow and potentials in the three coils. As noted in Figure 7.6a, the

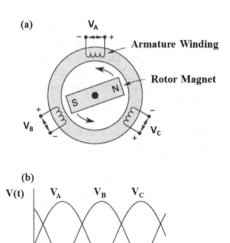

FIGURE 7.6 AC alternator construction and output waveform. (a) Generator/alternator construction. (b) Output of a typical AC alternator.

voltage thus developed across coil A is V_A, voltage developed across coil B is V_B, and the voltage developed across coil C is V_C.

The overall principle of operation and construction of a **single-phase AC alternator/generator** is similar to a **three-phase AC alternator/generator**, with the exception of the fact that the armature windings consist of one set of windings for the harnessing of single-phase AC.

While the complete representation of the AC sinusoidal voltage generated by single-phase AC generator is given by Eq. 7.5 (angle of the voltage is assumed to be "0"), the RMS and peak voltages would be stipulated by Eqs. 7.6 and 7.7, respectively. For the sake of reader's convenience, these equations are restated below:

$$V(t) = \pi n \frac{p}{60} NAB \sin\left(\frac{p}{2}\Omega\right)t \tag{7.5}$$

$$V_{DC} = V_{eff} = \frac{V_p}{\sqrt{2}} = \left(\frac{\pi}{2}\right)\cdot\left(\frac{n}{30}\right)\cdot\frac{pNAB}{(\sqrt{2})} = \frac{\pi np\,NAB}{(60)\cdot(\sqrt{2})} \tag{7.6}$$

$$V_m = V_p = \left(\frac{\pi}{2}\right)\cdot\left(\frac{n}{30}\right)\cdot pNAB = \frac{\pi np\,NAB}{60} \tag{7.7}$$

Example 7.3

A four-pole single-phase AC generator consists of windings constituting 90 series paths and is driven by a propane prime mover (engine). The effective or mean length of the armature is 20 cm and the cross-sectional radius of the armature is 5 cm. The armature is rotating at 2,000 rpm. Each armature pole is exposed to a magnetic flux of 1.5 T. The efficiency of this generator is 92% and it is rated 1.5 kW. Determine the following:

(a) The maximum voltage generated. (b) The RMS voltage generated. (c) The horsepower rating of the generator. (d) The horsepower output of the prime mover.

Solution:

The maximum voltage, V_m, generated by this alternator is given by Eq. 7.7.

$$V_m = V_p = \left(\frac{\pi}{2}\right)\left(\frac{n}{30}\right)\cdot pNAB = \frac{\pi np\,NAB}{60} \tag{7.7}$$

Given:
   $n = 2,000$ rpm
   $p = 4$
   $N = $ Number of series paths $= 90$
   $B = 1.5$ T

A = (Eff. diameter of the coil conductor) × (Eff. length of the coil)
  = (2 × 5 cm) × (20 cm) = (0.1 m) × (0.2 m) = 0.02 m²

a.

$$V_m = \frac{\pi n p \, NAB}{60} = \frac{(3.14) \cdot (2000 \text{ rpm})(4)(90)(0.02 \text{ m}^2)(1.5 \text{ T})}{60}$$

$$= 1,130 \text{ V}$$

b. The RMS voltage could be calculated by using Eq. 7.6, or simply by dividing V_m, from part (a), by the square root of 2 as follows:

$$V_{RMS} = V_{eff} = \frac{V_m}{\sqrt{2}} = \frac{1,130}{\sqrt{2}} = 799 \text{ V}$$

c. The horsepower rating of this generator is the power **output** rating specified in hp, premised on the stated **output capacity** of 1.5 kW. Therefore, application of the 0.746 kW/hp conversion factor yields:

$$P_{hp} = \frac{P_{kW}}{0.746 \text{ kW/hp}} = \frac{1.5 \text{ kW}}{0.746 \text{ kW/hp}} = 2.01 \text{ hp} \cong 2 \text{ hp}$$

d. The horsepower rating of the prime mover – or the propane fired engine – would need to offset the inefficiency of the AC generator. Therefore, based on the given 92% efficiency rating of the generator:

$$P_{hp \text{ - Prime Mover}} = \frac{P_{hp \text{ - Gen.}}}{Eff._{Gen.}} = \frac{2.01 \text{ hp}}{0.92} = 2.19 \text{ hp}$$

AC INDUCTION MOTOR

An induction motor is also referred to as an asynchronous motor; primarily, because a typical AC induction motor has a certain amount of "slip" (discussed in detail in the next section) and the shaft speed is less than the motor's synchronous speed. An AC induction motor can be considered as a special case of a transformer with a rotating secondary (rotor) and a stationary primary (stator). The electromagnetic field in the primary (stator) rotates at synchronous speed, n_s, as given by Eq. 7.9. The stator magnetic field, rotating at the synchronous speed, cuts through the rotor windings. This rotating or moving magnetic field – in accordance with Faraday's law – induces EMF and current through the rotor windings. Due to the fact that the rotor windings are inductive – and possess inductive reactance, X_L – the current induced in the rotor windings lags behind the induced voltage, resulting in a definite amount of slip.

$$n_s = \text{Synchronous speed} = \frac{120f}{p} = \frac{60\omega}{\pi p} \qquad (7.9)$$

In essence, induction motors allow the transfer of electrical energy from the primary (stator) windings to the secondary (rotor) windings. The primary or secondary windings – or the stator and the rotor – are separated by an air gap. In a wound rotor, the wire wound construction is similar to the rotor construction of AC alternators. However, with AC motors, a more common alternative to a wound rotor is a **squirrel cage rotor**, which consists of copper or aluminum bars – in lieu of insulated wire – embedded in slots of the cylindrical iron core of the rotor. See Figure 7.7. Construction of a common, TEFC, Totally Enclosed Fan Cooled, motor is illustrated in Figure 7.7. Major components of an AC induction motor, i.e. squirrel cage rotor, motor shaft, armature windings, cooling fan, and bearings, are labeled and highlighted in Figure 7.7.

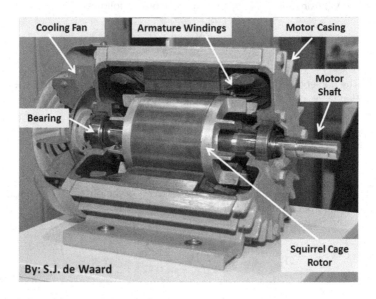

FIGURE 7.7   AC induction motor construction.

MOTOR SLIP

The rotational speed of rotor (secondary) of an induction motor – rotational speed of the rotor is the also the actual speed of the motor shaft – lags behind the synchronous speed, n_s. This difference in rotational speeds can range between 2% and 5% and is called **slip**. Slip can also be defined as the difference between synchronous speed and operating speed, at the same frequency, expressed in rpm or in percent or ratio of synchronous speed. Slip can be defined mathematically as represented by Eqs. 7.10 and 7.11.

$$\text{Slip} = \left(\frac{n_s - n}{n_s} \right) = \left(\frac{\Omega_s - \Omega}{\Omega_s} \right) \tag{7.10}$$

$$\% \text{ Slip} = \left(\frac{n_s - n}{n_s} \right) \cdot (100)\% = \left(\frac{\Omega_s - \Omega}{\Omega_s} \right) \cdot (100)\% \tag{7.11}$$

where

n_s = Synchronous speed of the AC induction motor, in rpm

= Rotational speed of the armature AC induction EMF, in rpm

n = Actual speed of the AC induction motor shaft, in rpm

= Rotational speed of the EMF in the rotor, expressed in rpm

Ω_s = Rotational speed of the armature AC induction EMF, in rad/s

= Synchronous speed of the AC induction motor, rad/s

Ω = Rotational speed of the rotor EMF, in rad/s.

= Actual speed of the rotor, in rad/s

Induction motors are made with slip ranging from less than 5% up to 20%. A motor with a slip of 5% or less is known as a **normal-slip motor**. A normal-slip motor is sometimes referred to as a **constant speed motor** because the speed changes very little from no-load to full-load conditions. Higher slip characteristics of motors are not always "undesirable." Since low slip is often accompanied by instantaneous imparting of a large amount of torque – which can be detrimental to the material and mechanical integrity of the shaft – to the motor shaft, motors with slip over 5% are often used for hard to start applications.

Typically, at the rated full load, slip ranges from more than 5% for small or special purpose motors to less than 1% for large motors. Speed variations due to slip can cause load-sharing problems when differently sized motors are mechanically connected. Slip can be reduced through various means. Due to the progressive decline in the cost and continuous technological improvements in the VFD, variable frequency drive, technology, they offer the most optimal solution for mitigating undesirable effects associated with slip.

A common four-pole motor with a synchronous speed of 1,800 rpm may have a no-load speed of 1,795 rpm and a full-load speed of 1,750 rpm. The rate of change of slip is approximately linear from 10% to 110% load, when all other factors such as temperature and voltage are held constant.

MOTOR TORQUE AND POWER

Torque generated or produced by a motor is equivalent to the amount of **external work** performed **by** the motor. In the absence of transactions involving other forms of energy, such as heat, the amount of work performed, or torque applied by a motor, is equal to the energy stored in the external system in the form of potential energy or kinetic energy. Therefore, torque, work, and energy are equivalent entities. This is further corroborated by the fact that torque, work, and energy can be quantified or measured in terms of equivalent units, i.e. ft-lbf, BTU, calories, therms, N-m, joules, kWh, etc.

On the other hand, power, as explained earlier in this text, is the rate of performance of work, rate of delivery of energy rate of application of torque, etc. In order to illustrate the difference between torque and power, let's consider the example of a crank wheel that is being turned at a certain rate. Suppose the wheel has a one-foot long crank arm and takes a force of one pound (lbf) to turn at steady rate. The torque required would be one-pound times one foot or one foot-pound, 1 ft-lbf. If we were to apply this torque for 2 seconds, the power delivered would be ½ ft-lbf/s. Now let's assume that we continue to apply the same amount of torque, i.e. 1 ft-lbf but twice as fast or for 1 second. Then we would have a case where the torque remains the same, or 1 ft-lbf, but the rate of delivery of torque would double to 1 ft-lbf/s.

Application of motors in mechanical or electromechanical systems often challenges engineers to assess the **amount** of torque involved. However, unlike power output or input of a motor which can be assessed through simple and safe measurement of voltage and current – with the help of help of held instruments, such as the handheld voltmeter or clamp-on ammeter – measurement of torques produced by large motors requires large, stationary, and rather unwieldy systems called dynamometers. See Figure 7.8. Therefore, if the motor power is known in hp or kW and if the actual shaft speed, **n** (rpm), is available, the torque can be calculated by using Eqs. 7.12 and 7.13.

$$\text{Torque}_{\text{(ft-lbf)}} = \frac{5,250 \times P_{\text{(horsepower)}}}{n_{\text{(rpm)}}} \tag{7.12}$$

$$\text{Torque}_{\text{(N-m)}} = \frac{9,549 \times P_{\text{(kW)}}}{n_{\text{(rpm)}}} \tag{7.13}$$

The torque computation equations stated above are derived from the basic torque, angular speed, and power equation stated as Eq. 7.14.

$$P = T \cdot \omega \text{ or, } T = \frac{P}{\omega} \tag{7.14}$$

where
   P = Power delivered by the motor, in hp or Watts
   T = Torque delivered by the motor, in ft-lbf or N-m or Joules
   ω = Actual rotational speed of the motor shaft, in rad/s

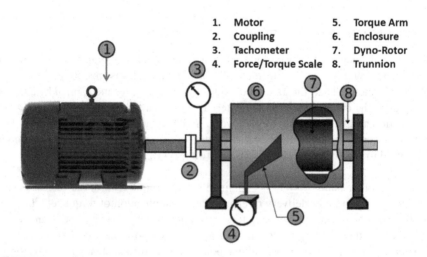

1.	Motor	5.	Torque Arm
2.	Coupling	6.	Enclosure
3.	Tachometer	7.	Dyno-Rotor
4.	Force/Torque Scale	8.	Trunnion

FIGURE 7.8   Motor dynamometer.

SINGLE-PHASE AND THREE-PHASE MOTOR LINE CURRENT COMPUTATION

While copious tables, charts, handbooks, and pocket cards include full-load current information for various (standard) motor sizes, at standard voltages, when accuracy is of the essence, one must **compute** correct current values pertaining to specific voltage, efficiency, power factor, and actual load. Therein rests the value of line current calculation equations for single-phase and three-phase motors. Those equations and their derivation are summarized below.

SINGLE-PHASE AC INDUCTION MOTOR CURRENT

$$S_{L,1\text{-}\phi} = \text{Full load apparent power on the line side of single phase motor}$$

$$= \frac{\text{Full Load Rating of Motor (in Watts)}}{(\text{Pf}) \cdot (\text{Eff.})} \tag{7.15}$$

or

$$\left| S_{L,1\text{-}\phi} \right| = \text{Magnitude of apparent power (drawn) on the line side of a single}$$

$$\text{phase motor} = \frac{P_{1\text{-}\phi}(\text{in Watts})}{(\text{Pf}) \cdot (\text{Eff.})} \tag{7.16}$$

Since the magnitude of single phase apparent power, $\left| S_{L,1\text{-}\phi} \right| = \left| V_L \right| \cdot \left| I_L \right|$,

$$\text{then } \left| I_L \right| = \frac{\left| S_{L,1\text{-}\phi} \right|}{\left| V_L \right|} \tag{7.17}$$

Therefore,

$$\text{Single Phase AC Line Current drawn, } \left| I_L \right| = \frac{\left| S_{L,1\text{-}\phi} \right|}{\left| V_L \right|} = \frac{P_{L,1\text{-}\phi} \text{ (in Watts)}}{\left| V_L \right| (\text{Pf}) \cdot (\text{Eff.})} \tag{7.18}$$

THREE-PHASE AC INDUCTION MOTOR CURRENT

$$S_{L,3\text{-}\phi} = \text{Full Load Apparent (3-}\phi\text{) Power on the Line Side}$$

$$= \frac{\text{Full Load (3-}\phi\text{) Rating of Motor (in Watts)}}{(\text{Pf}) \cdot (\text{Eff.})} \tag{7.19}$$

or

$$\left| S_{L,3\text{-}\phi} \right| = \text{Magnitude of Full Load Apparent Power on the Line Side of}$$

$$\text{a 3-phase motor} = \frac{P_{3\text{-}\phi}(\text{in Watts})}{(\text{Pf}) \cdot (\text{Eff.})} \tag{7.20}$$

Since the magnitude of 3-ϕ (three-phase) apparent power is given as: $\left|S_{L,3\text{-}\phi}\right| = \sqrt{3}\left(\left|V_L\right|\right)\cdot\left(\left|I_L\right|\right)$,   (7.21)

by rearranging this equation, the 3-ϕ phase line curent, $\left|I_L\right| = \dfrac{\left|S_{L,3\text{-}\phi}\right|}{\sqrt{3}\left(\left|V_L\right|\right)}$

Therefore,

The 3-ϕ phase line curent, $\left|I_L\right| = \dfrac{\left|S_{L,3\text{-}\phi}\right|}{\sqrt{3}\left(\left|V_L\right|\right)} = \dfrac{P_{L,3\text{-}\phi}\ (\text{in Watts})}{\sqrt{3}\left(\left|V_L\right|\right)(Pf)\cdot(Eff.)}$ (7.22)

where

$\left|I_L\right|_{1\text{-}\phi}$ = Single-phase RMS line current, measured in Amp
$\left|I_L\right|_{3\text{-}\phi}$ = Three-phase RMS line current, measured in Amp
$\left|V_L\right|$ = RMS line to line, or phase to phase, voltage, measured in volts
$\left|S_L\right|_{1\text{-}\phi}$ = Apparent power drawn by a single-phase motor, in VA or kVA, etc.
$\left|S_L\right|_{3\text{-}\phi}$ = Apparent power drawn by three-phase motor, in VA or kVA, etc.
$P_{1\text{-}\phi}$ = Real power demanded by a single-phase motor load, in W, or kW, etc.
$P_{3\text{-}\phi}$ = Real power demanded by a three-phase motor load, in W, or kW, etc.
Pf = Power factor of the motor, as specified on motor nameplate
Eff. = Motor efficiency, as specified on motor nameplate

Example 7.4

A three-phase, four-pole, AC induction motor, rated 150 hp, is operating at full load, 50 Hz, 480 V_{rms}, efficiency of 86%, power factor of 95%, and a slip of 2%. Determine the (a) motor shaft speed, in rpm, (b) torque developed, in ft-lbf, (c) line current drawn by the motor, and (d) the amount of reactive power, Q, sequestered in the motor under the described operating conditions.

Solution:

Given:
$P_{L,3\text{-}\phi}$ = Real power or rate of work performed by the motor = 150 hp
    = (150 hp)·(746 W/hp) = 111,900 W
p = Four poles
V_L = 480 V_{RMS}
Pf = 95% or 0.95
Eff. = 86% or 0.86
n_s = Synchronous speed, in rpm = ?
Slip, s = 2%
f = Frequency of operation = 50 Hz

a. **Shaft or motor speed:** Rearrange and apply Eq. 7.10:

$$\text{Slip} = s = \left(\frac{n_s - n}{n_s} \right)$$

And, by rearrangement: $n = n_s(1-s)$

Next, we must determine the synchronous speed of the motor by applying Eq. 7.9:

$$n_s = \text{Synchronous speed} = \frac{120f}{p} = \frac{(120) \cdot (50)}{4} = 1{,}500 \text{ rpm}$$

$$\therefore n = (1{,}500 \text{ rpm})(1 - 0.02) = 1{,}470 \text{ rpm}$$

b. **Torque developed, in ft-lbf:** There are multiple methods at our disposal for determining the torque developed. Formulas associated with two common methods are represented by Eqs. 7.12–7.14. Since the power is available in hp and the rotational speed in rpm, apply Eq. 7.12:

$$\text{Torque}_{(\text{ft-lbf})} = \frac{5{,}250 \times P_{(\text{horsepower})}}{n_{(\text{rpm})}} = \frac{5{,}250 \times 150 \text{ hp}}{1{,}470 \text{ rpm}} = 536 \text{ ft-lbf}$$

Note: The reader is encouraged to prove this result through application of Eq. 7.14.

c. **Line current drawn by the motor:** Full-load line current is the current the motor draws from the power source, or utility, in order to sustain the full load. Therefore, the determination of line current at full load requires that the full-load output of 150 hp be escalated in accordance with motor efficiency and power factor to compute the line current. Application of Eq. 7.22 takes this into account in the computation of three-phase motor line current, after the motor hp rating (or actual load) is converted into watts:

Therefore,

$$\text{The 3-}\phi \text{ phase line curent, } |I_L| = \frac{|S_{L,3\text{-}\phi}|}{\sqrt{3}\,(|V_L|)} = \frac{P_{L,3\text{-}\phi} \text{ (in Watts)}}{\sqrt{3}\,(|V_L|)(\text{Pf}) \cdot (\text{Eff.})}$$

Three-phase (total) real power was converted into watts under "Given" as $P_{L,3\text{-}\phi} = 111{,}900 \text{ W}$

Therefore,

$$\text{The 3-}\phi \text{ phase line curent, } |I_L| = \frac{111{,}900 \text{ W}}{\sqrt{3}\,(480 \text{ V}_{\text{RMS}})(0.95) \cdot (0.86)} = 165 \text{ A}$$

d. **Reactive power, Q:** There are multiple approaches available to us for determination of reactive power Q for this motor application. We will utilize the power triangle or apparent power component vector method:

$$S = P + jQ, \text{ and } S^2 = P^2 + Q^2, \text{ or, } Q^2 = S^2 - P^2 \text{ and } Q = \sqrt{S^2 - P^2}$$

Real power "P" was computed earlier as 111,900 W, and apparent power S can be assessed using Eq. 7.20:

$$\left|S_{L.3\text{-}\phi}\right| = \frac{P_{3\text{-}\phi}\ (\text{in Watts})}{(\text{Pf})\cdot(\text{Eff.})} = \frac{111,900\ \text{W}}{(0.95)\cdot(0.86)} = 136,965\ \text{VA}$$

Therefore,

$$Q = \sqrt{S^2 - P^2} = \sqrt{136,965^2 - 111,900^2} = \sqrt{6,237,665,433}$$
$$= 78,979\ \text{VAR}$$

Example 7.5

A four-pole induction motor operates on a three-phase, 240 V_{rms} line-to-line supply. The slip is 5%. The operating (shaft) speed is 1,600 rpm. What is most nearly the operating frequency?

Solution:

Given:
p = Four poles
n = Shaft or motor speed = 1,600 rpm
n_s = Synchronous speed, in rpm = ?
Slip, s = 5% or 0.05
f = Frequency of operation = ?

Since this case involves synchronous speed and slip, it requires the application of Eqs. 7.9 and 7.10. As apparent from examination of Eq. 7.9, the operating frequency can be determined by rearranging the equation if the synchronous speed n_s were known.

$$\text{Slip} = s = \left(\frac{n_s - n}{n_s}\right)\ \text{or,}\ n_s = \left(\frac{n}{1-s}\right) = \left(\frac{1,600}{1-0.05}\right) = 1,684\ \text{rpm} \qquad (7.10)$$

$$n_s = \text{Synchronous speed} = \frac{120f}{p}\ \text{or,}\ f = \frac{(p)\cdot(n_s)}{120} = \frac{(4)\cdot(1,684)}{120} = 56\ \text{Hz} \qquad (7.9)$$

SYNCHRONOUS MOTORS

Synchronous motors are AC induction motors that have no slip. In other words, in synchronous motors, the speed of rotor or motor shaft is the same as the rotational speed of the stator winding magnetic field. This is accomplished through the excitation of the rotor through the field windings such that the rotor is able to "**catch up**" to the rotational speed of the stator. This characteristic permits synchronous motors to induce **leading currents** in the branch circuit. Therefore, synchronous motors can serve as alternative means for power factor correction; albeit, application of power factor correcting capacitors is more economical and effective in most lagging power factor situations.

Example 7.6

A three-phase induction motor delivers 550 kW at a power factor of 82%. Determine the size of synchronous motor – in kVA – that should be installed to carry a load of 250 hp load and, at the same time, raise the (combined) power factor to 95%.

Solution:

Given:
P_I = Real power delivered by the 3-ϕ induction motor = 550 kW
P_s = Real power contributed by the synchronous motors = 250 hp =
   (250 hp) × (0.746 kW/hp) = 186.5 kW
Pf_i = Initial power factor = 82% = 0.82
Pf_f = Final power factor = 95% = 0.95

To solve this problem, as in many others, we will begin with the final state and work our way upstream. With the final combined power factor of 95% in mind, we will work our way back to the required synchronous motor specifications. The apparent power rating, **S**, of the synchronous motor that must be installed to achieve the combined "system" power factor of 95% (0.95), while contributing 250 hp toward the system's real power, P_T, requirement can be determined if we could assess the Q_T (kVAR) contributed by the synchronous motor in an effort to raise the combined power factor of the two motor system to 95% (0.95). This can be accomplished through assessment of final (combined) S_T and P_T values and application of the power triangle equation:

$$S_T = P_T + jQ_T, \text{ and } S_T{}^2 = P_T{}^2 + Q_T{}^2, \text{ or, } Q_T{}^2 = S_T{}^2 - P_T{}^2 \text{ and,}$$

$$Q_T = \sqrt{S_T{}^2 - P_T{}^2}$$

The combined real power of the induction motor and the synchronous motor:

$$P_T = 550 \text{ kW} + 186.5 \text{ kW} = 736.5 \text{ kW}$$

$$\text{Since } S_T \, Cos\theta = P_T, \; S_T = \frac{P_T}{Cos\theta},$$

And since θ_T, the final power factor angle $= Cos^{-1}(0.95) = 18.19°$,

$$S_T = \frac{P_T}{Cos\theta_T} = \frac{736.5 \text{ kW}}{Cos(18.19°)} = 775.26 \text{ kVA}$$

Therefore,

$$Q_T = \sqrt{S_T{}^2 - P_T{}^2} = \sqrt{775^2 - 736^2} = \sqrt{59,337} = 244 \text{ kVAR}$$

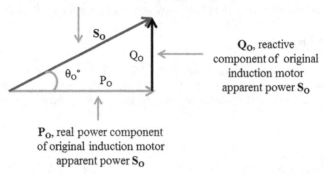

FIGURE 7.9   Power triangle.

Now, in order to determine Q_S, the reactive power contributed by the synchronous motor, we must subtract the original reactive power, Q_O, from the final, total, reactive power, Q_T. However, Q_O is unknown and can be determined through the power triangle as follows:

$$\text{Since } Tan(\theta_O) = \frac{Q_O}{P_O}, \text{ or, } Q_O = P_O\, Tan(\theta_O) = 550 \text{ kW } Tan(\theta_O)$$

$$\text{And, } \theta_O = Cos^{-1}(Pf_O) = Cos^{-1}(0.82) = 34.92°;$$

where Pf_O is the original power factor.

Therefore,

$$Q_O = 550 \text{ kW } Tan(34.92°) = 550 \text{ kW}(0.698) = 384 \text{ kVAR}$$

And, reactive power contributed by the synchronous motor would be:

$$Q_S = Q_O - Q_T = 384 \text{ kVAR} - 244 \text{ kVAR} = 140 \text{ kVAR}$$

Therefore,

$$S_S = P_S + jQ_S, \text{ and } S_S^2 = P_S^2 + Q_S^2, \text{ or, } S_S = \sqrt{P_S^2 + Q_S^2}$$

or

$$S_S = \sqrt{(186.5 \text{ kW})^2 + (140 \text{ kVAR})^2} = 233 \text{ kVA}$$

MOTOR STARTING METHODS FOR INDUCTION MOTORS

Often, the starting method for an induction motor depends on the type of motor. The five basic types of small induction motors are as follows:

1. Single-phase capacitor start
2. Capacitor run
3. Split phase
4. Shaded pole
5. Small polyphase induction motors

Because of the absence of a rotating EMF field in the armature, a single-phase induction motor requires separate starting circuitry to provide a rotating field to the motor. A starting circuit is needed to determine and "set" the direction of rotation due to the fact that the regular armature windings of a single-phase motor can cause the rotor to turn in either direction. Therefore, rotation direction is initiated in certain smaller single-phase motors by means of a "**shaded pole**" with a copper wire "**turn**" around a section of the pole. Then, the current induced in the shaded pole turn lags behind the supply current, creating a delayed magnetic field around the shaded part of the pole face. This imparts sufficient rotational field energy to start the motor in a specific direction. The shaded pole starting method is often employed in motors installed in applications such as small fans and pumps.

A common approach for staring larger single-phase motors is to incorporate a second stator (armature) winding that is fed with an "out-of-phase" current. The out-of-phase current may be created by feeding the winding through a **capacitor** or through an **impedance** that is different from the main winding. Often, the second "starting" winding is disconnected once the motor has accelerated up to normal steady-state speed. This is accomplished, commonly, through either a **centrifugal switch** or a **thermistor** – the thermistor heats up and increases its resistance, thereby reducing the current through the starting winding to a negligible level. Some designs maintain the starting winding in the circuit to supplement the regular motor torque.

Since three-phase, or polyphase, motors possess a rotating armature EMF, the direction of rotation is determined by phase sequence, ABC versus ACB. Hence, during start-up or commission of three-phase motors, if the motor rotation is incorrect, two of the three-phase conductors are "swapped" to reverse the direction of rotation. This reversing starter design – in **FVR, Full Voltage Reversing starters** – is premised on the "phase-swapping" principle to reverse the direction of rotation of motors, such as in conveyor or fan applications where direction reversal is required. Self-starting polyphase induction motors produce torque even when stationary. Common starting methods for larger polyphase induction motors are as follows:

1. Direct online starting
2. Reduced-voltage reactor or autotransformer starting
3. Star–delta starting
4. Application of solid-state soft starting systems
5. Use of VFDs for electronically controlled starting – and normal motor operation – through variation of frequency and voltage

In squirrel cage polyphase motors, the rotor bars are designed and shaped according to the desired speed-torque characteristics. The current distribution within the rotor bars varies as a function of the frequency of the induced current. In locked rotor scenarios – or when the rotor is stationary – the rotor current has the same frequency as the stator current, and the current tends to travel at the outermost parts of the cage rotor bars due to the **"skin effect."** The squirrel cage rotor bars are designed to meet the required speed-torque characteristics and to limit the inrush current.

DC MOTOR SPEED CONTROL

While DC motor speed control through armature or field windings was briefly discussed earlier in this chapter, specific approaches to DC motor shaft speed control can be categorized as follows:

1. **Armature Control:** The armature-based speed control technique involves changing the voltage across the armature through variation of parallel or series resistance, while holding the field voltage constant.
2. **Field Control:** The field control approach involves variation of the field voltage while holding the armature voltage constant. The field voltage is varied through the variation of series or shunt resistance.
3. **Electronic Control:** This approach involves the use of electronic controls for the variation of armature and/or the field voltage and current. Due to the fact that features like programmability, closed loop control, and automation are inherently available with electronic controls, this approach tends to provide better control over a wide range of speeds and torques.
4. **Combination of Basic Approaches:** In certain applications, combination of the three approaches described above is more suitable.

AC MOTOR SPEED CONTROL

A common and efficient means for controlling speed of AC motors is through the application of AC inverters, VFD or ASD (Adjustable Speed Drives). The premise of AC motor speed control is the ability to vary the frequency of AC power being supplied to the motor through a VFD. The shaft speed in such cases can be determined through the application of Eqs. 7.9 and 7.23.

$$n_s = \text{Synchronous speed} = \frac{120f}{p} = \frac{60\omega}{\pi p} \tag{7.9}$$

$$\text{Motor Shaft Speed, } n = (1-s)\cdot(n_s) = (1-s)\cdot\left(\frac{120f}{p}\right) \tag{7.23}$$

where
  n_s = Synchronous speed of the AC induction motor, in rpm
        = Rotational speed of the armature AC induction EMF, in rpm
  n = Actual speed of the AC induction motor shaft, in rpm
        = Rotational speed of the EMF in the rotor, expressed in rpm
  s = Motor slip

f = Frequency of the AC power source feeding the motor, in Hz
p = Number of poles, as stated on the motor nameplate

MOTOR CLASSIFICATIONS

There are numerous motor classifications, and these classifications tend to change and evolve over time as new applications for AC and DC motors emerge. Discussion and listing of all possible classifications and categories of AC and DC motors are beyond the scope of this text. However, some of the more common categories and classifications are stated below for reference.

    a. Self-commutated motors
    b. AC asynchronous motors (i.e. typical induction motor)
    c. AC synchronous motor
    d. Constant speed motors
    e. Adjustable speed motor
    f. General duty
    g. Special purpose

Classification based on the winding insulation type: Motor insulation class rating is determined by the **maximum allowable operating temperature** of the motor, which, subsequently, depends on the type/grade of insulation used in the motor. Seven mainstream insulation classifications and associated maximum allowable temperatures are listed below:

    I. Class A: **105°C**
    II. Class B: **130°C**
    III. Class F: **155°C**
    IV. Class H: **180°C**
    V. Class N: **200°C**
    VI. Class R: **220°C**
    VII. Class S: **240°C**

Aside from exercising proper care in specifying the correct insulation type for a motor when installing new motors – based on the operating environment of a motor – one must ensure that if a motor is rewound or repaired, the windings are replaced with **original classification of insulation**. For example, if the windings of a 100 hp **Class H** motor are replaced with **Class B** insulation windings, the motor will lose its capacity to operate in higher temperature environment.

TYPICAL MOTOR NAMEPLATE INFORMATION FOR LARGE THREE-PHASE INDUCTION MOTOR

Electric motors are typically labeled with a nameplate that displays pertinent specifications of that motor. While there are myriad possible specifications, ratings, and

TABLE 7.1

Motor Nameplate Example

Model No.	B200	TYPE:	Marine Duty, IEEE 45				
HP:	200	kW	150	RPM:	1760		
Volt	460	AMP	233	PH.	3	MAX. AMB:	40°C
Hz	60	S.F.	1.15	P.F.	84.5	Code: G	
NEMA NOM EFF:	96.2			DUTY:	CONT.	WT: 1052 kg	2314 lbs
FRAME:	447T	ENCL. Type: TEFC		Motor Type: TKKH		Insul. Type: NEMA: B	IP: 56

characteristics of motors, some of the more common ones are described and illustrated through a motor nameplate example in Table 7.1.

Interpretation of this nameplate information would be as follows:

Model No.: B200
Type: Marine Duty, IEEE-45
Power rating: 200 hp
Standard operating voltage: 460 V_{RMS}
Standard frequency of operation: 60 Hz
NEMA-based efficiency of the motor: 96.2%
Motor frame size: 447T. This NEMA designated frame classification often signifies the mechanical construction, mounting or installation of a motor
Power rating of the motor, in kW = 150 kW
AMP or full-load cCurrent rating of the motor: 233 A
Service factor of the motor: 1.15 or 115%
ENCL or NEMA rating of the motor enclosure: TEFC or, Totally Enclosed Fan Cooled
RPM: Full-load shaft or rotor speed: 1,760 rpm
PH or number of phases: 3
PF or power factor of the motor: 84.5%
Duty, or duration of operation: Continuous
TYPE: TKKH, high-efficiency motor
NEMA–B: Class B winding insulation
IP56: European enclosure rating, comparable to NEMA 4
Maximum allowable ambient temperature: 40°C
Weight of the motor: 1,052 kg or 2,413 lb

SELF-ASSESSMENT PROBLEMS AND QUESTIONS – CHAPTER 7

1. A gas-powered prime mover is rotating the rotor of a single-phase alternator at a speed of 1,200 rpm. The alternator consists of six-pole construction. The effective diameter of the coil is 0.15 m, and the length of the coil loop is 0.24 m. The coil consists of 20 turns. The magnetic flux density has been measured to be 1.2 T. Calculate the power delivered by this generator across a resistive load of 10 Ω.

2. A four-pole alternator/generator is producing electrical power at an electrical frequency of 50 Hz. (a) Determine the angular speed corresponding to the generated electrical frequency. (b) Determine the rotational (synchronous) speed of the armature/rotor. (c) Determine the angular velocity of the armature/rotor (rad/s).

3. A four-pole single-phase AC generator consists of windings constituting 80 series paths and is driven by a diesel engine. The effective or mean length of the armature is 18 cm and the cross-sectional radius of the armature is 5 cm. The armature is rotating at 1,800 rpm. Each armature pole is exposed to a magnetic flux of 1.0 T. The efficiency of this generator is 90% and it is rated 2 kW. Determine the following:

   (a) The maximum voltage generated. (b) The RMS voltage generated. (c) The horsepower rating of the generator. (d) The horsepower output of the prime mover.

4. A three-phase, four-pole, AC induction motor is rated 170 hp, is operating at full-load, 60 Hz, 460 V_{rms}, efficiency of 90%, power factor of 80%, and a slip of 4%. Determine the (a) motor shaft speed, in rpm, (b) torque developed, in ft-lbf, (c) line current drawn by the motor, and (d) the amount of reactive power, Q, sequestered in the motor under the described operating conditions.

5. A three-phase, four-pole, AC induction motor is tested to deliver 200 hp, at 900 rpm. Determine the frequency at which this motor should be operated for the stated shaft speed. Assume the slip to be negligible.

6. A three-phase induction motor delivers 600 kW at a power factor of 80%. In lieu of installing power factor correction capacitors, a synchronous motor is being considered as a power fact correction measure. Determine the apparent power size of the synchronous motor – in kVA – that should be installed to carry a load of 300 hp and raise the (combined) power factor to 93%. The source voltage is 230 V_{rms}.

8 Electrical Power Distribution and Control Equipment, and Safety- Related Devices

INTRODUCTION

This chapter explores power distribution equipment through the review of motor control center (MCC), disconnect switches, loop switches, motor starters, breakers, power switchgear, variable frequency drives (VFD), etc. In this chapter, pictures and diagrams are used to give the reader a "hands-on" feel of common electrical and electronic equipment. Of course, as with the rest of this text, the reader will have an opportunity to test their knowledge through self-assessment problems at the end of this chapter.

VOLTAGE CATEGORIES IN POWER DISTRIBUTION SYSTEMS

Before we delve into the specific voltage categories of electrical systems, let's examine the flow of power from the point of generation to the point of consumption through Figure 8.1.

As shown in Figure 8.1, after AC power is generated at a typical power plant in the United States, it is "stepped up" for transmission to voltages ranging from 138 to 765 kV. As explained earlier in this text, the higher the transmission voltage – based on the fact that $S = V \cdot I$ – the lower the current required to deliver a certain amount of power to the consumers. The end product delivered by the power distribution systems being AC power "S" or energy is expressed in kWh or kVAh. Note that many "progressive" utility (power) companies are beginning to bill the consumers on the basis of demand assessed in KVAs (i.e. apparent power "S") instead of kWs and energy in the form of KVA-h (i.e. S x·t, or energy) instead of kWh. This, as alluded to in the power factor discussion, is due to the fact that many utility companies are weary of "policing" the consumers' power factor. As shown in Figure 8.1, once the power arrives within reasonable proximity of the consumers, it is stepped down to lower voltages suitable for use by the consumers. As we describe the major categories of voltage systems in the sections below, the reader is cautioned against interpreting these categories in an absolute fashion. As explained below, the boundaries between the various categories vary, to some degree, depending on the entity, standard, or organization performing the interpretation.

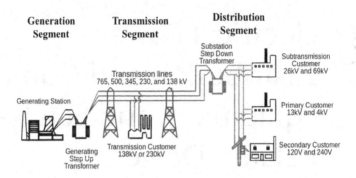

FIGURE 8.1   Electric power distribution and voltage levels. US DOE.

LOW-VOLTAGE SYSTEMS

The low-voltage category includes systems with voltages that **range from 50 to 1,000 V_{RMS} AC or 120 to 1,500 V DC**. This category has the following subcategories:

Extra low voltage: The voltage in this category is typically below 50 V_{RMS} AC or below 120 V DC. The extra low-voltage category is associated voltage, which typically can't harm humans, due to the low magnitude of potential difference. This conclusion is bolstered by NFPA 70E, which declares that electrical systems consisting of voltages 50 V or more in magnitude must comply with NFPA 70E shock hazard-related stipulations. This category applies to equipment and wiring widely used in bathrooms, showers, swimming pools, toys, and other electric devices, which might be in open contact with human.

Low voltage in power supplies for fluorescent lamps: This low-voltage category pertains to fluorescent lamp power supplies that use low DC voltage as source.

Low-voltage connectors: This low-voltage category is associated with low-voltage connectors and low-voltage plugs. Common examples include cigarette lighter 12 V plugs, low-voltage power adapters such as those used for charging rechargeable domestic and office cordless electronic, DC, equipment.

Low-voltage overhead power lines: This low-voltage category pertains to power lines that bring low voltage, 110–240 VAC power, to most homes and small commercial establishments.

LVDS voltage level: This low-voltage category pertains to LVDS systems. The term LVDS stands for Low-Voltage Differential Signaling. LVDS represents electrical digital signaling standard that pertains to high-speed digital communications. It specifies the electrical-level details for interface between inputs and outputs (I/O) on integrated circuit chips. Some common applications of LVDS standards are high-speed video, graphics, video camera data transfers, and general-purpose computer buses.

Microphone preamp voltage level: This category pertains to microphone preamplifiers used to amplify a microphone's low output voltage to a higher, more usable level.

MEDIUM-VOLTAGE SYSTEMS

Medium voltage is typically the voltage going from the substations to the service drops. The medium-voltage **category ranges from 1 up to 15 kV, or, 35 kV,** by some standards. According to ANSI/IEEE 1585-2002, medium voltage ranges from 1,000 V to 35 kV$_{RMS}$. NEMA 600-2003, on the other hand, refers to medium-voltage cables as "medium-voltage cables rated from 600 V to 69,000 V AC."

HIGH-VOLTAGE SYSTEMS

The high-voltage category, by most standards, ranges from 15,000 V (35 kV, by some standards) up to 230 KV. High voltage is typically the voltage going from substation to substation or, as shown in Figure 8.1, the voltage level at which power is typically transmitted in the United States from the point of generation to the point of consumption. Transmission-level voltages are usually considered to be 110 kV and above. Lower voltages such as 66 and 33 kV are usually considered sub-transmission voltages.

EXTRA HIGH-VOLTAGE (EHV) SYSTEMS

There is a general consensus among power experts that voltage ranging between **230 and 600 kV** could be aptly termed as extra high voltage, or **EHV.** However, some standards refer to voltages ranging from 230 to 600 kV as simply "High Voltage."

ULTRA-HIGH VOLTAGE SYSTEMS

There is a general consensus that **voltages exceeding 600 kV** could be referred to as ultra-high voltage, or **UHV.** Ultra-high voltage (UHV) transmission lines are rare, however. There are a very few of these in the world. A1.20 MV (1,200,000 V) UHV line, spanning from eastern to western part of India, when commissioned into operation is expected to be the highest voltage commercial power transmission line in the world.

POWER EQUIPMENT MEASUREMENT CATEGORIES

The measurement category system pertains to the classification of live electric circuits used in measurement and testing of installations and equipment.

There are four measurement categories that take into account the total continuous energy available at the given point of circuit and the occurrence of impulse voltages. Of course, this energy can be limited by circuit breakers or fuses and the impulse voltages by the nominal level of voltage. Figure 8.2 shows the four categories in relation to the typical power distribution components.

There are four categories, which are always stated with the designated voltage, for instance "CAT III, 150 V" or "CAT IV, 1,000 V." This has important safety implications for impulse voltages and insulation clearances.

CAT I: This category is applicable to instruments and equipment, which are not intended to be connected to the main supply. Because the available energy is very

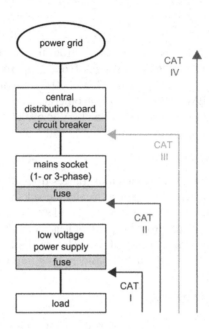

FIGURE 8.2 Power measurement categories.

limited, this category is typically not labeled on the equipment. Examples: Low-voltage electronic circuits, load circuits of typical test bench power supplies, etc.

CAT II: This category defines circuits which are intended for direct connection into main sockets. The energy in such installations is typically limited to below 100 A continuous. In CAT II systems, maximum available continuous power must be limited to 22 kVA or less through fuses or circuit breakers.

CAT III: This category pertains to circuits which can be connected to the main feeders. The energy in CAT III systems is limited by circuit breakers to less than 110 kVA with the current not exceeding 11 kA.

CAT IV: CAT IV systems include circuits which are connected directly to the source of power, or utility, for a given building. The level of energy in CAT IV systems is high, limited to an extent by the power transformer impedance. CAT IV systems, typically, carry higher arc flash hazard during energized work.

MCCS OR MOTOR CONTROL CENTERS

The term motor control center can be somewhat misleading. The application of MCCs is not typically limited to the "control of motors." While many of power controls housed in most MCCs are motor starters, often controls for lighting and other types of loads are included, as well. While our examination of MCCs will be limited to Allen-Bradley/Rockwell brand of MCCs and associated components, there are many other world-class brands available, i.e. Square-D, ABB, and Siemens.

As you tour an industrial or commercial facility, gray or blue cabinets such as the one depicted in Figure 8.3 can hardly escape notice. Those who are not engineers or

•20" (508mm) wide standard

•15" (381mm) or 20" (508mm) deep

•90" (2286mm) high standard,

– 71" (1790mm) high available

•Front mounted or back-to-back

FIGURE 8.3 Physical size specifications of a typical MCC.

technicians are likely to walk by one of these cabinets unaware of the electromechanical equipment performing critical functions behind the doors of those individual cubicles. Through examination of a series of MCC pictures and components, thereof, we will explore vital specifications of MCCs, the devices behind those MCC cubicle doors, how they function, and what they consist of.

An Allen-Bradley CENTERLINE MCC, pictured in Figure 8.3, is made up of one or more vertical sections. A standard section is 90" (2,286 mm) high, 20" (508 mm) wide, and either 15" (381 mm) or 20" (508 mm) deep for front-mounted configurations. Greater widths are sometimes supplied when larger equipment is required. Back-to-back configured MCCs are also available in 30" (762 mm) and 40" (1,016 mm) designs.

NEMA ENCLOSURE TYPES

Rockwell Automation offers a variety of NEMA-type enclosures to meet specific requirements. The standard enclosure is NEMA Type 1. This compares to the IEC enclosure IP40. NEMA Type 1 with gaskets, which is unique to MCCs, provides gasketing for unit doors. This compares to the IEC enclosure IP41. NEMA Type 12 provides gasketing for unit doors, bottom plates, and all cover plates (Figure 8.4).

MCCs in stainless steel NEMA Type 12 enclosures are also available for corrosive environment applications. This is comparable to the IEC enclosure IP54. For outdoor use, manufacturers offer NEMA Type 3R enclosures. This enclosure is essentially a metal shell around a NEMA Type 1 inner enclosure. This is comparable to the IEC enclosure IP44. For indoor or outdoor use, the manufacturers offer a NEMA Type 4 enclosure. This enclosure is essentially a stainless-steel shell around a NEMA Type 1 inner enclosure. This is comparable to the IEC enclosure IP65. All metallic parts (except stainless steel) are painted or plated before assembly, providing protection on all mating surfaces.

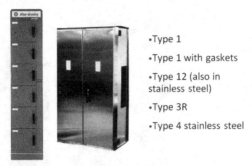

•Type 1

•Type 1 with gaskets

•Type 12 (also in stainless steel)

•Type 3R

•Type 4 stainless steel

FIGURE 8.4   MCC enclosure NEMA specifications.

MCC BUS BARS AND BUS CONNECTIONS

The motor starters, contactors, VFDs, and other devices housed in MCCs are mounted on draw-out rack assemblies. These devices or systems can be "drawn out" and removed for replacement or service after their disconnect switches have been turned off. This does not imply that the main power disconnect switch for power fed **into** the MCC must be disconnected or turned off while removing starters/cubicles for service.

If all cubicle assemblies were removed, the view would be similar to what is depicted in Figure 8.5. As apparent in Figure 8.5, behind the Micarta insulating panel, or equivalent, an array of bus bars is visible. Three visible bus bars are shown magnified on the right side of Figure 8.5. The three bus bars represent phases A, B, and C of three-phase AC. These bus bars are, essentially, **silver-plated copper conductors** forming a network of conductors carrying three-phase 480 VAC (or other specified/rated voltage) to all MCC cubicles. Power bus connections are, typically, made with at least two bolts. The vertical-to-horizontal bus connection is made with two bolts, as pictured in Figure 8.5. The horizontal bus splice connections are made with at least two bolts on each side of the splice (depending on bus size). The two-bolt

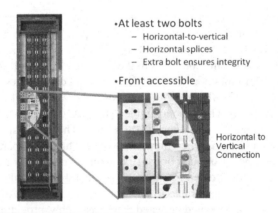

•At least two bolts
  − Horizontal-to-vertical
  − Horizontal splices
  − Extra bolt ensures integrity
•Front accessible

Horizontal to Vertical Connection

FIGURE 8.5   MCC bus bar network.

connection requires less maintenance and the extra bolt guards against the occurrence of "hot spots" (a result of loose connections) and arcing faults (a result of an open connection).

HORIZONTAL GROUND BUS

A 1/4" × 1" (6.4 mm × 25.4 mm) horizontal ground bus, rated 500 A, is supplied, as standard, with each vertical section. A 1/4" × 2" (6.4 mm × 50.8 mm) horizontal ground bus, rated 900 A, is optional. The bus can be unplated or supplied with optional tin plating. The horizontal ground bus can be mounted in the top or bottom horizontal wireway or top and bottom horizontal wireways of the MCC. See Figure 8.6.

VERTICAL GROUND BUS

Proper grounding is critical for safety of personnel and property. A substantial part of the NEC®, National Electrical Code, is dedicated safe grounding design, grounding equipment, and grounding methods. Hence, it is no surprise that grounding of MCCs and the connected equipment is facilitated through a network of ground buses in various segments of MCCs. Figures 8.6 and 8.7, for instance, show vertical and horizontal ground bus bars located inside the MCC cabinet.

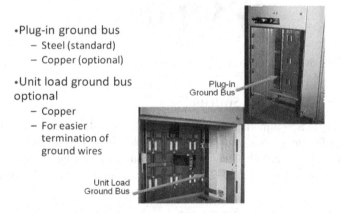

•Plug-in ground bus
 – Steel (standard)
 – Copper (optional)

•Unit load ground bus optional
 – Copper
 – For easier termination of ground wires

Plug-in Ground Bus

Unit Load Ground Bus

FIGURE 8.6   Vertical and horizontal ground bus bars in a typical MCC.

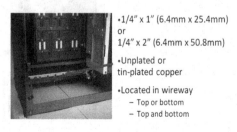

•1/4" x 1" (6.4mm x 25.4mm)
or
1/4" x 2" (6.4mm x 50.8mm)

•Unplated or tin-plated copper

•Located in wireway
 – Top or bottom
 – Top and bottom

FIGURE 8.7   Horizontal ground bus bars in an MCC.

Each standard vertical section is supplied with a steel vertical "plug-in ground bus" on the left side of the section. The vertical ground bus is bolted to the horizontal ground bus providing positive grounding for all plug-in units. The load ground wires, from various field locations, can be brought directly to the MCC.

INCOMING LUG COMPARTMENT

Power must be brought into the MCC. Therefore, the MCC needs to have some type of incoming compartment. There are three ways to accommodate the incoming power: incoming lug compartment, main fusible disconnect, and main circuit breaker.

Typically, an incoming lug compartment is used when the main disconnecting means for the MCC is located in switchgear near the MCC. See Figure 8.8. Incoming lug compartments are available for top or bottom entry of the power cables. The incoming lug compartment is designed so the cables are brought straight into the compartment. There is no power bus in the way, and therefore, there is no need for sharp cable bends. This configuration is available in sizes ranging from 600 to 3,000 A.

•Top or bottom
•Straight pull for cables

FIGURE 8.8   Source or utility connection, incoming power lugs.

MAIN FUSIBLE DISCONNECT SWITCH

Another way to accommodate the incoming cables is with a main fusible disconnect switch. See Figure 8.9. Main fusible disconnect switches are available for top or bottom entry of power cables. They are frame-mounted (non-plug-in) and hard-wired to the horizontal bus. For 600–2,000 A applications, a "Bolted Pressure Switch"

Main Fusible Disconnect Switch

•Top or bottom

•Frame mounted

•600A-2000A utilize "Bolted Pressure Switch"

•Visible blade

Bolted
Pressure
Switch

FIGURE 8.9   Anatomy of the main fusible disconnect switch.

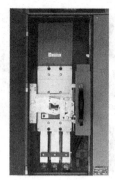

•Top or bottom

•Frame mounted

•Ground fault protection available

FIGURE 8.10 Use of a breaker instead of fusible disconnect switch for MCC power source feed.

is used. The bolted pressure switch features a contact system that tightly holds the blades during closure to provide a reliable current path and high withstand. All main disconnect switches provide visible blade indication.

Main Circuit Breaker

A main circuit breaker is another way that the incoming power cables can be accommodated. See Figure 8.10. Main circuit breakers are available for top or bottom entry of power cables and are frame-mounted (non-plug-in) and hard-wired to the horizontal bus. They are available in sizes up to 2,000 A. Ground fault protection is available for 600–2,000 A main circuit breakers.

There are pros and cons associated with fusible disconnect switches and breakers. When fuses clear (or "blow"), they must be replaced with exact equivalents. Incidents have been reported where improper fuse substitution has resulted in catastrophic failure of fuses, resulting in arc flash incidents. If breakers are applied as disconnecting and overcurrent protection means in a power distribution system, they simply need to be reset when they trip in response to overcurrent conditions. One possible disadvantage associated with breakers is that they don't offer as much flexibility in terms of current limiting and fast action as compared with some of the current limiting and fast acting fuses, i.e. Ferraz Shawmut class J and RK.

Stab Assembly

The stab assembly housing isolates each phase at the rear of the unit. See Figure 8.11. Since the power wires are isolated within the stab assembly, a fault barrier is effectively formed between the units and the vertical bus.

The tin-plated stabs, rated 240 A, are directly crimped to the power wires, minimizing any chance for a loose connection. The steel spring that backs the stab ensures a reliable high-pressure four-point connection on the vertical bus. The stabs are also free-floating and self-aligning, meaning they will position themselves for easy unit insertion.

•Housing
  – Isolates incoming
    phases acting as a
    fault barrier
•Stabs
  – Tin-plated
  – Rated 240A
  – Directly crimped
  – Steel spring backed
  – Free-floating and
    self-aligning

FIGURE 8.11   Stab assembly behind a typical MCC cubicle.

•Unit ground stab
  – Used with unit plug-in
    ground bus
  – Copper alloy (standard)
  – Solid copper (optional)

Standard
Plug-In
Ground
Stab

•Unit load ground connector
  – Used with unit load
    ground bus
  – Solid copper
  – For easier termination of load
    ground wire

Optional
Plug-In
Ground
Stab

Unit Load
Ground
Connector

FIGURE 8.12   Ground system bus stabs in a typical MCC.

UNIT GROUNDING PROVISIONS

A unit ground stab is provided on the back of each plug-in unit. See Figure 8.12. The ground stab engages the ground bus before the power stabs contact the vertical bus and disengages after the power stabs are withdrawn from the vertical bus. The standard unit ground stab is made of a copper alloy. An unplated or tin-plated solid copper unit ground stab is also available.

UNIT HANDLE

The unit handle is flange mounted and, therefore, stays in control of the disconnecting means at all times – whether the door is opened or closed.

The unit handle has positive status indication:

• Color-coded: **red** for **ON**, **green** for **OFF**.
• Labeled: ON and OFF (international symbols – **I** is used for **ON** and **O** for **OFF**).

The handle position is depicted in the ON and OFF positions (and TRIPPED position for circuit breakers). See Figure 8.13. The unit door is interlocked to the unit

Unit Handle

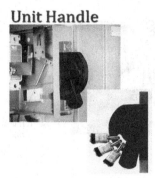

•Remains in control of disconnecting means

•Positive status identification

•Interlocked with door

•Padlockable

FIGURE 8.13   LOTO feature in the incoming power disconnect switch in a typical MCC.

handle. The door cannot be opened when the handle is in the ON position unless the operator "defeats" the mechanism using a screwdriver. The unit handle can be locked in the OFF position with up to three padlocks for LOTO, Lock Out Tag Out, purposes.

UNIT INTERLOCK

Plug-in units are supplied with a "unit interlock" that prevents the unit from being inserted into or withdrawn from the section while the handle is in the ON position. See Figure 8.14. When the handle is in ON position, the interlock mechanism moves upward engaging the unit support pan above the unit. The unit interlock can also be used to secure the unit in a service position to guard against accidental unit insertion. As shown in Figure 8.14, the interlock can be padlocked during servicing to prevent unit insertion even with the handle OFF.

•Prevents inserting or withdrawing unit with handle in ON position

•Padlockable

FIGURE 8.14   Mechanical safety interlock.

MOTOR STARTER

A motor starter is designed to control the flow of power to a motor, while, simultaneously, protecting the motor branch circuit against overcurrent conditions. Motor starter assemblies, such as the A-B Bulletin 500, are designed for up to 10 million

FIGURE 8.15   Rockwell FVNR starter.

operations. See Figure 8.15. Additional information on the design, specifications, and operation of some of the motor starter components will be discussed in greater detail in Chapter 10.

We will use the picture of a Rockwell starter, as captured in Figure 8.15, to examine the design and operation of a typical FVNR, Full-Voltage Non-reversing motor starter. Beginning at about 3'o clock, in the motor starter picture below, we notice the handle of breaker and operating handle. The handle of the starter is mechanically linked to the circuit breaker. The three terminals, with three emerging black wires, provide the continuity of the three-phase voltage to the top of the contactor.

A contactor, similar to a typical electrical relay, is simply an electromechanically controlled device that closes or opens contacts, to apply or disengage AC voltage, respectively, from the load. The load, in this case, is an electric motor.

The two wires visible directly in front of the contactor are **control wires** that energize the solenoid of the contactor. When the solenoid coil is energized, the core rod or plunger pushes the contact of the contactor shut and allows the flow of current and power downstream toward the motor via the solid-state overload protection device shown at about 8'o clock.

Notice the white terminal strips or terminal blocks visible along the bottom of the starter unit depicted in Figure 8.15. These terminal strips are called the "pull-apart terminal blocks." These pull-apart terminal blocks are, typically, used in plug-in units, and they represent a significant improvement in wire termination method. Prior to the advent of pull-apart terminal blocks, replacement, removal, and reinstallation of electrical required meticulous and arduous effort on the part of technicians to correctly terminate wires. The tedious task of disconnecting and connecting wires often resulted in miss-wiring and workmanship defects. The pull-apart terminal blocks have a front half and a rear half that detach for easy unit removal. The back half of the terminal block is factory wired, and the front half is where the customer terminates field wires.

PILOT DEVICES

Pilot devices are depicted in Figure 8.16. Typical set of pilot devices include status indicating lamps and control switches. The pilot device set shown in Figure 8.16 consists of one motor/load status indicating light at the top (typically, red or green).

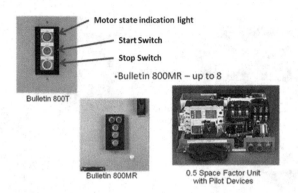

FIGURE 8.16   Motor starter pilot devices.

The two push buttons located below the indicating light are momentary (spring loaded) push button-type switches. The middle switch, typically green in color, is the "START" switch and the bottom switch, typically red in color, functions as a "STOP" switch. The control sequence and algorithm associated with the pilot devices is explained in the wiring diagram discussion, in Chapter 10.

FVR, FULL VOLTAGE REVERSING STARTER AND VARIETY OF OTHER POWER CONTROL DEVICES

As discussed earlier, when "on the fly," instantaneous, reversal of a three-phase AC motor is desired, a second contactor is incorporated into the motor starter cubicle to reverse the three-phase AC sequence from ABC to ACB and vice versa. Of course, the two starters in such cases are interlocked to ensure that simultaneous connection of the two opposite-phase sequences is avoided.

An FVR starter is pictured on the top right side of Figure 8.17. See labels "Contactor No. 1" and "Contactor No. 2" in Figure 8.17. An FVNR starter – with a single contactor – is depicted to the left of the FVR starter in Figure 8.17 for contrast. The bottom half of the picture in Figure 8.17 shows two other power control devices: (1) A soft start or two-speed starter unit and (2) a full-voltage lighting contactor unit.

VARIABLE FREQUENCY DRIVES – UP TO 250 HP

In addition to the power control devices discussed above, engineers can design or specify the integration of VFDs into MCC systems. When VFDs are housed into an MCC cabinet, it is advisable to maintain **electrical shielding** and adequate physical separation between power control components and more sensitive electronic systems like the VFDs and PLCs, Programmable Logic Controllers. In the absence of such caution, **electromagnetic interference**, or **EMI**, can adversely affect the operation of sensitive electronic systems.

Smaller VFDs are often, specifically, designed to fit into plug-in units or "**rack-out assemblies**." The VFD pictured on the left side of Figure 8.18 is a smaller Bulletin 1305 drive. The smaller drives range from 3 hP at 240 V to 5 hp at 480 V. The VFD

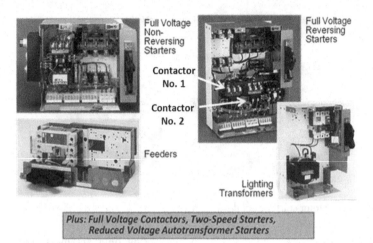

FIGURE 8.17   Motor starter pilot devices and comparison of FVNR and FVR starters.

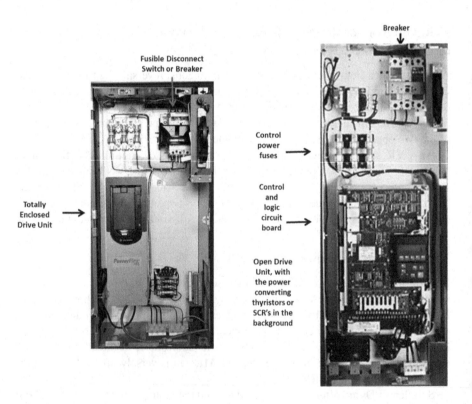

FIGURE 8.18   VFD, variable frequency drives.

pictured on the right side of Figure 8.18 is a larger Bulletin 1336 PLUS drive. Some of the larger VFDs, such as the A-B 1336, fall in the following size categories:

- 1/2–30 hP at 480 V, plug-in design
- 40–125 hP at 480 V, frame-mounted design
- 150–250 hP at 480 V, roll-out design

Smart Motor Controllers (SMCs) – Up to 500 A

The Allen-Bradley, like other power and controls manufacturers, offers various smart motor controllers that allow motor loads to be operated at multiple, **discrete**, speeds to take advantage of energy savings in accordance with the **fan laws** or **affinity laws**. This essentially means that if the motor is operated at 50% of the normal full-load speed (rpm), the power or energy consumption of the motor drops to 12.5%.

The smart motor controller featured in Figure 8.19 is an SMC-50 smart motor controller by Rockwell Automation. It is a solid-state motor controller that is premised on three-phase, solid-state, silicon-controlled rectifiers (SCRs, or thyristors).

Motor Soft Starts

Soft starters are typically employed in power control systems to limit excessive initial inrush of current associated with motor start-up and provide a gentler ramp up to full speed. Contrary to VFDs, soft starters are only used during start-up of motors. See Allen-Bradley soft start pictured in Figure 8.20.

Smart Motor Controllers (SMCs) - Up to 500A

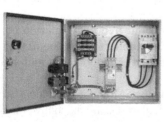

SMC-3

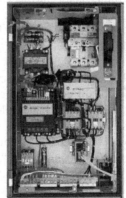

SMC Dialog Plus

FIGURE 8.19 SMC, Rockwell-Automation smart motor controllers.

FIGURE 8.20   Rockwell-Automation soft start.

In pump applications, a soft start can be installed to avoid pressure surges. Soft starts can facilitate smoother starting, prevent jerking and stressing of mechanical drive components, for example in conveyor belt systems that are loaded with bulk materials, intermittently. Using soft starts, fans or other systems with belt drives can be started slowly to avoid belt slipping. In essence, a soft start **limits the inrush current, improves stability** of the power supply, and **reduces transient voltage drops** that may affect other loads and the overall power.

Among "non-electronic" methods for "soft-starting" are means such as the installation of a **series reactor** (coil or inductance, in general). The presence of a series reactance limits motor starting current.

Comparison between SMCs, VFDs, and Soft Starts

The key differences between the application of VFDs versus SMCs are as follows:

1. VFDs have the ability to vary the armature frequency – and the motor speed – in practically **continuous**, gradual, or **analog** fashion. Smart motor controllers, on the other hand, allow variation of the motor at multiple **discrete**, preset, levels or steps.
2. While VFDs tend to cost more than the SMCs, the continuous adjustability of the VFDs can provide "finer," higher resolution, control of AC motors. This could provide the end user better process control, in addition to the energy savings opportunity.
3. SMCs and soft start systems tend to be smaller, in physical size, than VFDs.
4. VFDs offer the greatest energy savings for fans and pumps.
5. Soft starts tend to present the smallest footprint as compared to SMCs and VFDs.

PLC AND I/O CHASSIS

Even though the focus of discussion – through the exploration of Allen-Bradley MCCs – has been attainment of a better understanding of power or motor control devices, we will probe the hardware aspects of PLCs in this section; primarily, because Allen-Bradley ControlLogix and CompactLogix PLCs can be integrated into a typical Allen-Bradley MCC. See Figure 8.21. Of course, smaller PLCs, i.e. the Micro-Logix series, could be specified to be included in a section of an MCC, just as well.

However, once again, care should be exercised to protect the PLC and associated I/O from potential EMI emitted from power components in close proximity. The left picture of a PLC shown in Figure 8.21 depicts a large ControlLogix PLC and associated I/O blocks. In Figure 8.21, adjacent to the CPU, Central Processing Unit – the PLC power supply is shown mounted to the "**backplane**" of the PLC. A backplane, often, consists of a circuit board with rigid connectors that the I/O modules, CPU module, and the power supply can be plugged into. A PLC, not unlike a PC (Personal Computer), operates off of low DC voltage, 5–10 V. This low DC voltage is generated by the PLC power supply. The power supply, in turn, is powered by AC voltage.

Often, PLCs are integrated into a single cabinet with peripheral control equipment such as VFDs for automatic speed control and the control of higher current field devices through relays and contactors. Figure 8.22 depicts one such scenario.

METERING UNITS

It is common for engineers to specify and incorporate power monitoring equipment into an MCC. Examples of typical power monitors are pictured in Figure 8.23. These digital power monitors tend to be more versatile; they are often designed to capture and display electrical parameters like three-phase voltage, current, kWs (or real power **P**), kVAs (or apparent power **S**), kVARs (or reactive power **Q**), power factor, frequency, etc.

The rear of a multifunctional, digital, power monitor is displayed in Figure 8.24. The wires bringing the current signal from line current measuring CTs (current transformers) would be terminated at terminals I_1, I_2, and I_3. The voltage signal coming from the voltage or potential transformers would be terminated at points

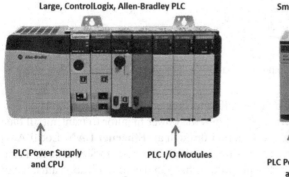

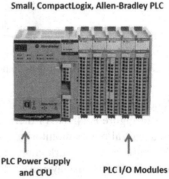

Large, ControlLogix, Allen-Bradley PLC

Small, CompactLogix, Allen-Bradley PLC

PLC Power Supply and CPU

PLC I/O Modules

PLC Power Supply and CPU

PLC I/O Modules

FIGURE 8.21 Allen-Bradley PLCs.

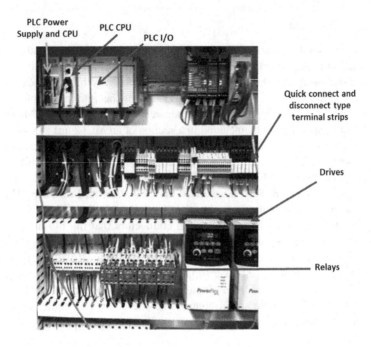

FIGURE 8.22   Complete control system in single control cabinet.

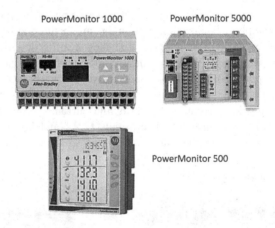

FIGURE 8.23   MCC compatible power monitoring equipment

labeled V_1, V_2, and V_3. Also, note the Ethernet connection port in the top right corner of Figure 8.24. This Ethernet port allows the power monitor to transmit all data to a central power monitoring work station through an **Ethernet LAN**, Local Area Network. Such network connectivity allows multiple monitors to feed data to a central location for display and control purposes. The digital communication attribute of digital power monitors – and the connectivity to LAN – also permits users to broadcast the data for off-site monitoring through **routers** and **modems**.

FIGURE 8.24 Rear termination points on an Allen-Bradley power monitor.

MAIN SWITCH YARD AND MEDIUM-VOLTAGE SWITCHGEAR

During the course of our discussion of power distribution equipment in this text we made mention of "Main Switch Yard." For illustration purposes, a picture of the main switch yard of a facility is shown in Figure 8.25. Main switch yards are compounds, on site at the consumer's facility, where the utility (power company) "lands" the power into the customer's facility and transforms it from high voltage to medium voltage, i.e. 100–13 kV.

MEDIUM-VOLTAGE SWITCHGEAR – LOOP SWITCH

Switching within facilities, at the medium voltage level, is accomplished through loop switches and breakers. A typical, three-phase, 13 kV loop switch is pictured in Figure 8.26. Some of the key functional components of the loop switch, i.e. the operating mechanism, viewing ports, and switch status (open or closed), are annotated on Figure 8.26.

POWER FACTOR CORRECTING CAPACITORS AND STEP-DOWN TRANSFORMER, MAIN SWITCH YARD

Power factor correcting capacitors, installed on the power company side of the main switch yard are pointed out in Figure 8.27. Also annotated are the large 100–13 kV step-down transformers on the power company side of the switch yard.

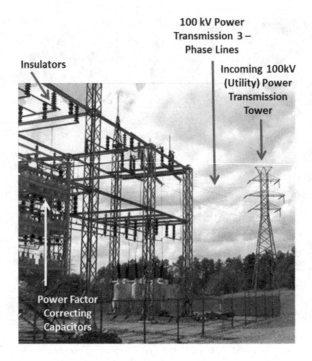

FIGURE 8.25   Main Switch Yard; 100 kV.

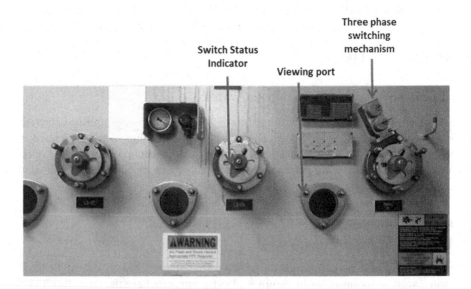

FIGURE 8.26   13 kV loop switch.

| Step down transformer (100kV to 13kV), on utility side | Power factor correcting capacitors, on utility side |

FIGURE 8.27 Power factor correcting capacitors and main step-down transformer in a Main Switch Yard.

CIRCUIT BREAKERS

A circuit breaker is an overcurrent protection device that is an automatically operated. A circuit breaker, basically, is an electrical switch designed to protect an electrical circuit from damage caused by overload or short circuit. Its core function is to detect a fault condition and interrupt current flow. Unlike a fuse, which operates once and then must be replaced, a circuit breaker can be reset (either manually or automatically) to resume normal operation. The most common and basic type of circuit breaker is the type that is often applied in residential dwelling breaker panels. Circuit breakers are categorized based on the following features or functions:

  a. Voltage
  b. Current
  c. Principle of operation

Circuit breakers for large currents or high voltages are equipped with sensing and transducing devices that detect fault current and operate the trip or circuit opening mechanism. The term "trip," essentially means "opening" of an electrical circuit such that the current ceases to flow. The trip function is typically executed by a solenoid that releases a latch that keeps the breaker closed (ON). In order to maintain the failsafe properties – typically required in control systems – the tripping solenoid is often energized by a separate battery. High-voltage circuit breakers are self-contained with CTs, protection relays, and an internal control power source.

Upon detection of a fault, or fault current, contacts within the circuit breaker open to interrupt the circuit. Often, mechanically stored energy – in springs or compressed air – is used to separate and force open the contacts, thus interrupting the flow of current.

Low-Voltage Circuit Breakers

As the name implies, low-voltage breakers are applied in domestic, commercial, and industrial applications, operated at less than 1,000 V_{AC}. The design of low-voltage breakers is premised on thermal or thermal-magnetic operation. Low-voltage power circuit breakers can be mounted in multi-tiers in low-voltage switchboards or switchgear cabinets. Low-voltage circuit breakers are also made for DC applications, i.e. electrical protection of subway lines.

A sectional diagram of a common 10 A DIN (derived from German term Deutsches Institut für Normung) rail-mounted thermal-magnetic miniature circuit breaker is depicted in Figure 8.28.

Where

1. Actuator lever – Used to manually trip and reset the circuit breaker. The legend on the breaker typically indicates the status of the circuit breaker (ON or OFF).
2. Actuator mechanism – Forces the contacts to "close" or "open."
3. Contacts – Allow the load current to flow when **closed** interrupt the flow when **open**.
4. Terminals.
5. Bimetallic strip for thermal operation or tripping of the breaker under prolonged overload conditions.
6. Calibration screw – Allows the manufacturer to precisely adjust the trip current of the device after assembly.
7. Solenoid.
8. Arc divider or extinguisher.

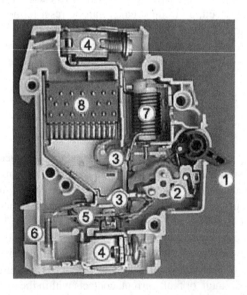

FIGURE 8.28 Sectional diagram of a common rail-mounted thermal-magnetic miniature circuit breaker. Dorman Smith circuit breaker, annotated by: Ali, UK.

MEDIUM-VOLTAGE CIRCUIT BREAKERS

Medium-voltage circuit breakers are classified based on the medium (or substance) used, within, to extinguish the arc. Some of the medium-voltage circuit breakers types are discussed below:

Vacuum circuit breakers: The vacuum circuit breakers – rated up to 3,000 A – interrupt the current by creating and extinguishing the arc in a near vacuum environment, within a sealed container. The vacuum circuit breakers are generally applied for voltages up to about 35,000 V; they tend to have longer life expectancies, and they offer longer service spans between overhauls than do air circuit breakers.

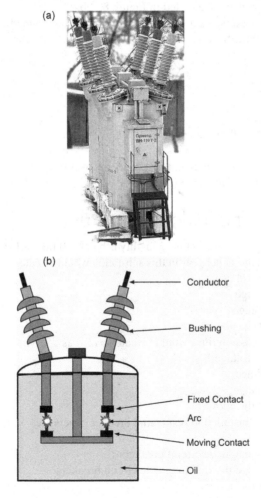

FIGURE 8.29 (a) Oil circuit breaker MKP-110 (110 kV) on traction substation 110 kV 50 Hz/3.3 kV DC of railway, Tolyatti city, Russia. By: Vivan. (b) OCB schematic.

Air circuit breakers: Air circuit breakers are rated, in current, up to 10,000 A. Their trip characteristics are often fully adjustable including configurable trip thresholds and delays. They are often electronically controlled.

SF_6 **circuit breakers**: The SF6 medium-voltage circuit breakers extinguish the arc in a chamber that is filled with **sulfur hexafluoride gas**.

HIGH-VOLTAGE CIRCUIT BREAKERS

High-voltage circuit breakers are often applied in electrical power transmission networks. High-voltage breakers are almost always solenoid-operated, with current-sensing protective relays operated through CTs. A high-voltage circuit breaker, rated 110 kV, is depicted in Figure 8.29a. This breaker is designed for 50 Hz operation and is equipped with dielectric oil-based arc quenching system. The schematic of the OCB, oil circuit breaker, is depicted in Figure 8.29b.

High-voltage breakers are broadly classified based on the medium used to extinguish the arc. Technically, any of the following can be used for arc quenching:

- Bulk oil
- Minimum oil
- Air blast
- Vacuum
- SF_6, sulfur hexafluoride gas

However, due to environmental and cost concerns over insulating oil spills, most new breakers use SF_6 gas to quench the arc.

SELF-ASSESSMENT QUESTIONS – CHAPTER 8

1. A substation in a manufacturing facility is being fed from a 13 kV transformer secondary. This switchgear in this substation would be categorized as:
   A. Medium voltage
   B. Low voltage
   C. High voltage
   D. None of the above
2. Power transmission lines would be categorized as:
   A. Medium voltage
   B. Low voltage
   C. High voltage
   D. None of the above
3. The breakers installed in residential breaker panels are:
   A. OCBs
   B. Thermal magnetic circuit breakers
   C. Low-voltage thermal magnetic circuit breakers
   D. None of the above
   E. Both B and C

4. The vacuum circuit breakers tend to offer longer service spans between overhauls than do air circuit breakers.
   A. True
   B. False
5. The SF_6-type high-voltage circuit breakers are not preferred due to environmental concerns.
   A. True
   B. False
6. MCCs are not designed to accommodate PLCs and VFDs.
   A. True
   B. False
7. The bus bars in MCCs are commonly constructed out of:
   A. Aluminum
   B. Silver-plated copper
   C. Silver
   D. Iron
8. Pilot devices on MCCs:
   A. Control circuit breakers
   B. Indicate the status of MCC
   C. Indicate the status of motor/load
   D. Include "Start" and "Stop" controls
   E. Both (C) and (D).
9. Power control cubicles in MCCs are fixed and cannot be removed while the main fusible disconnect switch of the MCC is ON.
   A. True
   B. False
10. A control transformer in a given MCC compartment:
    A. Steps down the voltage for control circuit operation
    B. Provides power for MCC cabinet lighting
    C. Is seldom needed
    D. Serves as an isolation transformer
    E. Both (C) and (D)

9 National Electric Code® and Arc-Flash Electrical Safety Standards

INTRODUCTION

The discussion of NEC®, National Electrical Code, in this chapter is simply an introduction. The coverage of the NEC® in this chapter is not intended to provide the reader intermediate-, advanced-, or expert-level knowledge that is required and expected of a NEC®-trained and practicing electrical Professional Engineer, specializing in power system design. Notwithstanding the foregoing clarification regarding the limited depth of NEC® knowledge provided in this chapter, you will find that this chapter on NEC® and electrical safety opens the door and allows you to appreciate the complexity and depth of the NEC®. And, most of all, if you find yourself leading a group of electrical engineers and electricians in a "2:00 AM triage situation," trying to troubleshoot and reinstate an important piece of equipment back into operation, the knowledge and familiarity they will gain in this chapter should prepare them better to comprehend the code-related jargon used by electrical engineers and electrical technicians.

The NFPA (National Fire Protection Agency) promulgates NEC® revisions every 3 years. The reader is reminded that the objective of this text is not to provide precise code content and references at the time this text is read. Instead, this text aims to give the reader **general** references of NEC® articles that have, **traditionally**, addressed the minimum requirements associated with equipment and appurtenances, i.e. conductors, conduits, raceways, fuses, breakers, grounding systems, etc. The objective of NEC coverage in this book is to dispel phobia and mystic professionals experience during their first encounter with this, rather intimidating, code.

Electrical or non-electrical engineering professionals often wonder if they should invest in the *NEC®* book. If so, what format and version would be most suitable? While the most appropriate answer to this question lays in the depth and extent of study intended, most non-electrical engineers might find the **NEC® handbook**, in hardcopy or electronic form, more beneficial. The handbook version of the NEC® is replete with copious pictures, diagrams, and illustrative example problems that aptly facilitate quicker comprehension of code and associated concepts. Of course, the electronic version of the *NEC Handbook* provides the added advantage of being able to carry the code in your portable laptop PC and be able to conduct electronic search of NEC articles, terms, and tables expeditiously.

In keeping with the approach utilized in the rest of this text, as we get introduced to certain commonly applied codes, we will take our knowledge and comprehension to the next level through example problems, end-of-chapter problems, and the solutions at the back of the text.

NEC® ARTICLES

With the exception of Article 90, all article numbers of the code correspond to the chapter numbers. In that, the first article of Chapter 1 begins with Article 100, the first article of Chapter 2 is Article 200, etc. up to Chapter 9. Chapter 9 comprises mostly tables vital to the implementation of the code.

Listed below are major NEC® articles that are introduced in this chapter, followed by a few details that explain the significance of the articles. Later, in this chapter, we will illustrate the significance of some of these NEC® articles through examples.

- Art. 90 – Introduction to NEC® and Outline
- Art. 100 – Definitions, Including Enclosure Ratings
- Art. 110.6 – Conductor Sizes, AWG, and Circular Mils
- Art. 110.16 – Arc-Flash Regulations
- Art. 110.26 – Clearances and Working Space Requirements
- Art. 210 – Load Configurations and Voltages in Branch Circuits
- Art. 210.9 – Autotransformers
- Art. 210.20 – Branch Circuit Ampacity Determination and Overcurrent Protection
- Art. 240 – Overcurrent Protection
- Art. 240.50–240.101 – Circuit Breaker and Fuse Types
- Art. 250 – Grounding
- Art. 310 – Conductor Insulation Rating
- Art. 310.15 – Conductor Ampacity
- Art. 358–392 – Conduit and Cable Trays
- Art. 408.13–408.35 – Panel Boards
- Art. 430.6 – Ampacity and Motor Rating Determination. Table 430.250
- Art. 500 – Hazardous or Classified Locations, Classes I, II, and III; Divisions 1 and 2
- Chapter 6 – Special Equipment
- Chapter 7 – Special Conditions
- Chapter 8 – Communications Equipment
- Chapter 9 – Tables

ART. 90 – INTRODUCTION TO NEC® AND OUTLINE

Article 90 of the NEC® is a basic introduction to the intention of the codes. This article states the purpose of the code provision of uniform and practical means to safeguard people and equipment from electrical hazards. These safety guidelines are not meant to describe the most convenient or efficient installations and don't guarantee good service or allow for future expansion. The NEC® articles of code are designed to provide a standard for safety that protects against electrical shock and thermal effects, as well as dangerous overcurrents, fault currents, overvoltage, etc.

The NEC® comports, for the most part, with the principles for safety covered in Section 131 of the International Electrotechnical Commission Standard for electrical installations. International Electrotechnical Commission, or IEC, is the world's leading organization that prepares and publishes **International Standards** for all electrical, electronic, and related technologies.

Art. 100 – Definitions

Professionals who are not electrical engineers or electricians are likely to find this article one of the most useful articles. In that, this article contains definitions of terms that are essential in the understanding and interpretation of the code. As an introduction, a few of the terms described in Article 100 are listed and explained below.

Ampacity: Ampacity is defined as "The **maximum** current, in amperes, that a conductor can carry continuously under the conditions of use without exceeding its temperature rating."

Note: Ampacity does vary depending on several factors. Appropriate NEC® tables and derating rules must be applied to determine the correct ampacity. See Article 310 for additional explanation.

Bonded (Bonding): Bonding is defined as equipment or objects "Connected to establish electrical continuity and conductivity." In other words, bonding constitutes permanent joining of metallic parts to form an electrically conductive path that ensures electrical continuity and the capacity to conduct electrical current safely.

Note: This is not the same as grounding, but bonding jumpers are essential components of the bonding system, which is an essential component of the grounding system. Furthermore, note that the NEC does not authorize the use of the earth as a bonding jumper because the resistance of the earth is more than 100,000 times greater than that of a typical bonding jumper.

Branch Circuit: Branch circuit is defined as "The circuit conductors between the final overcurrent device protecting the circuit and the outlet(s)." As we will discuss and illustrate later, a branch circuit, as defined by the NEC®, can be the entire circuit that lays between the overcurrent fuses and a motor load, in a typical motor branch circuit.

Continuous Load: A continuous load is defined by the NEC® as "A load where the maximum current is expected to continue for 3 hours or more."

Note: The maximum running current referred to in this definition is exclusive of the **starting current**.

Feeder: A feeder is defined by the NEC® as "All circuit conductors between the service equipment, the source of a separately derived system, or other power supply source and the final branch-circuit overcurrent device."

Ground: A ground is defined by the NEC® as simply "The earth." On the other hand, the term **Grounded (Grounding)** is defined as "Connected (Connecting) to ground or to a conductive body that extends the ground connection."

Note: Simply driving an electrode into the earth does not constitute grounding a circuit. The ground must be made with the source or supply in mind, as the flow of electrons – or current – always tries to **return** to the source.

In Sight From (Within Sight From, Within Sight): The NEC® defines this as "Where this Code specifies that one equipment shall be 'in sight from,' 'within sight from,' or 'within sight of,' and so forth, another equipment, the specified equipment is to be visible and not more than 15 m (50 ft) distant from the other." In other words, lockable disconnecting means must be provided in sight of the load, unless the specific installation meets the criteria stated under Article 430.102 (B) (2) (a) or (b).

Labeled: The NEC® defines this as "Labeled. Equipment or materials to which has been attached a label...acceptable to the authority having jurisdiction (AHJ) and concerned with product evaluation...."

Note: The reader is advised to read the entire original definition in order to attain full understanding of this term.

Neutral Conductor: The NEC® defines this as "The conductor connected to the neutral point of the system that is intended to carry current under normal circumstances."

Overcurrent: The NEC® defines overcurrent as "Any current in excess of the rated current of equipment or the ampacity of a conductor. It may result from overload, short circuit or ground fault."

Overload: The NEC® defines overload as "Operation of equipment in excess of normal, full-load rating, or of a conductor in excess of rated ampacity that, when it persists for a sufficient length of time, would cause damage or dangerous overheating. A fault, such as short circuit or ground fault, is not an overload."

Raceway: The NEC® defines raceway as "An enclosed channel of metal or nonmetallic materials designed expressly for holding wires, cables, or bus-bars."

Note: The reader is advised to read the entire original definition in order to attain full understanding of this term.

Short-Circuit Current Rating: The NEC® defines this term as "The prospective symmetrical fault current at a nominal voltage to which an apparatus or system is able to be connected without sustaining damage exceeding defined acceptable criteria."

Voltage, Nominal: The NEC® defines nominal voltage as "A nominal value assigned to a circuit or system for the purpose of conveniently designating its voltage class (e.g., 120/240 volts, 480/277 volts, 600 volts). The actual voltage at which a circuit operates can vary from the nominal within a range that permits satisfactory operation of equipment."

Art. 110.6. – Conductor Sizes, AWG, and Circular Mils

Overall Article 110 lays out the requirements for electrical installations. Article 110.6, specifically, addresses conductor sizes. This article clarifies the fact that the conductor sizes are expressed in AWG, American Wire Gage, or in circular mils. For copper, aluminum, or copper clad aluminum conductors up to size 4/0 AWG, the code identifies the conductor sizes in AWG. Conductors larger than 4/0 AWG are sized in circular mils, beginning with 250,000 mils; formerly referred to as 250 MCM. Where first "M" stands for 1,000. The unit **kcmil** was adopted 1990. Therefore, the more contemporary identification of a 250,000 cmil conductor would be 250 **kcmil**. Note: 1 mil is equal to 1/1,000th of an inch and a circular mil (cmil) is a unit of area equal to the area of a circle with a diameter of one mil. See diagram below. In Canada and the United States, the NEC uses the circular mil to define wire sizes larger than 0000 AWG.

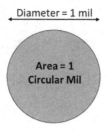

Diameter = 1 mil

Area = 1
Circular Mil

ARTICLE 110.16 – ARC-FLASH HAZARD WARNING

This article addresses stipulation associated with potential hazard of arc-flash incidents initiated by electrical faults. The essence of the **Code**, in this article is that switchboards, panelboards, industrial control panels, meter socket enclosures, and motor control centers in other than dwelling units, which are likely to require examination, adjustment, servicing, or maintenance while energized, shall be field marked to warn qualified persons of potential electric arc-flash hazards. The marking or label shall be located so as to be clearly visible to qualified persons before examination, adjustment, servicing, or maintenance of the equipment.

Two alternative and mutually exclusive approaches for identification of arc-flash and shock hazard risks are (1) The Category Method and (2) The Incident Energy Method. The Category Method can only be applied when the piece of electrical equipment in question fully comports with the exact specifications listed in the respective AC and DC tables provided in the NFPA 70E code. When the specifications and installation parameters of the actual electrical equipment **do not** match the voltages and conditions specified in Article 130 tables, as is often the case, one must adopt the incident energy method. Examples of labels, generated or manually filled, are depicted in Figure 9.1. Look for additional discussion on the topic of arc flash later in this chapter.

ART. 110.26 – CLEARANCES AND WORKING SPACE REQUIREMENTS

This article of the Code states: "Access and working space shall be provided and maintained about all electrical equipment to permit ready and safe operation and maintenance of such equipment." The essence of this article is that one cannot install electrical equipment in just **any** available physical space. Equipment must be installed such that it can be accessed, safely, at all times by maintenance personal.

ART. 210 – BRANCH CIRCUITS

This article applies to branch circuits that supply power to motor and non-motor loads. Article 430 must be consulted for situations where branch circuits feed motors only. The *NEC® Handbook* contains myriad illustrations of different types of branch circuits and associated code requirements.

ART. 210.9 – CIRCUITS DERIVED FROM AUTOTRANSFORMERS

This article stipulates: "Branch circuits shall not be derived from autotransformers unless the circuit supplied has a grounded conductor that is electrically connected to

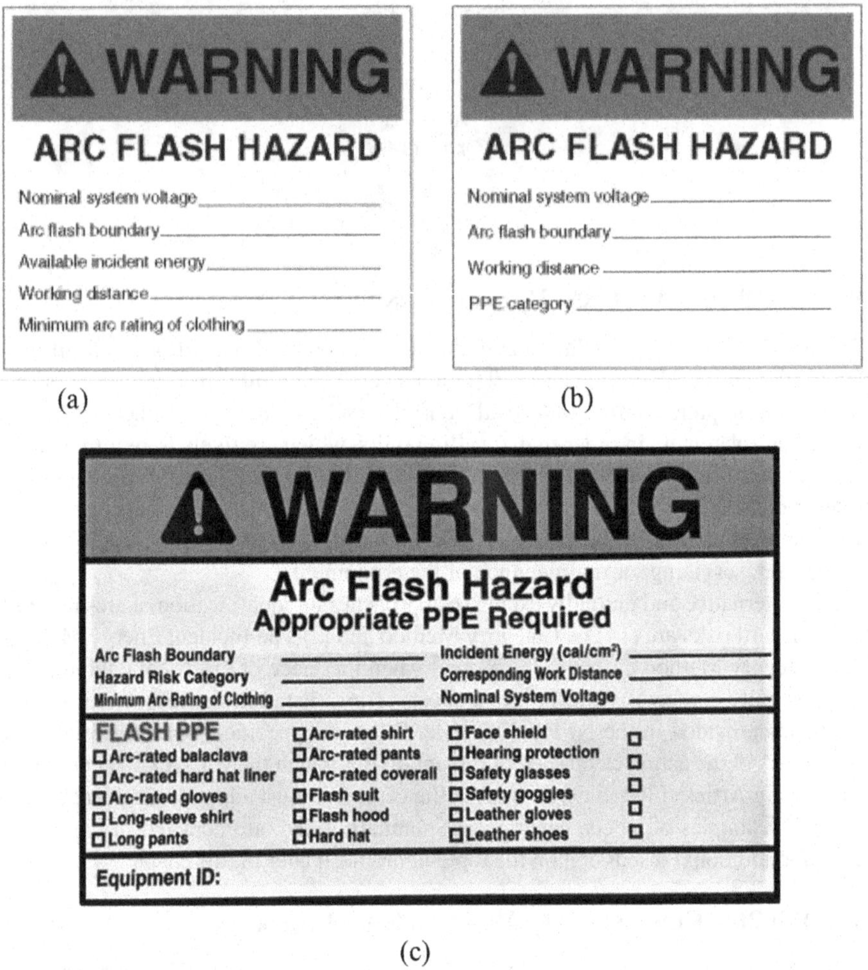

(a)

(b)

(c)

FIGURE 9.1   (a) Standard arc-flash label when Incident Energy Method is used. (b) Label used with Category Method. (c) Arc-flash label filled manually.

a grounded conductor of the system supplying the autotransformer." In addition to explanation of this code, the handbook shows circuit diagrams of autotransformers connected in various configurations.

ART 210.20 – OVERCURRENT PROTECTION

According to 210.20, an overcurrent device that supplies continuous and non-continuous loads must have a rating that is not less than the sum of 100% of the non-continuous load plus 125% of the continuous load, calculated in accordance with Article 210. Because grounded/neutral conductors are generally not connected to the terminals of an overcurrent protective device, this requirement for sizing conductors subject to continuous loads does not apply.

Art. 240 – Overcurrent Protection

This article of the *Code* provides the requirements for selecting and installing overcurrent protection devices (OCPDs). On the one hand, an OCPD protects a circuit by opening the circuit when current reaches a value that would cause an excessive temperature rise in the conductors. If one were to apply the rising water analogy, current rises like water in a tank, and at a certain level, the OCPD shuts off the faucet. On the other hand, an OCPD protects equipment by opening the circuit when it detects a short circuit or ground fault. All electrical equipment must have a short-circuit current rating that permits the OCPDs to clear short circuits or ground faults without a catastrophic failure.

Art. 240.50–240.101 – Circuit Breaker and Fuse Types

Articles, ranging from Art. 240.50–240.101, address requirements associated with selection, specification and installation of various OCPDs such as circuit breakers and fuses. Once again, in addition to explanation of the code, the handbook shows diagrams and pictures of various types of fuses and a breaker tripping unit.

Art. 250 – Grounding and Bonding

Overall, this article addresses general requirements for grounding and bonding of electrical installations, types and sizes of grounding and bonding equipment, methods of grounding and bonding, and situations when guards, isolation, or insulation may be substituted for grounding.

Art. 310 – Conductors and General Wiring

This NEC® article addresses general requirements for conductors, their types, insulations, markings, mechanical strengths, and ampacity ratings. Table 310.15 lists ampacities of various conductors and is, possibly, the most frequently used page in the *Code*. See additional discussion below on conductor ampacity.

Art. 358 – Electrical Metallic Tubing: Type EMT

This article of the *Code* addresses the application, installation, and construction specifications for electrical metallic tubing (conduit) and associated tubing.

Art. 408 – Switchboards and Panelboards

This article of the *Code* covers switchboards and panelboards associated with equipment operating at 600 V or less.

Art. 500.5 – Classifications of Locations

This article addresses hazardous locations and classifications, thereof. Hazardous locations are classified depending upon the presence of the following in concentrations or quantities that could increase the likelihood of combustion:

- Flammable gases
- Flammable liquids

- Flammable vapors
- Vapors produced by combustible liquids
- Combustible dust
- Combustible fibers or flyings

Art. 600 – Electric Signs and Outlet Lighting

This article covers the installation of conductors, equipment, and field wiring for electric signs retrofit kits and outline lighting regardless of voltage.

Art. 700 – Emergency Systems

This article applies to the electrical safety of the installation, operation, and maintenance of emergency systems. These systems could consist of circuits and equipment intended to supply, distribute, and control electricity for illumination, actual energy usage by large loads, or both, to required facilities when normal electrical supply or system is interrupted.

Art. 800 – Communication Systems

This article covers communication circuits and communications equipment.

Chapter 9 – Tables

The tables in Chapter 9 are part of the mandatory requirements of the NEC®. Tables 1 through 10 deal with conductors and raceways. The last four tables provide parameters for power limitations for Class 2 and 3 power-limited circuits and for power-limited fire alarm circuits.

Ampacity of Conductors – Table 310-15

In order to provide a measure of familiarity with the most frequently used section of the NFPA, NEC, an older version of NEC Table 310-15 has been included under Tables 9.1 and 9.2. Closer examination of this table shows that the ampacities of various conductors are listed under two separate sections. The left section represents copper conductors and the right section represents the aluminum or copper clad aluminum conductors. These two sections are divided further into three columns each. These three columns list the ampacities of various commercially available conductors under three temperature categories: 60°C (140°F), 75°C (167°F), and 90°C (194°F). Note that these three separate temperature columns lump different types of commercially available insulations. The temperature and insulation columns are selected in accordance with specific application, environment, and ambient temperature. The bottom section of Table 9.2 represents the conductor **ampacity correction factors** based on the ambient temperatures above or below 30°C.

TABLE 9.1
Partial NEC® Table 310.15(B)(16)

Partial Ampacity Information from NEC® Table 310.15(B)(16) – Allowable Ampacities of Insulated Conductors Rated Up to and Including 2,000V, 60°C through 90°C (140°F through 194°F), for Not More Than Three Current-Carrying Conductors in Raceway, Cable, or Earth (Directly Buried), Based on Ambient Temperature of 30°C (86°F)[a].

	Temperature Rating of Conductor; Ref: NEC® Table 310.104(A)						
	60°C	75°C	90°C	60°C	75°C	90°C	
	(140°F)	(167°F)	(194°F)	(140°F)	(167°F)	(194°F)	
Size AWG or kcmil	Types TW, UF	Types RHW, THHW, THW, THWN, XHHW, USE, ZW	Types TBS, SA, SIS, FEP, FEPB, MI, RHH, RHW-2, THHN, THHW, THW-2, THWN-2, USE-2, XHH, XHHW, XHHW-2, ZW-2	Types TW, UF	Types RHW, THHW, THW, THWN, XHHW, USE	Types TBS, SA, SIS, THHN, THHW, THW-2, THWN-2, RHH, RHW-2, USE-2, XHH, XHHW, XHHW-2, ZW-2	Size AWG or kcmil
	Copper			Aluminum or Copper-Clad Aluminum			
18	-	-	14	-	-	-	-
16	-	-	18	-	-	-	-
14[b]	15	20	25	-	-	-	-
12[b]	25	25	30	15	20	25	12[b]
10[b]	30	35	40	25	30	35	10[b]
8	40	50	55	35	40	45	8
6	55	65	75	40	50	55	6
4	70	85	95	55	65	75	4
3	85	100	115	65	75	85	3
2	95	115	130	75	90	100	2
1	110	130	145	85	100	115	1
1/0	125	150	170	100	120	135	1/0
2/0	145	175	195	115	135	150	2/0

(Continued)

TABLE 9.1 (Continued)
Partial NEC® Table 310.15(B)(16)

Partial Ampacity Information from NEC® Table 310.15(B)(16) – Allowable Ampacities of Insulated Conductors Rated Up to and Including 2,000V, 60°C through 90°C (140°F through 194°F), for Not More Than Three Current-Carrying Conductors in Raceway, Cable, or Earth (Directly Buried), Based on Ambient Temperature of 30°C (86°F).

Temperature Rating of Conductor; Ref: NEC® Table 310.104(A)

Size AWG or kcmil	60°C (140°F) Types TW, UF	75°C (167°F) Types RHW, THHW, THW, THWN, XHHW, USE, ZW	90°C (194°F) Types TBS, SA, SIS, FEP, FEPB, MI, RHH, RHW-2, THHN, THHW, THW-2, THWN-2, USE-2, XHH, XHHW, XHHW-2, ZW-2	60°C (140°F) Types TW, UF	75°C (167°F) Types RHW, THHW, THW, THWN, XHHW, USE	90°C (194°F) Types TBS, SA, SIS, THHN, THHW, THW-2, THWN-2, RHH, RHW-2, USE-2, XHH, XHHW, XHHW-2, ZW-2	Size AWG or kcmil
	Copper			Aluminum or Copper-Clad Aluminum			
3/0	165	200	225	130	155	175	3/0
4/0	195	230	260	150	180	205	4/0
250	215	255	290	170	205	230	250
300	240	285	320	195	230	260	300
350	260	310	350	210	250	280	350

Note: Included for general illustration purposes only. Use actual NEC® tables for work.

[a] Refer to NEC® Table 310.15(B)(2)(a) for the ampacity correction factors where the ambient temperature is other than 30°C (86°F). Refer to NEC® Table 310.15(B)(3)(a) for more than three current-carrying conductors.

[b] Refer to NEC® Article 240.4(D) for conductor overcurrent protection limitations.

TABLE 9.2

Partial Table 310.15(B)(2)(a) Ambient Temperature Correction Factors Based on 30°C (86°F)

For Ambient Temperatures Other Than 30°C (86°F), Multiply the Allowable Ampacities Specified in the Ampacity Tables by the Appropriate Correction Factor Shown Below

Ambient Temperature (°C)	Temperature Rating of Conductor			Ambient Temperature (°F)
	60°C	75°C	90°C	
10 or less	1.29	1.2	1.15	50 or less
11–15	1.22	1.15	1.12	51–59
16–20	1.15	1.11	1.08	60–68
21–25	1.08	1.05	1.04	69–77
26–30	1	1	1	78–86
31–35	0.91	0.94	0.96	87–95
36–40	0.82	0.88	0.91	96–104
41–45	0.71	0.82	0.87	105–113
46–50	0.58	0.75	0.82	114–122
51–55	0.41	0.67	0.76	123–131

Note: For illustration purposes only. Use actual NEC® tables for work.

The methods associated with the application of NEC® articles are illustrated through example problems listed below. However, the NEC® articles cited are subject to change, without notice, due to periodic code revisions and were meant to be valid at the time of development of these problems. Therefore, these NEC® article references may not be current and should not be used for actual practice of electrical engineering design. Instead, the reader should refer to the most current NEC® articles in their practice of engineering.

Example 9.1

Overcurrent Protection:

Applicable Code/Codes: Articles 210.19 (A) (1), 210.20 (A), and 310.15. Tables 310.15(B)(2)(a) and 310.15(B)(16).

The branch circuit in the exhibit below consists of two (2) 8 A continuous loads. Overcurrent protection in the branch circuit is provided through a 20 A circuit breaker. (a) Verify the size/specifications of the circuit breaker and the 12 AWG conductor, assuming conductor temperature at 60°C (140°F) or less. (b) If the ambient temperature were to rise to 50°C, how would the conductor size be impacted?

Conductor and OCPD verification:

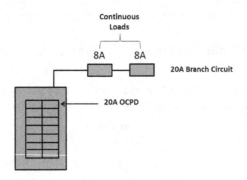

Solution:

a. In accordance with Article 210.20 (A), which stipulates:

    Branch-circuit conductors and equipment shall be protected by over-current protective devices that have a rating or setting that complies with 210.20(A) through (D).

    (A) Continuous and Non-continuous Loads. Where a branch circuit supplies continuous loads or any combination of continuous and non-continuous loads, the rating of the overcurrent device shall not be less than the non-continuous load plus 125 percent of the continuous load.

NEC® 2017

The overcurrent protection should be rated $= 1.25 \times$ continuous load

$$= 1.25 \times (8 \times 2 \text{ A}) = 20 \text{ A}.$$

∴The 20 A circuit breaker as an overcurrent protection device is adequate for the given branch circuit.

    In accordance with Article 210.19 (A) (1): "Branch circuit conductors shall have an ampacity not less than maximum load to be served... (and) shall have an allowable ampacity not less than the non-continuous load plus 125% of the continuous load."

∴ The conductor ampacity for the given branch circuit

$$= 1.25 \times \text{Continuous Load} + 1.00 \times \text{Non-Continuous Load}$$

$$= 1.25 \times (16 \text{ A}) + 1.00 \times (0) = 20 \text{ A}$$

According to Table 310.15(B)(16), for 60°C operation, with Type TW or UF insulation, AWG 12 conductor carries an allowable ampacity of 20 A for conductors that are insulated, rated for 0–2,000 V operation, in situations with no more than three (3) current-carrying conductors in raceway, cable earth (directly buried); under ambient temperature (not exceeding) 30°C (85°F); with no required/applicable derating.

    ∴ Selection of AWG 12 would be adequate for this scenario.

b. Ambient temperature rise and conductor size:

    According to Article 310.15, Tables 310.15(B)(2)(A) and 310.15(B)(16), when ambient deviates from 30°C to 50°C, a derating multiplier of 0.58 must be applied to adjust the ampacity of the conductor.

∴ The adjusted or derated ampacity of AWG #12 conductor, in this case, would be:

$$= 0.58 \times 20 \text{ A} = 11.6 \text{ A}$$

Since the derated ampacity of AWG #12, for this application, falls below the 20 A capacity mandated by 210.19 (A) (1), AWG #12 would no longer be adequate. Therefore, AWG #10, which is the next size above AWG#12, must be considered. According to Table 310.15(B)(16), for 60°C operation, with Type TW or UF insulation, AWG#10 conductor carries an allowable ampacity of 30 A. Then, if the 50°C adjustment rating of 0.58 is applied to 30 A ampacity of AWG #10, the derated ampacity would be:

$$= 0.58 \times 30 \text{ A} = 17.4 \text{ A}$$

Since the 17.4 A derated ampacity of AWG #10 still falls short of the 20 A requirement, it would **not** meet the code. So, the next larger size conductor, AWG #8, with an ampacity of 40 A should be considered.

If the 50°C adjustment rating of 0.58 is applied to the 40 A ampacity of an AWG # 8 conductor, the derated ampacity would be:

$$= 0.58 \times 40 \text{ A} = 23.2 \text{ A}$$

The 23.2 A derated ampacity of AWG #10 exceeds the 20 A requirement; therefore, it would meet the code.

Example 9.2

Overcurrent Protection and Minimum Conductor Size

Applicable Code/Codes: 210.20 (A), 240.6(A), 210.19 (A) (1), 310.15 and Table 310.15 (B) (16).

Determine the size of overcurrent protection device and the minimum conductor size for the following scenario assuming that no derating applies:

- Four (4) current-carrying copper conductors in a raceway.
- Operating temperature and OCPD terminal rating: 60°C
- Insulation: THWN
- Load: (calculated) 26 Amp, continuous.

Solution:

SIZE OF THE OCPD

In accordance with Article 210.20 (A), overcurrent protection should be rated $= 1.25 \times$ continuous load $= 1.25 \times (26 \text{ A}) = 32.5 \text{ A}$.

According to Article 240.6(A), standard ampere rating above 30 A is 35 A.

∴ The minimum standard size or rating of the **OCPD device should be 35 A.**

MINIMUM CONDUCTOR SIZE

In accordance with Article 210.19 (A) (1), the branch circuit conductors shall have an ampacity not less than maximum load to be served… (and) shall have an allowable ampacity not less than the non-continuous load plus 125% of the continuous load.

∴ The conductor ampacity for the given branch circuit = 1.25 × continuous load + 1.00 × non-continuous load = 1.25 × (26 A) + 1.00 × (0) = 32.5 A. **Select AWG 8 which has an ampacity of 40 A.**

Example 9.3

Appliance Load – Dwelling Unit(s):

Applicable Code/Codes: 220.53
Determine the feeder capacity needed for a 120/240 VAC, fastened-in-place, appliance load in a dwelling unit for the following appliances:

- Water heater, rated: 4,000 W, 240 V; load: 4,000 VA (PF = 1)
- Kitchen disposal, rated: 0.5 hp, 120 V; load: 1176 VA (PF & Eff. ≪100%)
- Dishwasher, rated: 1,200 W, 120 V; load: 1,200 VA (PF = 1)
- Furnace motor, rated: 0.25 hp, 120 V; load: 696 VA (PF & Eff. ≪100%)
- Attic fan motor, rated: 0.25 hp, 120 V; load: 696 VA (PF & Eff. ≪100%)
- Water heater, rated: 0.5 hp, 120 V; load: 1176 VA (PF = 1)

Solution:

Total Load = 4,000 VA + 1,176 VA + 1,200 VA + 696 VA + 696 VA + 1,176 VA

= 8,944 VA

Since the total load consists of more than four (4) appliances, according to Article 220.53, a demand factor of 75% is permissible.

∴ The size of the service and feeder conductors may be based on net load:

= 0.75 × 8,944 VA

= 6,708 VA

Example 9.4

Outlets in Dwelling Unit(s):

A 120 V dwelling branch circuit supplies four outlets, one of which has four receptacles. What is the total volt-ampere load?

Solution:

According to Article 220.14(I), a single outlet is counted as a 180 VA load. So, the three outlets, out of the four, would constitute a load of:

$$= (3) \cdot (180 \text{ VA}) = 540 \text{ VA}$$

Each of the receptacles, in the fourth outlet – with four receptacles – according to Article 220.14(I), would constitute a load of 90 VA. Therefore, the total load contribution from the four receptacles would be:

$$= (\text{numbers of receptacles})(90 \text{ VA/receptacle})$$

$$= (4) \cdot (90 \text{ VA}) = 360 \text{ VA}$$

∴ The total load is $= 540 \text{ VA} + 360 \text{ VA} = \textbf{900 VA}$

ARC FLASH

While some background information on the subject of arc flash is presented under Article 110.16 of the NEC®, the NEC® is not a core source for information on arc-flash regulations. The subject of arc flash is comprehensively addressed by **NFPA 70E**. NFPA 70E, similar to the NEC®, is revised every 3 years. Like the NEC®, NFPA 70E is available in print or in electronic format. Arc flash is being introduced in this text primarily due to the importance and gravity of arc-flash hazard in the electrical work environment. The introduction of arc flash in this text is at the basic level and **does not** enable the reader to adequately practice arc-flash safety.

Basic facts related to arc flash are listed below:

- Arc flash is the result of a rapid release of energy due to an arcing fault between a **phase bus bar and another phase bus bar, neutral or a ground**.
- Arc faults are typically limited to systems with the bus voltage in **excess of 50 V**.
- An arc fault is similar to the arc obtained during **electric welding**.
- The massive energy discharged during and arc fault phenomenon has the **capacity to burn bus bars, vaporize the copper, and cause an explosive volumetric increase.**
- An arc blast is estimated to cause explosive expansion of gas or air **67,000** times the nominal volume.
- Temperature in ARC plasma ranges from 5,000°F to 35,000°F.
- Approximately 5–10 arc-flash incidents (explosions) are recorded per day, in the United States. Average medical cost associated with remediation is estimated to exceed $1.5 million per incident. Total cost, including litigation, is estimated to be $8–10 million per incident.
- OSHA, Occupational Safety and Health Administration, carries the arc-flash regulation enforcement responsibility.
- The essence of relative magnitude or intensity of arc flash can be understood and appreciated through Eq. 9.1:

$$\textbf{Arc Flash Energy, or Incident Energy} = (\textbf{V}) \cdot (\textbf{I}) \cdot (\textbf{t}) \qquad (9.1)$$

In Eq. 9.1, the arc-flash energy, or incident energy, is the energy released by an arc-flash fault. This energy can be measured in kWh, Watt – sec, Joules, calories, BTUs, therms, etc. In arc-flash analysis, arc-flash classifications, and arc-flash PPE (personal protective equipment) ratings, arc-flash energy "**intensity**" term "**Calories/cm^2**" is commonly used. In Eq. 9.1, "**V**" is the rms (root mean square) voltage, "**i**" is the rms (root mean square) fault current, and "**t**" is the duration of the arc fault, in seconds. The arc fault current magnitude, typically, ranges in thousands of Amp and the fault duration is typically in milliseconds. As obvious, this equation stipulates that arc fault energy is directly proportional to the voltage, current, and fault duration. Note that, over the time span or duration of the fault, the voltage change is negligible in comparison with the order of magnitude rise in the current level. Therefore, mostly, the magnitude of the fault current is responsible for the enormity of the arc-flash energy.

In general, compliance with NFPA 70E and OSHA involves the following:

1. Facility owner, manager, or employer must be able to demonstrate exitance of an NFPA 70E-based safety process and program with defined responsibilities.
2. As a minimum, utilize the "Category Method" to assess the arc flash and shock hazard, identify the required PPE for performance of energized work– followed by incident energy analysis as warranted – for all electrical equipment rated or operating at 120 V (nominal) or greater.
3. Produce and affix warning labels on electrical equipment. Note that the arc-flash labels must be generated – as stated in item 2 above – by the equipment owners, not the manufacturers.
4. Provision of proper PPE for workers, as prescribed by proper arc-flash risk analysis.
5. Initial training, and subsequent refresher training, for workers on:
   a. The hazards of arc flash
   b. The proper interpretation of arc-flash risk and PPE labels
   c. Proper use of arc-flash PPE and voltage rated tools and gloves.

PHYSICAL AND THERMAL BACKGROUND OF ARC FLASH

This section is devoted to the exploration of the physical, chemical, and thermal aspects of arc-flash phenomena. This discussion is intended to enhance technicians' and engineers' appreciation of the reasoning behind arc-flash hazard analysis requirements and the need for arc-flash PPE. Pictures from two different arc-flash simulations are depicted in Figures 9.2 and 9.3.

The picture in Figure 9.2 depicts a simulated arc-flash incident, conducted at 250 V, 13.1 kA (13,100 A). The energy intensity for this simulated 13,100 A fault is estimated to be 1.48 cal/cm^2. The picture in Figure 9.3 depicts a simulated arc-flash incident, conducted at 250 V, 44 kA (44,000 A). The energy intensity for this simulated 44,000 A fault is estimated to be 8.48 cal/cm^2, almost **six times the fault energy** released by the lower current fault. Note that the voltage in both simulations was 250 V. This observation supports the directly proportional relationship between fault current and the arc-flash fault energy as stipulated in the fault energy equation, Eq. 9.1.

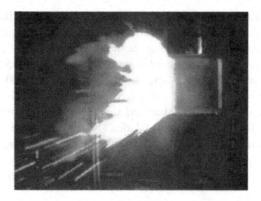

FIGURE 9.2   Simulated arc-flash incident, conducted at 250 V, 13.1 kA (13,100 A). Energy intensity: 1.48 cal/cm².

FIGURE 9.3   Simulated arc-flash incident, conducted at 250 V, 44 kA (44,000 A). Energy intensity: 8.48 cal/cm².

Some of the hazards associated with an arc-flash incident are evident from pictures in Figures 9.2 and 9.3. These hazards are as follows:

- **Extreme Heat** – Energy contained in explosive arc-flash events raises the temperature thousands of degrees; to the extent that it not only melts copper components but vaporizes them. Immense heat energy contained in vaporized copper can burn through or ignite regular work clothes. In addition, human skin coming in contact with vapors at such high temperatures would result in third-degree burns. Temperature in ARC Plasma can range from 5,000°F to 35,000°F. Proper PPE, including hood, balaclava, arc-rated coveralls, and jacket can provide a measure of protection against the radiated heat and the heat contained in the vapors. See the PPE section below.
- **Brilliant Flash** – Without adequate tinting and shielding, the intense light that accompanies the release of immense explosive energy can result in retina damage and can cause blindness.

- **High UV Emission** – The ultraviolet light that accompanies arc-flash events can damage human epithelial (skin) layer. Once again, proper arc-rated clothing can provide a measure of protection.
- **Shock Wave** – Vaporization of copper accompanies 67,000-fold expansion of air and gases. The intense instantaneous expansion of gases is tantamount to explosion of ordinance and results in a shock wave. The shock wave can launch workers off their feet resulting in broken limbs and bones. The explosive energy-laden shock waves have the capacity to subject anterior of a human body to immense pressure, potentially, fracturing ribs, puncturing, and collapsing lungs. Best practices associated with **safe posture** can provide a measure of protection against effect of a shock wave.
- **Loud Noise** – Due to the fact that expanding vaporized copper is an explosive event, it generates a loud report. This is the reason why hearing protection is required by NFPA 70E.
- **Projectiles** – As visible in Figures 9.2 and 9.3, energy contained in arc-flash events projects components outward in solid or molten form. Note the bright streaks in the picture of the simulated arc faults. The projectiles launched in arc-flash events can result in injuries due to shrapnel penetration.
- **Electrical Shock** – The explosion associated with arc-flash events dislodges components, insulation, and conductors, thus exposing personnel, in close proximity, to live energized components. This exposure can result in lethal electrical shock hazard.

ARC-FLASH PPE

Having gained some appreciation of the arc-flash hazards and causes thereof, let's consider some of the PPE required, in general, by NFPA 70E. The sets of PPE introduced here pertain to situations when (1) incident energy is 1.2–12 cal/cm^2 and (2) when incident energy is greater than 12 cal/cm^2. Arc-flash PPE for these two categories as listed below:

PPE, FOR INCIDENT ENERGY OF 1.2–12 CAL/CM^2

- Arc-rated clothing with an arc rating equal to or greater than the estimated incident energy
- Long-sleeve shirt and pants or coverall or arc-flash suit (SR)
- Arc-rated face shield and arc-rated balaclava or arc-flash suit hood (SR)
- Arc-rated outerwear (e.g., jacket, parka, rainwear, hard hat liner) (AN)
- Heavy-duty leather gloves, arc-rated gloves, or rubber-insulating gloves with leather protectors (SR)
- Hard hat
- Safety glasses or safety goggles (SR)
- Hearing protection
- Leather footwear

PPE, FOR INCIDENT ENERGY GREATER THAN 12 CAL/CM²

- Arc-rated clothing with an arc rating equal to or greater than the estimated incident energy
- Long-sleeve shirt and pants or coverall or arc-flash suit (SR)
- Arc-rated arc-flash suit hood
- Arc-rated outerwear (e.g., jacket, parka, rainwear, hard hat liner) (AN)
- Arc-rated gloves or rubber insulating gloves with leather protectors (SR)
- Hard hat
- Safety glasses or safety goggles (SR)
- Hearing protection
- Leather footwear

where
    SR means selection of one in group is required.
    AN means as needed.

An example of a PPE system, similar to what is required for a typical situation pertaining to greater than 12 cal/cm², is depicted in Figure 9.4.

As alluded to earlier, PPE requirement can be assessed on the basis of NFPA 70E "category" method or through NFPA 70E-based incident energy analysis. The label depicted on the left side of Figure 9.5 shows the format of the arc-flash labels

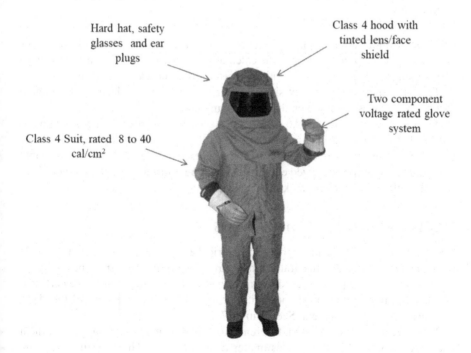

Hard hat, safety glasses and ear plugs

Class 4 hood with tinted lens/face shield

Two component voltage rated glove system

Class 4 Suit, rated 8 to 40 cal/cm²

FIGURE 9.4   Example of NFPA 70E Class 4 PPE.

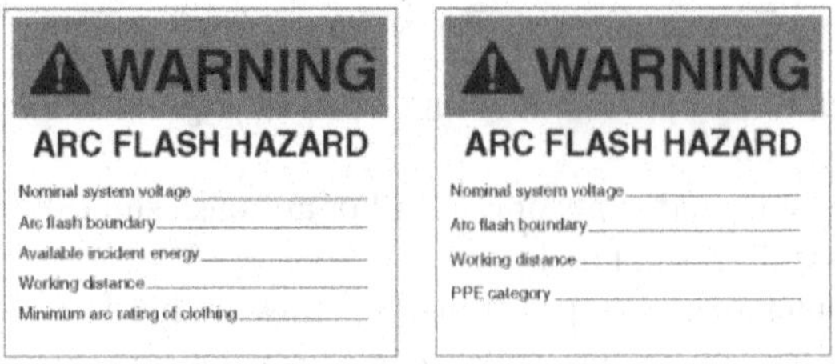

FIGURE 9.5 Labels based on arc-flash risk analysis, NFPA 70E.

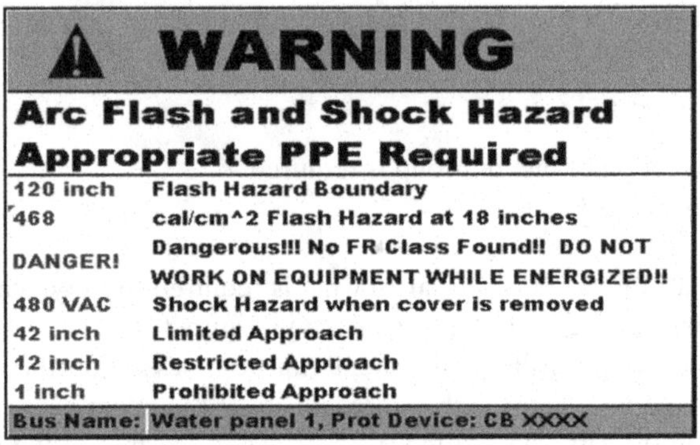

FIGURE 9.6 Example of a label pertaining to higher than 40 cal/cm² arc-flash hazard, based on NFPA 70E.

and type of information displayed as a result of incident energy-based arc-flash risk analysis. Especially noteworthy is the calculated "incident energy" and the "level of PPE" required when transacting with the pertaining electrical equipment.

The label shown on the right side of Figure 9.5 pertains to electrical equipment that has been assessed based on the category method.

The label shown in Figure 9.6 pertains to electrical equipment that has the potential for drawing enough fault current to exceed the energy intensity level of 40 cal/cm². When fault energy intensity of 40 cal/cm² is exceeded, as stipulated on the label in Figure 9.6, work on energized equipment is **not permitted**. "Robotic methods" are typically utilized to perform racking and switching in such situations.

ELECTRICAL SAFETY CERTIFICATIONS

Certifications and certification labels on electrical equipment are intended to provide the end user, installer, or integrator a measure of assurance that the labeled equipment is safe when applied or used as prescribed by the manufacturer. The certification labels can be found at the bottom, rear, or sides of non-custom, off-the-shelf, standard electrical equipment. See Figure 9.7.

Various certifications exist in different parts of the world. A few, more prominent ones, are mentioned in this text as matter of introduction. These certifications are listed below:

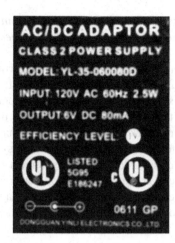

FIGURE 9.7   Examples of a safety certification labels – ETL and UL listings.

- **UL®**, Underwriters Laboratories, United States. UL® tests equipment to be certified for safety, either at laboratories owned and operated by UL or at laboratories owned and operated by its sub-contractors, such as **ETL**CM.
- **ETL**CM, Intertek Listing. The ETL Listed Mark is proof of product compliance (electrical, gas, and other safety standards) to North American safety standards. Following authorities having jurisdiction in 50 states and Canada accept the ETL Listed Mark as proof of product safety: UL, ANSI (American National Standards Institute), CSA (CSA Group), ASTM (formerly known as: American Society for Testing and Materials), NFPA, and NOM (Norma Official Mexicana).
- **NOM Mark**: NOM is a mark of product safety approval for virtually any type of product exported into Mexico.
- **ULC**, Underwriters Laboratories of Canada.
- **IEC**, International Electrotechnical Commission. The IEC plays an important role in developing and distributing electrical standards. IEC was instrumental in developing and distributing standards for units of measurement, particularly the gauss, hertz, and weber. IEC first proposed a system of standards, the Giorgi System, which ultimately became the SI, or Système International d'unités (in English, the International System of Units).
- **IP Rating**: IP rating stands for **International Protection** rating. Sometimes interpreted as **Ingress Protection** rating, the IP rating consists of the letters **IP** followed by two digits and an optional letter. As defined in IEC 60529 60529, it classifies the degrees of protection provided against the intrusion of solid objects (including body parts like hands and fingers), dust, accidental contact, and water in electrical enclosures. See Table 9.2 for correspondence between the American **NEMA® ratings** and their counterparts in the **IP** realm. Also see Example 9.5 for illustration of enclosure ratings application.

- **CE Certification:** The CE Mark stands for *Conformité Européenne*,
  a French term that can be literally translated into English as *European Conformity*.
- **Safety Compliance Statement** from the manufacturer, engineering
  firm, general contractor, turn-key installer, or system integrator is often
  required on custom-engineered systems or equipment. This requirement
  must be clarified and agreed upon – at the quotation/bidding phase of the
  project – between the project manager, end user/customer, plant safety manager, engineering firm, general contractor, turn-key installer, and the system integrator (if applicable). Absence of such documented agreement on
  Safety Compliance Statement requirement, at the very outset of a custom-
  engineered project, can result in system commissioning and start-up delays,
  in addition to penalties, unforeseen costs, potential breach of contract,
  future business prospects, etc.

OTHER, COMMON SAFETY CERTIFICATIONS

Examples of other electrical equipment safety certifications are listed in Table 9.3.

TABLE 9.3
Examples of Other Electrical Equipment Safety Certifications

c-UL-us	UL listed industrial control equipment, certified for United States and Canada. See UL File E65584.
	UL listed for Class I, Division 2, Group A, B, C, D hazardous locations, certified for United States and Canada. See UL File E194810.
CSA	CSA-certified process control equipment. See CSA File LR54689C.
	CSA-certified process control equipment for Class I, Division 2, Group A, B, C, D hazardous locations. See CSA File LR69960C.
FM	FM-approved equipment for use in Class I, Division 2, Group A, B, C, D hazardous locations
CE	European Union 2004/108/EC EMC Directive, compliant with: • EN 61326-1; meas./control/lab., industrial requirements • EN 61000-6-2; industrial immunity • EN 61000-6-4; industrial emissions • EN 61131-2; programmable controllers (Clause 8, Zones A & B)
RCM	Australian Radiocommunications Act, compliant with EN 61000-6-4; industrial emissions
Ex	European Union 94/9/EC ATEX Directive, compliant with: • EN 60079-15; potentially explosive atmospheres, protection "n" • EN 60079-0; general requirements • II 3 G Ex nA IICT4 Gc • DEMKO13ATEX1325026X
IECEx	IECEx System, complaint with: • IEC 60079-15; potentially explosive atmospheres, protection "n" • IEC 60079-0; general requirements • II 3G Ex nA IIC T4 Gc • IECEx UL 14.0008X

(Continued)

TABLE 9.3 (*Continued*)

Examples of Other Electrical Equipment Safety Certifications

KC	Korean Registration of Broadcasting and Communications Equipment, complaint with Article 58-2 of Radio Waves Act, Clause 3
EAC	Russian Customs Union TR CU 020/2011 EMC Technical Regulation
	Russian Customs Union TR CU 004/2011 LV Technical Regulation
EtherNet/IP	ODVA conformance tested to EtherNet/IP specifications

ADDITIONAL INFORMATION ON NEMA® AND IP RATINGS

The digits or numeral in the NEMA or IP code indicate conformity with the conditions summarized in Table 9.4. The first digit indicates the level of protection that the enclosure provides against the ingress of solid foreign objects. For example, an electrical socket rated IP22 is protected against insertion of fingers and will not be damaged or become unsafe during a specified test in which it is exposed to vertically or nearly vertically dripping water. IP22 or IP2X are common minimum requirements for the design of electrical accessories for indoor use.

TABLE 9.4

NEMA® and IP Electrical Enclosure Ratings Descriptions, Comparison, and Correspondence

NEMA® Type	Definition	IEC Equivalent
1	General purpose. Protects against dust, light, and indirect splashing but is not dust-tight; primarily prevents contact with live parts; used indoors and under normal atmospheric conditions.	IP10
2	Drip-tight. Similar to Type 1 but with addition of drip shields; used where condensation may be severe (as in cooling and laundry rooms).	IP11
3 and 3S	Weather-resistant. Protects against weather hazards such as rain and sleet; used outdoors on ship docks, in construction work, and in tunnels and subways.	IP54
3R	Intended for outdoor use. Provides a degree of protection against falling rain and ice formation. Meets rod entry, rain, external icing, and rust-resistance design tests.	IP14
4 and 4X	Watertight (weatherproof). Must exclude at least 65 GPM of water from 1-in. nozzle delivered from a distance not less than 10 ft for 5 minutes. Used outdoors on ship docks, in dairies, and in breweries.	IP66
5	Dust-tight. Provided with gaskets or equivalent to exclude dust; used in steel mills and cement plants.	IP52
6 and 6P	Submersible. Design depends on specified conditions of pressure and time; submersible in water; used in quarries, mines, and manholes.	IP67
7	Hazardous. For indoor use in Class I, Group A, B, C, and D environments as defined in the NEC®.	N/A

(Continued)

TABLE 9.4 (*Continued*)
NEMA® and IP Electrical Enclosure Ratings Descriptions, Comparison, and Correspondence

NEMA® Type	Definition	IEC Equivalent
8	Hazardous. For indoor and outdoor use in locations classified as Class I, Groups A, B, C, and D as defined in the NEC®.	N/A
9	Hazardous. For indoor and outdoor use in locations classified as Class II, Group E, F, or G as defined in the NEC®.	N/A
10	MSHA. Meets the requirements of the Mine Safety and Health Administration, 30 CFR Part 18 (1978).	N/A
11	General purpose. Protects against the corrosive effects of liquids and gases. Meets drip and corrosion-resistance tests.	N/A
12 and 12K	General purpose. Intended for indoor use, provides some protection against dust, falling dirt, and dripping noncorrosive liquids. Meets drip, dust, and rust resistance tests.	IP52
13	General purpose. Primarily used to provide protection against dust, spraying of water, oil, and noncorrosive coolants. Meets oil exclusion and rust resistance design tests.	IP54

NEMA® versus IP Enclosure Ratings

Example 9.5

Electrical specifications for a factory call for a junction box that must be submersed into a water tank. (a) Determine the NEMA rating of junction box for the US installations. (b) Determine the IP rating of the junction box for the European installations.

Solution:

a. Examination of the NEMA–IP rating table in this chapter shows that NEMA 6 enclosure is rated as:
   Submersible. Design depends on specified conditions of pressure and time; submersible in water; used in quarries, mines, and manholes.
   Therefore, a **NEMA 6 enclosure should be specified** for the US installation.
b. Since the European installation would be exposed to the same worst-case conditions, US NEMA 6's European counterpart, **IP 67 should be specified**.

Common Acronyms Associated with Electrical Standards Organizations

A few organizational acronyms, commonly used in the field of electrical engineering, are listed below with respective background information and website addresses to facilitate further exploration of the roles of these organizations by the reader:

NEMA: National Electrical Manufacturers Association; www.nema.org

NEMA, created in the fall of 1926 by the merger of the Electric Power Club and the Associated Manufacturers of Electrical Supplies, provides a forum for the standardization of electrical equipment, enabling consumers to select from a range of safe, effective, and compatible electrical products.

ANSI: American National Standards Institute; www.ansi.org

The ANSI is a private, nonprofit organization that administers and coordinates the US voluntary standardization and conformity assessment system.

IEC: International Electrotechnical Commission.

IEC is the authoritative worldwide body responsible for developing consensus global standards in the electrotechnical field. IEC is the European counterpart to NEMA and ANSI.

IEEE: Institute of Electrical and Electronic Engineers: www.ieee.org

The IEEE is a nonprofit, technical professional association for Electrical and Electronics Engineers. IEEE makes vital contributions in the electrical engineering realm, in many ways. Some of IEEE's more prominent contributions are as follows:

- Publishing of texts and reference materials that promote education and training on various electrical and electronic technologies and subjects.
- Development of Arc-Flash Hazard Calculation formulas through IEEE 1584 committee.
- Facilitation of the development of protocols that promote communication between various electronic devices.

International Society of Automation (ISA): www.isa.org

Founded in 1945, the ISA is a leading, global, nonprofit organization that is setting the standard for automation by helping members and other professionals solve difficult technical problems. ISA is based in Research Triangle Park, North Carolina. ISA develops standards, certifies industry professionals, provides education and training, publishes books and technical articles, and hosts conferences and exhibitions for automation professionals.

RIA – Robotics Industries Association: www.robotics.org

RIA publishes information to help engineers, managers, and executives apply and justify robotics and flexible automation. The RIA website includes a proprietary search engine algorithm that makes it easy to find and compare leading companies, products, and services. Robotics Online is dedicated to news, articles, and information specifically for the robotics industry.

COMMON ELECTRICAL/ELECTRONIC SAFETY DEVICES

This section introduces the reader to electrical and electronic devices commonly employed in automated control systems and process controls, in general, for safety-related actions and events. We will first familiarize the reader to each of the devices, individually. Then, we will discuss integration of the devices in an automated manufacturing system example.

Similar to the pictorial tour approach utilized with the MCC and power distribution system discussion, we will explore Rockwell Automation® safety control devices to gain familiarization with what these devices look like and gain a better understanding about their functions.

SAFETY E-STOP DEVICES

E-stop switches are, functionally, passive mechanical devices similar to the light switches in offices and homes. Rockwell/A-B® offer Series 800T/800E Push-Buttons and Self-Monitoring Contact Block-based E-stop switches. The emergency stopping function is implemented through a combination of two components: (1) Push–pull operator and (2) contact block.

The description and specification of these two components are as follows:

- E-Stops Operator
  - Available in 30 mm and 22 mm sizes
  - Metal and plastic construction
  - Meet EN418 and IEC 60947-5-5 standards
  - Push–pull, push–pull/twist release, illuminated, or key-operated devices
- Self-Monitoring Contact Blocks
  - For use with 800T and 800E E-Stops
  - If contact block becomes separated from E-stop, monitoring circuit automatically opens and shuts down the controlled process. This feature essentially eliminates contact separation concerns from improper installation, damage, or high-vibration applications.

Contact Blocks   **Operators**

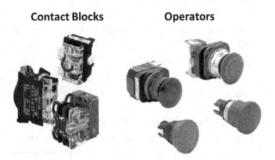

FIGURE 9.8   Rockwell® Safety E-stop switch operators and contact blocks.

SAFETY LIGHT CURTAIN SYSTEM

Even though safety light curtains often perform emergency stopping function, similar to the emergency stop switches introduced above, they are powered, active, somewhat automated, and substantially more sophisticated than the E-stop switches. See Figures 9.9 and 9.10.

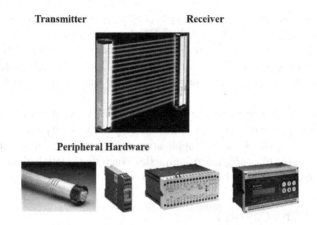

FIGURE 9.9   Rockwell® safety light curtain.

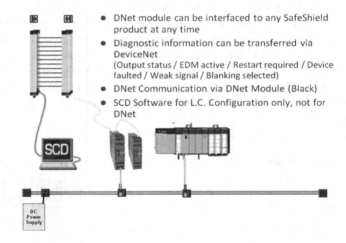

FIGURE 9.10   SafeShield DeviceNet: Rockwell® Safety Light Curtain and Device-Net interface.

Some of the key components of a safety light curtain system are listed in Figure 9.9. As depicted in Figure 9.9, the transmitting "column" transmits an array of invisible (infrared) light beams. When the path between the transmitter and the receiver is clear and unobstructed, the transmitted beams are received by the receiver. This, in most applications of light curtains, constitutes the norm. If, however, the light beams are interrupted by equipment or personnel, typically, an emergency stop command is generated, thus shutting down the protected system. The interface cable pictured in Figure 9.9, as simple and unsophisticated as it appears, constitutes a reliable and robust approach to electrical/control connections. Electrical connections, a decade or so ago, had to be made one wire and one terminal at a time. With hundreds of connections required in mid- to large-size control systems, the old wire to connector method often resulted in miss-wiring, loose connections, unreliability, and delayed system start-ups.

Other important features and components associated with the light curtains are included in Figure 9.10. Rockwell's® DNet module (hardware) is an interface module and is pictured in Figure 9.10. The software or protocol that permits the safety devices, safety Programmable Logic Controllers (PLCs), and other Rockwell® control devices to communicate with each other is referred to as DeviceNet®.

The laptop PC shown in Figure 9.10 allows control engineers and technicians to configure or "program" the light curtain to respond to safety events in a desired fashion, In addition, the Rockwell® application software loaded on the laptop allows the control engineer to **configure "diagnostics"** such that safety incidents and other associated events can be troubleshot promptly.

AAC (Area Access Control)

Area access control (AAC) system depicted in Figure 9.11 offers a simpler, less costly, less complex alternative to light curtains. Typical AAC system consists of a small single beam transmitter that transmits an infrared beam across the protected opening, to a pair of prism-shaped reflectors. The reflectors return the beam across the opening to a receiver. When the beam is interrupted by person-nel or equipment, an emergency stop is triggered by the AAC system. As obvious from a comparison between Figures 9.10 and 9.11, the protected opening area of cross-section covered by a light curtain is greater than the two-dimensional area covered by the AAC.

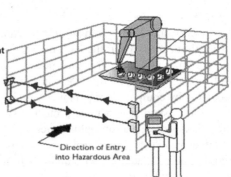

- Long Scan Range up to 70M (230Ft.)
- Die Cast Aluminum Housing
- Two Ranges available
  - -.5M to 18M (19.5" to 59")
  - -15M to 70M (4' to 230')
- Easy Installation
- Heated Front Screen, I.e. can be used in outdoor applications
- Fast response time 22ms
- 24vd/115vac standard / 230vac (special order)
- Built in monitored safety relays
  - -2 NO/1NC / 2A Max switching Current
- IP 65 enclosure rating
- Operating temp. -25°C to 50°C
- PG connector IP 67

Direction of Entry into Hazardous Area

FIGURE 9.11 Rockwell® AAC system.

CABLE PULL SWITCHES

Cable pull safety switches are designed to trip an emergency stop circuit when the cord, cable, or chain connected to the pull switch is pulled. Rockwell® cable pull switches incorporate a reset button on the front of the switch.

Some of the cable pull safety switches are equipped with a tension-viewing window to facilitate set-up of the cable tension. Among other important features incorporated in the Rockwell® cable pull switches is a function that latches out the contacts electrically and mechanically in accordance to EN 418. The reason is that if a person were to pull an ordinary cable pull switch as they were being dragged into a machine, the system would stop, but it would not prevent the system from being restarted at the other end by an operator, who did not see the switch being pulled. Thus, the operator could start the machine again, dragging the other subject back into the machine. In order to reset the lifeline 4 switch, an operator has to physically go up to the switch and reset the device by moving the designated lever into the run position. This allows for inspection of the area before the machine is restarted. The switch is yellow in accordance to EN 60204-1 which stipulates that all E-stops have a red button and yellow background.

KEY INTERLOCK SOLENOID SWITCHES

The next family of safety switches offered by Rockwell® are solenoid locking switches. See switches depicted in Figure 9.12. These switches are used to prevent access to a hazardous area until the hazardous motion has been contained. An example would be to have a gate locked until a set of cutting shears have come to rest, after which time a voltage is applied to the coil of the solenoid, releasing the key, allowing access to the area.

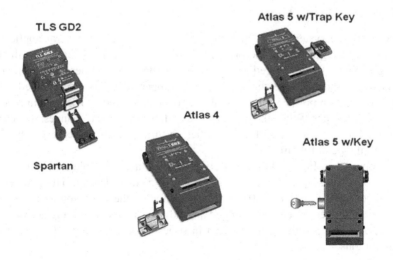

FIGURE 9.12   Rockwell® key interlock solenoid safety switch system.

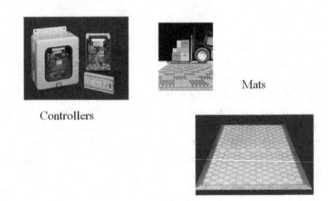

FIGURE 9.13   Rockwell® GaurdMat switch system.

GuardMat™ Safety Mats

Guardmaster offers a full range of standard mats and controllers for all applications. The GuardMat™ system is a tripping device. When a person steps on the mat, or equipment rolls onto it, their presence is detected and the safety output opens, shutting down the machine. See Figure 9.13.

The design of the mat includes two sheets of hardened rolled steel which are separated by small insulators. The separated sheets of metal have an approximate 24 V potential difference across them. When pressure exceeding 70 psi is applied to the top sheet, the sheets make contact, creating a short-circuit condition, changing the resistance, which is detected by the GaurdMat controller. The controller, in turn, executes emergency shutdown through system interlocks and safety control circuits.

GuardEdge™ Safety Edges

The GuardEdge system is a trip device which lends itself to several applications as depicted in Figure 9.14. The principle behind this system is that when the protruding rubber strip – referred to as profile – is depressed, emergency stop signal is issued and the machine shuts down. The rubber profile is impregnated with a conductive carbon powder, which creates a conductive rubber. The system reacts to change in resistance. When the profile is pressed, the resistance changes; this change is detected by the controller. The controller, in turn, executes emergency shutdown through safety control circuits.

As apparent in Figure 9.14, a variety of profiles are available to match specific applications. These profiles can be made up to 50 m in length. The profiles can also be bent to a radius of 200 mm to accommodate the most non-linear applications. The safety edges can be wrapped around corners, incorporating active corner connections. The profiles can be wired in series or parallel, depending on system configurations.

Profiles/Rails and Controllers

FIGURE 9.14 Rockwell® GaurdEdge system.

SAFETY PLCS

Traditionally, National Fire Protection Association's "Electrical Standard for Industrial Machinery" (NFPA79) has required hard wiring and the use of electromechanical components for safety circuits. With the advent, subsequent development, and enhanced reliability of safety PLCs, electromechanical safety circuits continue to be replaced by safety PLCs. Some of the more significant advantages offered by safety PLCs are as follows:

*   Safety PLC-based control circuits require fewer wires and terminations as compared to electromechanical relay and contactor-based safety control circuits.
*   Safety PLC-based control circuits require fewer "moving parts," fewer contactors, timers, and relays, if any, as compared to electromechanical relay and contactor-based safety control circuits.
*   Safety PLC-based control circuits are programmable. Therefore, their logic and functionality can be modified through modification of application program or code. While with electromechanical relay and contactor-based safety control circuits, significant hardware and wiring modifications are necessary for accomplishment of functional modification.

    Functionally and logically, safety PLCs are similar to regular PLCs. Notwithstanding the similarities, there are some notable differences between regular PLCs and safety PLCs. These differences are as follows:
*   Safety PLCs are color-coded **red** to signify the fact that they are "control reliable" and safety rated.
*   Safety PLCs employ **robust diagnostics and operational verification**. Standard PLC inputs provide **no** internal means for testing the functionality of the input circuitry. By contrast, safety PLCs have an internal "output" circuit associated with each input for the purpose of periodic testing and verification of the input circuitry. Simulated highs (1's) and lows (0's) are presented, automatically, to the inputs to verify their functionality.

- A safety PLC has **redundant microprocessors,** Flash and RAM memory that are continuously monitored by a **"watchdog" circuit** and a synchronous detection circuit. Regular PLCs are typically equipped with **one microprocessor.**
- A regular PLC has one output switching device, whereas a safety PLC's digital output logic circuit contains a test point after each of two safety switches located behind the output driver and a third test point downstream of the output driver. Each of the two safety switches is controlled by a unique microprocessor. If a failure is detected at either of the two safety switches, the operating system of a safety PLC will **automatically acknowledge the anomaly** and will default to a known state, thus facilitating an orderly equipment shutdown.

Typical specifications of safety PLCs are listed in Figure 9.15.

GuardLogix 5580 and Safety Partner
SIL CL3, PLe

GuardLogix 5580
SIL CL2, PLd

Compact GuardLogix 5380
SIL CL2, PLd

- GuardPLC 2000 and 1200 shipping since August
  – TÜV Certified (Entire System) - IEC 61508 SIL 3, DIN VDE 19250 AK6, EN 954-1 Category 4,
  – UL Listed
  – Primary Target Market - Machinery Safety
- GuardPLC 2000 - 6 I/O slot, modular design
  – 24 Input / 16 Output digital
  – 8 Channel Analog Input & Output (12 bit resolution)
  – 2 Channel HSC (100kHz, 24 bit)
- Guard 1200 - packaged design
  – 20 Inputs / 8 Outputs + 2 HSC Inputs (100khz, 24 bit)
- Communications
  – Proprietary GuardPLC Ethernet + ASCII
  – Peer to Peer Safety Communications
- RSLogix Guard Software (2 versions)
  – Lite and Professional Versions
  – Windows NT/2000
  – RSLogix "Look & Feel"
  – IEC 1131 Function Block Programming
  – User Defined Function Block Capabilities (1755-PCS)

FIGURE 9.15   Rockwell® safety PLC system.

Safety Relays

A regular relay consists of an electromagnetic coil or solenoid and associated contacts. The solenoid functions in the same manner as described earlier in this this text; it operates or controls multiple normally open or normally closed contacts. When a relay is off, its coil is de-energized, its normally open contacts are OPEN, and its normally closed contacts are CLOSED. Conversely, when a relay is turned on, the coil is energized, the normally open contacts are CLOSED and the normally closed contacts are OPEN. A regular relay is shown applied in an alarm circuit in Figure 9.16.

In the schematic diagram of the regular relay shown in Figure 9.16, a relay labeled CR1 (Control Relay 1) is being used to annunciate the status of an **alarm switch** to the left. The coil of CR1 is connected to neutral "N" on the right side. The other side of CR1 is connected to 110 V only when the alarm switch is operated. As shown in the circuit diagram, the control relay CR1 has two normally open contacts, depicted as two parallel lines and two normally closed contacts shown with a diagonal bar across the parallel lines. When the alarm switch is operated, or closed, the CR1 coil is energized. This closes the normally open CR1 contacts and opens the normally closed contacts. The two normally open contacts are being used to turn on the **alarm horn** and the **red alarm light**. One of the normally closed contacts is being used to maintain the **green light** on, as long as the safety switch is **not** closed. The second normally closed contact is left unconnected, as a spare contact for possible future use.

Safety relays operate somewhat similar to regular relays, with the following exceptions:

1. Safety relays are equipped with redundant coils and contacts.
2. Safety relays are equipped with diagnostics features.

Manufacturers, such as Rockwell®, also offer solid-state relays for applications involving high cycle rates. Safety relays offered by Rockwell® are shown in Figure 9.17, along with respective features and functions.

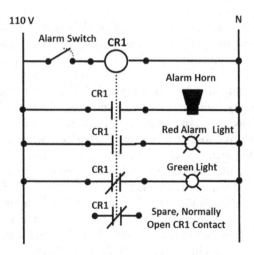

FIGURE 9.16   Regular relay applied in an alarm circuit.

Software Configurable Safety Relay

Guardmaster Safety Relays

Single-function Safety Relays

FIGURE 9.17   Rockwell® safety relays.

Now that we have gained a measure of familiarity with electrical control and safety devices, let's explore typical application of some of these devices through a tour of an automated manufacturing cell depicted in Figure 9.18. If we begin at approximately 11:00'o clock and move clockwise, we notice a cabinet housing the **safety PLC** and other safety devices such as safety relays, safety contactors, controllers for various safety devices, i.e. safety mat, safety edge, safety laser scanners, etc. The control cabinet housing the **safety** control devices is painted yellow to distinguish it from typical, gray, control cabinets for regular, **non-safety**, control systems.

At about 1:00'o clock, the common **E-stop switch** is shown mounted on the right side of the processing machine opening. In the diagram, the E-stop switch is labeled as "Safety Button."

The **safety limit switch** pointed out in the middle of the automated cell diagram limits inadvertent rotation of the robot about its major axis, beyond a safe point. Such application of a safety limit switch, typically, serves as a back-up to a software-based rotational limit.

Trapped key safety switch is shown installed on one of the man doors at about 3:00 o'clock. A **safety cable pull switch** is shown spanning the length of the conveyor. A **safety guard** is shown applied just outside a short conveyor section, at about 6:00'o clock, for the purpose of preventing **authorized** personnel in the area from coming too close to moving parts, i.e. the conveyor rollers.

A safety laser scanner is shown mounted on the frame that sustains the two gantry robots. This laser scanner is not clearly visible in Figure 9.18. Therefore, a picture of a Rockwell/Allen-Bradley, multi-zone, laser scanner is shown in Figure 9.19. Safety laser scanner represents a sophisticated and relatively new approach to three-dimensional protection that is not available through two-dimensional safety light curtains and other less sophisticated safety devices. However, not unlike light curtains, safety laser scanners must be programmed and configured for desired function.

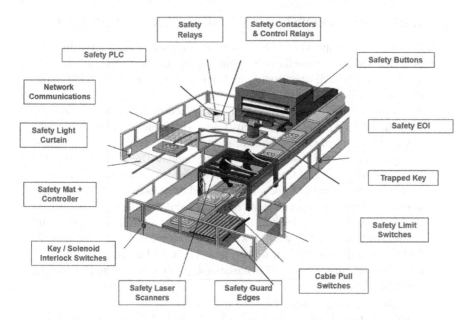

FIGURE 9.18   Rockwell® safety equipment applied to an automated manufacturing cell.

FIGURE 9.19   Rockwell® SafeZone™ safety laser scanners.

SELF-ASSESSMENT PROBLEMS AND QUESTIONS – CHAPTER 9

1. A given circuit is meant to carry a continuous lighting load of 16 A. In addition, four loads designed for permanent display stands are fastened in place and require 2 A each when operating. What is the rating of the OCPD on the branch circuit?

2. A three-phase, four-wire feeder with a full-sized neutral carries 14 A continuous and 40 Amp non-continuous loads. The feeder uses an overcurrent device with a terminal or conductor rating of 60°C. What is the minimum copper conductor size? Assume no derating applies. Use Tables 9.1 and 9.2.

3. Electrical specifications for a brewery company call for a fusible disconnect switch enclosure that must be able withstand occasional splashing of water during periodic wash downs required by the local health codes. This design will be applied in breweries in the United States as well as Europe. The water flow from the 1-in wash down nozzles is expected to be less than 60 GPM from a distance of 11 ft for less than 4 min. (a) Determine the NEMA rating of enclosure for the US installations. (b) Determine the IP rating of enclosure for the European installations.

4. Overcurrent protection and conductor ampacity:

   Applicable Code/Codes: Articles 210.19 (A) (1), 210.20 (A), and 310.15 and Tables 9.1 and 9.2 of this text (Note: This is not a current table and is only reproduced for exercise and illustration purposes).

   The branch circuit in the exhibit below consists of three continuous loads. Overcurrent protection in the branch circuit is provided through a 20 A circuit breaker. (a) Determine the size of copper conductor based on the ampacities given in Tables 9.1 and 9.2 assuming conductor temperature is at 75°C or less. Assume 75°C operation and selection are allowed. (b) Verify the size/specifications of the circuit breaker. Assume no derating applies. (c) If the ambient temperature were to rise to 50°C, how would the conductor size be impacted?

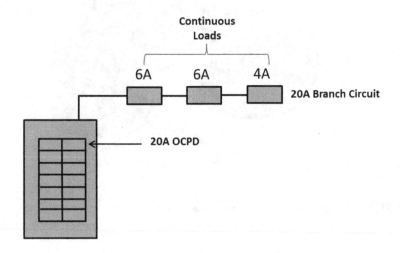

5. A US appliance manufacturer is planning to market a new appliance in Mexico. The most appropriate safety certification for this appliance would be:

A. UL
B. ULC
C. ETL
D. NOM

6. Assume that the alarm switch in the control circuit depicted below is opened after being closed for a prolonged period of time. Which of the following conditions would best describe the status of the annunciating lights and the horn when the switch is opened?

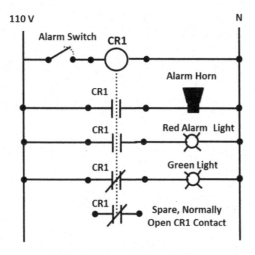

A. Alarm horn will turn off
B. Red light will turn off
C. Green light will turn on
D. All of the above

10 Electrical Drawings and PLC Relay Ladder Logic Program

INTRODUCTION

In this chapter, the coverage of electrical design drawings, and the application of National Electrical Code (NEC®) in the associated design work, is **not** intended to provide the reader expert-level knowledge that is required and expected of an NEC®-trained and practicing electrical Professional Engineer or electrical power system designer. The purpose of this chapter is to merely **introduce** the reader to established electrical power distribution drawings, controls drawings, and associated best practices and standards.

Engineering professionals or non-engineer readers will find this chapter helpful in equipping them with enough knowledge to be able to stay abreast of discussion at hand when those electrical and control drawings are spread across the table in the process of troubleshooting and reinstating an important piece of equipment back into operation. In addition, through a brief introduction to Programmable Logic Controller (PLC) relay ladder logic, we will introduce the reader to the programming technique utilized by most control engineers to control electrical and mechanical systems with PLC-based control systems. Of course, as before, we will illustrate the concepts and practices discussed, through examples, end-of-chapter problems, and the solutions at the back of the text.

ELECTRICAL DRAWINGS

Electrical engineers, electrical designers, and electrical CAD (Computer-Aided Design) operators can draw pertinent, applicable, references from the Electrical Engineering Drawing Standards listed in Table 10.1. Of course, the litany of standards listed in Table 10.1 are but a few to draw from. There are copious other standards applicable to the formulation of electrical drawings in the United States and abroad. In addition, CAD drawing software, i.e. AutoCAD Electrical, come incorporated with universally recognized electrical drawing symbols and standards.

Three common types of electrical drawings are discussed in this chapter. These include a one-line power distribution schematic, a wiring diagram, and electrical control drawings. The objective of this chapter is to inculcate basic understanding of electrical symbols, electrical drawing conventions, and electrical design. In addition, the reader will be shown how NEC and other methods for application of the code are employed in the electrical power distribution system design.

TABLE 10.1
Electrical Drawing Standards

Name	Description
ISO 01.100.25	Electrical and Electronics Engineering Drawings
NASA	NASA Engineering Drawing Standards
NEMA ICS 19-2002 (R2007)	NEMA Diagram, Device Designations and Symbols for Industrial Controls and Systems
STS 904	Hunter Water Corporation Standard Technical Specifications for Preparation of Electrical Engineering Drawing
IEEE Std 100-1996	IEEE Standard Dictionary of Electrical and Electronic Terms, Sixth Edition
IEEE Std 315™	IEEE Std 315™, IEEE Standard Graphic Symbols for Electrical and Electronics Diagrams (Including Reference Designation Letters).
IEEE Std 280-1985	American National Standard for Mathematical Signs and Symbols for Use in Physical Sciences and Technology
IEEE Std 280-1985	IEEE Standard Letter Symbols for (R1997) Quantities Used in Electrical Science and Electrical Engineering
ANSI Y32.9	American National Standard Graphic Symbols for Electrical Wiring and Layout Diagrams Used in Architecture and Building Construction.

ONE-LINE SCHEMATIC DIAGRAM

The one-line schematic diagram is also, simply, referred to as a "one-line drawing." This drawing is called a "one-line" drawing because it depicts electrical circuits and design through representation of just one of the three phases. This is predicated on a reasonable assumption that all three phases on the three-phase loads and sources are substantially identical in current, voltage, impedance, and other important considerations. As obvious, one important benefit derived from one-line representation of electrical circuits is that a greater number of circuits can be captured on one drawing. In other words, one can examine a large portion of the overall electrical system at one glance. This facilitates a quicker and more effective comprehension of a large segment of the electrical system being examined without flipping from one drawing to another. This is of considerable value in a triage situation when troubleshooting a system that is down. A simplified one-line schematic is shown in Figure 10.1. This drawing pertains to an MCC (motor control center)-based power distribution system. Due to the extensity of information captured in this diagram, certain segments and annotation are somewhat illegible. Therefore, those segments of the schematic that are examined in greater detail in this chapter are excerpted and duplicated in Figure 10.2.

As we examine the one-line schematic, and other drawings, we will note the more conventional symbols and nomenclature, as well as, certain practices adopted by the electrical engineer/designer of these drawings that deviate from the more universally accepted methods. As we examine the top portion of the one-line schematic in Figure 10.1, the first piece of information we notice is the rating of the MCC power distribution system:

480 V Bus, 3 ϕ, 4W, 600A, 60 Hz

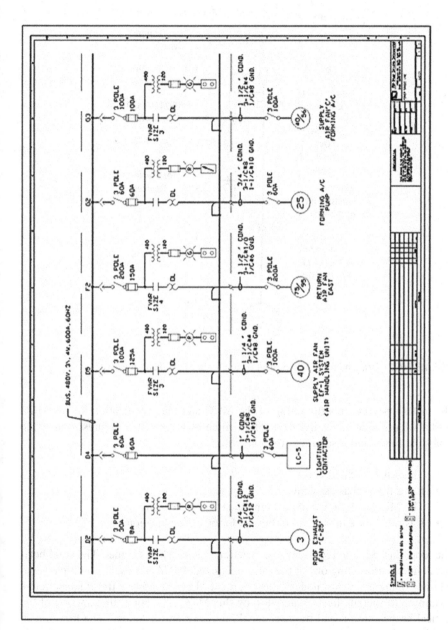

FIGURE 10.1   One-line schematic for a power distribution system.

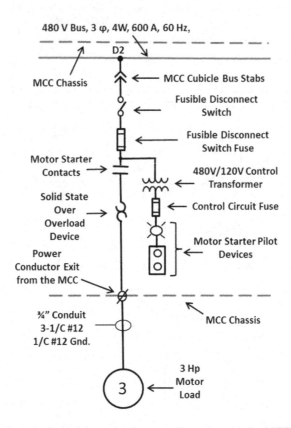

FIGURE 10.2 Branch circuit D2 excerpt from one-line schematic for MCC.

This notation represents the rating of the MCC and the specification for the MCC bus bars. This notation encircles one of the three-phase bus bars – represented by the long solid line – and it stipulates the following:

- This MCC is rated 480 V
- It is a three-phase system
- The MCC is designed to accommodate four wire loads
- It is rated for a maximum of 600 A, for 60 Hz application

The long, dashed, line represents the outer chassis or MCC cabinet. The solid horizontal line, representing one of the three energized 480 V phases, has multiple vertical lines "hanging" below nodes. These vertical "drops" represent the branch circuits pertaining to specific loads, catered to by this MCC. The dot at the junction of the vertical branch circuit and the horizontal bus represents a "node." The first branch circuit is identified by a unique identifier "D2," at the bus bar, on the line side. The other branch circuits are identified as D4, D5, and so on. Note the chevron symbols directly below each branch circuit drop. These symbols represent the MCC cubicle bus stabs, pointed out earlier in the MCC pictures. Focusing on branch circuit D2, and following this circuit down to the load, the next component we notice is the fusible disconnect switch. Refer to Figure 10.2 for a detailed excerpt of branch circuit D2. The three-phase, or three pole, "switch" component of the fusible disconnect switch is followed by the associated overcurrent protection fuse. The branch circuit bifurcates after the overcurrent protection fuse into the motor pilot device control circuit and the remaining branch circuit leading to the 3-hp motor in the field. The function and operational logic of the motor pilot device control circuit will be described in the wiring diagram section. As we follow the remaining branch circuit, leading to the motor in the field, the next entity we see is the solid-state overload device. The branch circuit exits the MCC chassis on the load side of the overload device, via the terminal strip in the cubicle.

As labeled in Figure 10.2, the power in branch circuit D2 flows from the MCC to the 3-hp motor, through a bundle of three energized conductors, accompanied by one ground conductor. This bundle of conductors and the conduit are specified on the one-line schematic and interpreted, as follows:

- **¾" Conduit:** The electrical designer of this power distribution system selected a ¾" (ID) conduit, or pipe, to house the four conductors specified for this branch circuit.
- **3 – 1/C #12:** The electrical designer has selected or specified a three, single (1/C), AWG #12 conductor to supply three-phase AC power to the 3-hp motor.
- **1/C #12 Gnd.** The electrical designer has specified one (1/C), AWG #12 conductor to serve as ground for the grounded three-phase AC service to the 3-hp motor.

The description of branch circuit symbols and notations provided above should be sufficient for understanding the symbols, nomenclature, and notations used for other branch circuits in the one-line schematic depicted in Figure 10.1 with the exception of branch circuit D4. Branch circuit D4 differs from all other branch circuits in that it represents a **lighting load**. Note the square block labeled "LC-5" used to represent the lighting load instead of the circles used for the motor branch circuits.

Examination of Branch Circuit D2

Having introduced the symbols, nomenclature, and labels employed in the representation of power distribution branch circuits, we are better poised to explore and examine the design criteria and specifications incorporated in this one-line schematic. We will verify the following design specifications – pertaining to **branch circuit D2** (Figures 10.1 and 10.2) – in order to give the reader an appreciation of the design process an electrical engineer might go through in designing electrical power distribution systems:

1. Current-carrying conductor size
2. Ground conductor size
3. Conduit size
4. Overload setting

5. Starter size
6. Overcurrent protection fuse size
7. Fusible disconnect switch size

Current-Carrying Conductor Size

As discussed in the coverage of NEC®, in Chapter 9 of this text, a conductor expected to support a certain continuous load must be rated 125% of the full-load Amp (FLA). So, the next step would be to determine the FLA expected to be drawn by the 3-hp motor. As discussed earlier, the code requires that the FLA information must be based on the NEC (NFPA 70) Table 430.250, with some exceptions. For illustration purposes we will base our discussion and analysis on the information available in the Buss® table, shown in Figure 10.3, with some verification through NEC®.

BUSS® SYSTEM 300 MOTOR PROTECTION GUIDE
"NO-DAMAGE" "TYPE 2" SHORT-CIRCUIT PROTECTION OVERLOAD OR BACK-UP OVERLOAD PROTECTION

115VAC (120V), 1 ph. (LPN-RK-SP or LPJ-SP)

HP	Motor FLA	Overload 1.15 S.F.	Overload All Other	Back-up 1.15 S.F.	Back-up All Other	Switch/Fuseholder Size	Min NEMA Starter	Min Copper Conductor†	Min Conduit Size‡
1/6	4.4	5	5	5.6	5.6	30	00	14	1/2
1/4	5.8	7	6.25	7	7	30	00	14	1/2
1/3	7.2	9	8	9	9	30	0	14	1/2
1/2	9.8	12	10	15	12	30	0	14	1/2
3/4	13.8	15	15	17.5	17.5	30	0	14	1/2
1	16	20	17.5	20	20	30	0	14	1/2
1½	20	25	25	25	25	30	1	12	1/2
2	24	30	25	30	30	30	1	10	1/2

200V (208V), 3 ph. (LPN-RK-SP or LPJ-SP)

HP	Motor FLA	Overload 1.15 S.F.	Overload All Other	Back-up 1.15 S.F.	Back-up All Other	Switch/Fuseholder Size	Min NEMA Starter	Min Copper Conductor†	Min Conduit Size‡
1/2	2.3	2.6	2.5	3	2.8	30	00	14	1/2
3/4	3.22	4	3.5	4.5	4	30	00	14	1/2
1	4.14	5	4.5	5.6	5	30	00	14	1/2
1½	5.98	7	6.25	7.5	7	30	00	14	1/2
2	7.82	9	8	10	9	30	0	14	1/2
3	11	12	12	15	15	30	0	14	1/2
5	17.5	20	20	25	25	30	1	12	1/2
7½	25.3	30*	25*	35	30*	60	1	8	1/2
10	32.2	40	35	45	40	60	2	6	3/4
15	48.3	60	50	70**	60	60	3	4	1
20	62.1	75	70	80	75	100	3	3	1
25	78.2	90	80	100	90	100	3	1	1¼
30	92	100	100*	125	110	200	4	1/0	1¼
40	120	150	125	150	150	200	4	1/0	1¼
50	150	175	150	200	175	200	5	3/0	1½
60	177	200*	200*	225	225	400	5	4/0	2
75	221	250	250	300	300	400	5	300	2
100	285	350	300	400	350	400	6	500	3
125	359	400*	400*	450	450	600	6	2-4/0	2-2
150	414	500	450	600	500	600	6	2-300	2-2

230V (240V), 1 ph. (LPS-RK-SP or LPJ-SP)

HP	Motor FLA	Overload 1.15 S.F.	Overload All Other	Back-up 1.15 S.F.	Back-up All Other	Switch/Fuseholder Size	Min NEMA Starter	Min Copper Conductor†	Min Conduit Size‡
1/6	2.2	2.5	2.5	2.8	2.8	30	00	14	1/2
1/4	2.9	3.5	3.2	4	3.5	30	00	14	1/2
1/3	3.6	4.5	4	4.5	4.5	30	00	14	1/2
1/2	4.9	5.6	5.6	6.25	6	30	00	14	1/2
3/4	6.9	8	7.5	9	8	30	00	14	1/2
1	8	10	9	10	10	30	00	14	1/2
1½	10	12	10	15	12	30	0	14	1/2
2	12	15	12	15	15	30	0	14	1/2
3	17	20	17.5	25	20	30	1	12	1/2
5	28	35	30*	35	35	60	2	8	1/2
7½	40	50	45	50	50	60	3	6	3/4
10	50	60	50	70**	60	60	3	4	3/4

460V (480V), 3 ph. (LPS-RK-SP or LPJ-SP)

HP	Motor FLA	Overload 1.15 S.F.	Overload All Other	Back-up 1.15 S.F.	Back-up All Other	Switch/Fuseholder Size	Min NEMA Starter	Min Copper Conductor†	Min Conduit Size‡
1/2	1	1.25	1¼	1.25	1.25	30	00	14	1/2
3/4	1.4	1.6	1.6	1.8	1.8	30	00	14	1/2
1	1.8	2.25	2	2.25	2.25	30	00	14	1/2
1½	2.6	3.2	2.6	3.5	3	30	00	14	1/2
3	4.8	5.6	5	6	5.6	30	0	14	1/2
7½	11	12	12	15	15	30	1	14	1/2
10	14	17.5	15	17.5	17.5	30	1	14	1/2
15	21	25	20	30	25	30	2	10	1/2
20	27	30*	30*	35	35	60	2	8	1/2
25	34	40	35	45	40	60	2	6	3/4
30	40	50	45	50	50	60	3	6	3/4
40	52	60*	60*	70*	70	100	3	4	1
50	65	80	70	90	75	100	3	3	1
60	77	90	80	100	90	100	4	1	1¼
75	96	110	110	125	125	200	4	1/0	1¼
100	124	150	125	175	150	200	4	2/0	1½
125	156	175	175	200	200	200	5	3/0	1½
150	180	225	200*	225	225	400	5	4/0	2
200	240	300	250	300	300	400	5	350	2½

* Fuse reducers required.   ** 100A switch required.
† THWN connected to 60°C terminations up to #1. AWG to 75°C terminations for 1/0 and larger. Consult equipment manufacturer for listed termination temperature rating. Higher equipment termination temperature ratings may allow smaller conductor and conduit size.
‡ Based on 3 conductors for 3 ph. circuits and 2 conductors for 1 ph. circuits.

MPG

FIGURE 10.3   Bus® table used for electrical design verification.

As we focus on the circled/highlighted section of the Buss® table in Figure 10.3, we can spot most of the code-compliant design parameters associated with the 3-hp load in branch circuit D2. According to this table, a fully loaded three-phase motor, operating at 480 V, will demand 4.8 A. In addition, according to this table, in order to comply with the code, the three current-carrying conductors must be a minimum of AWG #14. If we follow the NEC® requirement that the conductor ampacity must be a minimum of 125% (1.25) times the motor FLA, each of the three current-carrying conductors must be capable of carrying a minimum of:

$$(1.25) \times (4.8 \text{ A}) = 6 \text{ A}$$

Examining Table 10.2, we see that an AWG #14 has an ampacity of 20 A, for THHW and THWN insulation (and other types of insulation) at 60°F terminal or conductor temperature. Note, however, that the 20 A ampacity is predicated on other conditions and stipulations, i.e. overcurrent protection restrictions. Further discussion on additional stipulations, exceptions, and code implications is outside the scope and context of this text. For simplicity, acknowledging the fact that AWG #14 has the ampacity to carry 20 A, we establish that AWG #14 – as stated in the BUS® table (Figure 10.4) – is adequate for this motor branch circuit.

Nevertheless, as we refer back to the schematics in Figures 10.1 and 10.2, we notice the electrical engineer/designer in this case, selected size AWG #12 copper conductors for this circuit, which is rated 25 A, and it clearly exceeds the minimum requirement. This decision by electrical engineer/designer **could** serve as an example of how some engineers/designers and engineering firms subscribe to a "**hard deck**" on the low end of conductor sizes. In other words, in some cases, the design engineer and the firm, as result of "**in-house best practice**," don't design power circuits with conductors smaller than AWG #12. However, in this case, the more compelling reason for the designer selection of an AWG #12 conductor is, perhaps, the fact that the load – 3-hp Roof Exhaust Fan Motor – is located at a distant location on the roof. The long conductor run to the roof – due to higher resistance – would result in excessive voltage drop. See NEC, Article 215 (2) (A) (1)(a) and (b) & Informational Note 2. So, prudently, the designer selected a larger conductor (AWG #12) to prevent excessive voltage drop.

Ground Conductor Size

Next, let's examine the ground conductor selected for branch circuit D2. As apparent from examination of Bus table in Figure 10.3, this table does not specify the ground conductor size for various branch circuits. One could specify the **ground conductor** size to be the same as the **current-carrying conductors**, as done by the designer of branch circuit D2, in schematics on Figures 10.1 and 10.2, where the designer-specified AWG #12 conductors for the three, 3-hp motor, current-carrying conductors as well as the ground conductor. However, NEC prescribes the use of NEC® Table 250.122 for establishing the size of the ground conductor, based primarily on the overcurrent protection device rating. According to NEC® Table 250.122, selection of an AWG #14 would have been sufficient in branch circuit D2. Note that the purpose of the ground conductor is to carry fault current and to serve as a "mechanism" for safely

TABLE 10.2
Partial NEC® Table 310.15(B)(16)

Partial Ampacity Information from NEC® Table 310.15(B)(16) – Allowable Ampacities of Insulated Conductors Rated Up to and Including 2,000 V, 60°C through 90°C (140°F through 194°F), for Not More Than Three Current-Carrying Conductors in Raceway, Cable, or Earth (Directly Buried), Based on Ambient Temperature of 30°C (86°F)[a]

Size AWG or kcmil	Temperature Rating of Conductor; Ref.: NEC® Table 310.104(A).						Size AWG or kcmil
	60°C (140°F)	75°C (167°F)	90°C (194°F)	60°C (140°F)	75°C (167°F)	90°C (194°F)	
	Types TW, UF	Types RHW, THHW, THW, THWN, XHHW, USE, ZW	Types TBS, SA, SIS, FEP, FEPB, MI, RHH, RHW-2, THHN, THHW, THW-2, THWN-2, USE-2, XHH, XHHW, XHHW-2, ZW-2	Types TW, UF	Types RHW, THHW, THW, THWN, XHHW, USE	Types TBS, SA, SIS, THHN, THHW, THW-2, THWN-2, RHH, RHW-2, USE-2, XHH, XHHW, XHHW-2, ZW-2	
	Copper			Aluminum or Copper-Clad Aluminum			
18	-	-	14				-
16	-	-	18				-
14[b]	15	20	25	-	-	-	-
12[b]	25	25	30	15	20	25	12[b]
10[b]	30	35	40	25	30	35	10[b]
8	40	50	55	35	40	45	8
6	55	65	75	40	50	55	6
4	70	85	95	55	65	75	4
3	85	100	115	65	75	85	3
2	95	115	130	75	90	100	2
1	110	130	145	85	100	115	1

(Continued)

TABLE 10.2 (Continued)
Partial NEC® Table 310.15(B)(16)

Partial Ampacity Information from NEC® Table 310.15(B)(16) – Allowable Ampacities of Insulated Conductors Rated Up to and Including 2,000 V, 60°C through 90°C (140°F through 194°F), for Not More Than Three Current-Carrying Conductors in Raceway, Cable, or Earth (Directly Buried), Based on Ambient Temperature of 30°C (86°F)[a]

Temperature Rating of Conductor; Ref.: NEC® Table 310.104(A).

Size AWG or kcmil	60°C (140°F) Types TW, UF	75°C (167°F) Types RHW, THHW, THW, THWN, XHHW, USE, ZW	90°C (194°F) Types TBS, SA, SIS, FEP, FEPB, MI, RHH, RHW-2, THHN, THHW, THW-2, THWN-2, USE-2, XHH, XHHW, XHHW-2, ZW-2
		Copper	
1/0	125	150	170
2/0	145	175	195
3/0	165	200	225
4/0	195	230	260
250	215	255	290
300	240	285	320
350	260	310	350

	60°C (140°F) Types TW, UF	75°C (167°F) Types RHW, THHW, THW, THWN, XHHW, USE	90°C (194°F) Types TBS, SA, SIS, THHN, THHW, THW-2, THWN-2, RHH, RHW-2, USE-2, XHH, XHHW, XHHW-2, ZW-2	Size AWG or kcmil
		Aluminum or Copper-Clad Aluminum		
	100	120	135	1/0
	115	135	150	2/0
	130	155	175	3/0
	150	180	205	4/0
	170	205	230	250
	195	230	260	300
	210	250	280	350

Note: Included for general illustration purposes only. Use actual NEC® Tables for work.

[a] Refer to NEC® Table 310.15(B)(2)(a) for the ampacity correction factors where the ambient temperature is other than 30°C (86°F). Refer to NEC® Table 310.15(B)(3)(a) for more than three current-carrying conductors.

[b] Refer to NEC® Article 240.4(D) for conductor overcurrent protection limitations.

de-energizing circuits with electrical faults. For example, if the insulation of one of the energized phases in branch circuit D2 were to atrophy, or break down due to other reasons, the 460 V potential could come in contact with the motor chassis and raise its potential to 460 V. With the motor grounded properly, such an electrically unsafe condition would be rectified through a **phase to ground fault** clearing the fuse. Once again, the designer could be credited with the more appropriate decision of specifying a larger, AWG#12, ground conductor due to the long run of all the four wires to the remotely located motor. The discourse presented in this paragraph shows the complexities and intricacies associated with correct interpretation of the code, with the paramount interest of the safety of public and electrical workers in mind.

Conduit and Conduit Size

As stated on schematics in Figures 10.1 and 10.2, the conduit specified by the engineer is ¾" ID. Before we determine if the ¾" conduit is adequate for this branch circuit, we need to recount the purpose and general characteristics of an electrical conduit. Electrical conduits can be rigid or flexible. Electrical conduits are similar to pipes used to convey fluids. Conduits can be constructed out of metal or PVC (polyvinyl chloride). The key function of an electrical conduit is to protect the electrical conductors against mechanical damage, exposure to adverse environmental ambient factors, i.e. moisture, corrosive liquids, solids, or gases. Unlike pipes applied in fluid system, conduits used in electrical systems cannot be filled completely. As apparent from Chapter 9 in this text, and pertinent tables therein, electrical conduits are permitted to be filled only partially. Two ostensible reasons for this constraint are: (1) It is, physically, difficult to pull too many wires through a conduit and associated elbows and fittings, and (2) a certain minimum, cross-sectional, open space is necessary in an electrical conduit to facilitate ventilation of any heat dissipated in the conductors due to I^2R losses.

The ¾" conduit specified for branch circuit D2, on schematics in Figures 10.1 and 10.2, is oversized based on the Bus® table in Figure 10.4. As noted earlier, the Bus® table identifies a ½" conduit for the 3-hp branch circuit D2 in combination with the AWG #14 conductors. The ¾" conduit specified on the drawings is sized based on the fact that four AWG #12 conductors were selected by the electrical designer. Another rationale behind oversizing of conduits can be found in the electrical installation craft and challenges therein. If you observe electricians "pulling" conductors through conduits, you come away with appreciation of physical challenges associated with pulling wires or cables through conduits. Obstructions offered by conduit elbows, unions, and conduit wall friction, in general, warrant application of substantial force for pulling the wires. However, in order to understand the process of selection of conduits, using the NEC®, we will compare both alternatives: a ½" conduit and the ¾" conduit.

According to NEC®, Chapter 9, Table 4, a ½" **Rigid Metallic Conduit (RMC),** with greater than two conductors, can only be filled up to 40% of its internal area of cross-section. The available internal area of cross-section, according to NEC®, Chapter 9, Table 4 is 81 mm².

According to NEC® Chapter 9, Table 5, the total area of cross-section that the three AWG #14 current-carrying conductors would occupy, would be:

$$= 3 \times 8.968 = 26.9 \text{ mm}^2$$

Based on NEC®, Chapter 9, Table 4, 40% fill would require that the minimum internal area of the cross-section of the selected conduit must be:

$$\text{Minimum Area of Cross-section} = \frac{26.9}{0.4} = 67.26 \text{ mm}^2$$

Since this computed minimum area of cross-section of 67.26 mm² is well within the available internal area of cross-section of 81 mm² offered by a ½″ RMC, **a ½″ conduit for D2 branch circuit would have been adequate, had AWG #14 conductors been used.**

However, since the designer chose AWG #12 for branch circuit D2, according to NEC Chapter 9, Table 5, the total area of cross-section that the three AWG #12 current-carrying conductors would occupy, would be:

$$= 3 \times 11.68 = 35 \text{ mm}^2$$

Based on NEC®, Chapter 9, Table 4, 40% fill would require that the minimum internal area of the cross-section of the selected conduit must be:

$$\text{Minimum Area of Cross-section} = \frac{35}{0.4} = 87.6 \text{ mm}^2$$

Since this computed minimum area of cross-section of 87.6 mm² is well within the available internal area of cross-section of 141 mm² offered by a ¾″ RMC conduit, a ¾″ conduit for D2 branch circuit would be adequate with the AWG #12 conductors specified.

Overload Protection Setting

Overload protection devices are included in motor and branch circuits to protect motors, motor control apparatus, and motor branch circuit conductors against overheating due to overloads. A solid-state motor overload device offers flexibility in setting of overload set points. And motor overload device setting can be determined using the Bus® table or NEC® Article 430.32. In the case of branch circuit D2, according to NEC 430.32 (1) (for separate overload device, for motors other than those with service factor of 1.15 and marked operating temperature of 40°C), the overload device should be set at:

$$= 1.15 \times \text{Full Load Amp}$$

$$= 1.15 \times 4.8 = 5.52 \text{ A, or simply 5 A}$$

According to the Bus® table, overload current setting, for this specific case, should be 5 A. So, once again we see the Bus® table and the NEC® yielding, practically, the same results for design specification.

Starter Size

The motor branch circuit starter size is, generally, based on NEMA standards. According to the Bus® table, in Figure 10.4, the starter should be "**Size 0**." However, once again, we see that the designer of the branch circuit selected a size that is greater than the minimum requirement. The designer of the 3-hp motor branch circuit incorporated a NEMA "Size 1" starter in the schematics shown in Figures 10.1 and 10.2.

Overcurrent Protection Fuse Size

As with the motor overload protection device, the selection of motor overcurrent device can be accomplished by using the Bus® table or NEC®. In the case of branch circuit D2, according to NEC 430.52 and Table 430.52 (for a polyphase motor protected by a time-delay fuse), the short-circuit fuse must be rated as:

$$= 1.75 \times \text{Full Load Amp}$$

$$= 1.75 \times 4.8 = 8.4 \text{ A}$$

According to NEC® Article 240.6 (A), when overcurrent calculation yields an ampere specification that is non-standard, the next higher size may be used. Therefore, a standard 10 A fuse could have been selected. Instead, the designer decided to exceed the requirements once again and chose the standard 8 A fuse. As long as the 8 A fuse does not cause nuisance trips or fuse clearing, it should be acceptable.

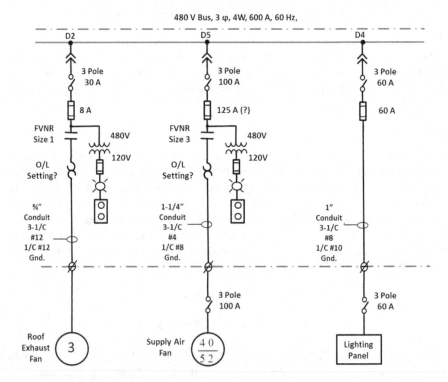

FIGURE 10.4   Excerpt one-line schematic for a power distribution system

Fusible Disconnect Switch Size

The fusible disconnect switch specification can be determined through the Bus® table in Figure 10.4. According to the Bus® table, at the point of intersection of the row representing 3-hp motor and the column titled: "Switch or Fuse Holder Size," we see that a 30 A disconnect switch is needed to meet the code. As shown on the one-line diagram in Figures 10.1 and 10.2, in compliance with the code, a 30 A disconnect switch is specified by the designer. This is also in compliance with NEC® Article 430.101 and NEC® Table 430.251 (B).

So, all in all, it appears that design of branch circuit D2 is in compliance with the code, and in some aspects, it exceeds the code. However, if one compares the design of branch circuit D2 with the rest of the branch circuits, it becomes apparent that, unlike the other branch circuits, D2 lacks a safety disconnect switch at the motor. And, as obvious from the label on the motor load, this motor pertains to a roof exhaust fan and is, possibly, located on the roof out of line of sight from the MCC disconnect switch. According to NEC® Article 430.102 (B) (2), the disconnecting means should be located in sight of the motor, baring qualification for exceptions stated under this article. It is unknown if this particular installation qualified for the exceptions. Nevertheless, it is always desirable, in the interest of safety, to have a lockable disconnect switch at each motor load.

Example 10.1

Answer the following questions pertaining to the branch circuit shown in the schematic diagram below:

a. What is the maximum current the power distribution system for this branch circuit rated for?
b. What is the turns ratio of the control transformer shown in the motor branch circuit?
c. What is the full-load current in this circuit?
d. What should the solid-state overload device be set for at commissioning of this system?

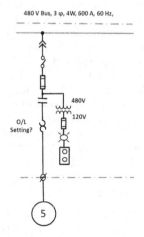

480 V Bus, 3 φ, 4W, 600 A, 60 Hz,

O/L Setting?

480V

120V

5

Solution:

   a. The maximum current the power distribution system for this branch cir-
      cuit rated for is stated in the specification stated at the very top of the
      schematic diagram, as 600 A, within the caption:

                        "480 V Bus, 3 ϕ, 4 W, 600 A, 60 Hz."

      Therefore, the answer is: **600 A.**
   b. The turns ratio of the control transformer shown in the motor branch cir-
      cuit would be based on the voltage transformation stated on the control
      transformer. As obvious from the schematic diagram, the control trans-
      former has a 480 V primary and a 120 V secondary. Apply Eq. 10.1,
      introduced earlier in this text:

$$\text{Turns Ratio} = \frac{N_P}{N_S} = \frac{V_P}{V_S} \tag{10.1}$$

$$\text{Turns Ratio} = \frac{N_P}{N_S} = \frac{V_P}{V_S} = \frac{480 \text{ V}}{120 \text{ V}} = \frac{4}{1} = 4:1$$

   c. What is the full-load current in this circuit?
         This question can be answered through multiple approaches, includ-
      ing the calculation method based on Eq. 10.2, provided the motor effi-
      ciency ($\eta_{\text{motor efficiency}}$) and the power factor are known.

$$|S| = \frac{P(\text{Watts})}{(\text{pf}) \cdot (\eta_{\text{motor efficiency}})} = |V| \cdot |I|, \text{ or}$$

$$|I| = \frac{P(\text{Watts})}{(\text{pf}) \cdot (\eta_{\text{motor efficiency}}) \cdot (|V|)}, \text{ or,} \tag{10.2}$$

$$|I| = \frac{P(\text{hp}) \cdot (746 \text{ W/hp})}{(\text{pf}) \cdot (\eta_{\text{motor efficiency}}) \cdot (|V|)}$$

      Since the efficiency and the power factor are not given, as introduced
      earlier in this text, we will use the Buss® table introduced earlier in this
      chapter. As stated in schematic diagram – in the circled motor symbol –
      the motor's full-load rating is **5 hp.** As specified at the top of schematic
      diagram, the motor is being powered by a **480 V, three-phase** source
      (**480 V Bus, 3 ϕ,** 4 W, 600 A, 60 Hz). Therefore, according to the Buss®
      table, and as highlighted (circled) below – under the 460 V(480 V), three-
      phase, section – **the motor full-load current would be 7.6 A.**
   d. Using the Buss® table above, Figure 10.5a, for **three-phase, 5 hp** motor,
      operating at a **480 V,** the overload device should be set at **8.0 A.**
         Note: If NEC tables were used here, the 115% multiplier by NEC®
      would result in an overload setting of 8.7 A.

Examination of Branch Circuit D5

While we will not examine branch circuit D5 to the degree we examined D2, it is
important to point out one apparent oversight on the part of the designer. This oversight

BUSS® SYSTEM 300
MOTOR PROTECTION GUIDE
"NO-DAMAGE" "TYPE 2" SHORT-CIRCUIT PROTECTION
OVERLOAD OR BACK-UP OVERLOAD PROTECTION

115VAC (120V), 1 ph. (LPN-RK-SP or LPJ-SP) / 200V (208V), 3 ph. (LPN-RK-SP or LPJ-SP)

HP	Motor Full Load Amps	Overload 1.15 S.F.	Overload All Other	Back-up 1.15 S.F.	Back-up All Other	Switch/Fuseholder Size	Min NEMA Starter	Min Copper Cond.	Min Conduit
1/6	4.4	5	5	5.6	5.6	30	00	14	1/2
1/4	5.8	7	6.25	7.5	7	30	00	14	1/2
1/3	7.2	9	8	9	9	30	00	14	1/2
1/2	9.8	12	10	15	12	30	0	14	1/2
3/4	13.8	15	15	17.5	17.5	30	0	14	1/2
1	16	20	17.5	20	20	30	0	14	1/2
1½	20	25	20	25	25	30	1	12	1/2
2	24	30	25	30	30	30	1	10	1/2
1/2	2.3	2.6	2.5	3	2.8	30	00	14	1/2
3/4	3.22	4	3.5	4.5	4	30	00	14	1/2
1	4.14	5	4.5	5.6	5	30	00	14	1/2
1½	5.98	7	6.25	7.5	7	30	00	14	1/2
2	7.82	9	8	10	9	30	0	14	1/2
3	11	12	12	15	15	30	0	14	1/2
5	17.5	20	20	25	25	30	1	12	1/2
7½	25.3	30*	25*	35	30*	60	1	8	1/2
10	32.2	40	35	45	40	60	2	6	3/4
15	48.3	60	50	70**	60	60	3	4	1
20	62.1	75	70	80	75	100	3	3	1
25	78.2	90	80	100	90	100	3	1	1¼
30	92	100	100*	125	110	200	4	1/0	1¼
40	120	150	125	150	150	200	4	1/0	1¼
50	150	175	150	200	175	200	5	3/0	1½
60	177	200*	200*	225	225	400	5	4/0	2
75	221	250	250	300	300	400	5	300	2
100	285	350	300	400	350	400	5	500	3
125	359	400*	400*	450	450	600	6	2-4/0	2-2
150	414	500	450	600	500	600	6	2-300	2-2

230V (240V), 1 ph. (LPS-RK-SP or LPJ-SP) / 460V (480V), 3 ph. (LPS-RK-SP or LPJ-SP)

HP	Motor Full Load Amps	Overload 1.15 S.F.	Overload All Other	Back-up 1.15 S.F.	Back-up All Other	Switch/Fuseholder Size	Min NEMA Starter	Min Copper Cond.	Min Conduit
1/6	2.2	2.5	2.5	2.8	2.8	30	00	14	1/2
1/4	2.9	3.5	3.2	4	3.5	30	00	14	1/2
1/3	3.6	4.5	4	4.5	4.5	30	00	14	1/2
1/2	4.9	5.6	5.6	6.25	6	30	00	14	1/2
3/4	6.9	8	7.5	9	8	30	00	14	1/2
1	8	10	9	10	10	30	00	14	1/2
1½	10	12	10	15	12	30	0	14	1/2
2	12	15	12	15	15	30	0	14	1/2
3	17	20	17.5	25	20	30	1	12	1/2
5	28	35	30*	35	35	60	2	8	1/2
7½	40	50	45	50	50	60	2	6	3/4
10	50	60	50	70**	60	60	3	4	3/4
1/2	1	1.25	1¼	1.25	1.25	30	00	14	1/2
3/4	1.4	1.6	1.6	1.8	1.8	30	00	14	1/2
1	1.8	2.25	2	2.75	2.25	30	00	14	1/2
1½	2.6	3.2	2.6	3.5	3	30	00	14	1/2
2	3.4	4	3.5	4.5	4	30	00	14	1/2
5	7.6	9	8	10	9	30	0	14	1/2
10	14	17.5	15	17.5	17.5	30	1	14	1/2
15	21	25	20	30	25	30	2	10	1/2
20	27	30*	30*	35	35	60	2	8	1/2
25	34	40	35	45	40	60	2	6	3/4
30	40	50	45	50	50	60	3	6	3/4
40	52	60*	60*	70	60	100	3	4	1
50	65	80	70	90	75	100	3	3	1
60	77	90	80	100	90	100	4	3	1¼
75	96	110	110	125	125	200	4	1/0	1¼
100	124	150	125	175	150	200	4	2/0	1½
125	156	175	175	200	200	200	5	3/0	1½
150	180	225	200*	225	225	400	5	4/0	2
200	240	300	250	300	300	400	5	350	2½

* Fuse reducers required. ** 100A switch required.
† THWN connected to 60°C terminations up to #1. AWG to 75°C terminations for 1/0 and larger. Consult equipment manufacturer for listed termination temperature rating. Higher equipment termination temperature ratings may allow smaller conductor and conduit size.
‡ Based on 3 conductors for 3 ph. circuits and 2 conductors for 1 ph. circuits.

MPG

FIGURE 10.5a Bussman® Motor Protection Guide.

pertains to the overcurrent protection fuse specification of 125 A, as it appears on the one-line drawing in Figures 10.1 and 10.2.

See the circled data, in Figure 10.5b Bus® table, for FLA associated with a 40-hp motor. In the case of branch circuit D5, which supports a 40-hp (52 FLA, FLA, as stated under the column adjacent to the HP column) motor, according to NEC 430.52 and Table 430.52 (for a polyphase motor protected by a time-delay fuse), the short-circuit fuse must be rated as:

$$= 1.75 \times \text{Full Load Amp}$$

$$= 1.75 \times 52 = 91 \text{ A}$$

According to NEC® Article 240.6 (A), when overcurrent calculation yields an ampere specification that is non-standard, the next higher size may be used. Therefore, standard 100 A fuses should have been selected. Instead, the designer chose a 125 A overcurrent protection fuse. In the absence of possible qualifying exceptions, this

BUSS® SYSTEM 300
MOTOR PROTECTION GUIDE
"NO-DAMAGE" "TYPE 2" SHORT-CIRCUIT PROTECTION
OVERLOAD OR BACK-UP OVERLOAD PROTECTION

Circuit	HP	Motor Full Load Amps	Fuse Size And Type Protection: Over-load: 1.15 S.F. Or Greater Or Rated Not Over 40°C	Fuse Size And Type Protection: Over-load: All Other Motors	Fuse Size And Type Protection: Back-up: 1.15 S.F. Or Greater Or Rated Not Over 40°C	Fuse Size And Type Protection: Back-up: All Other Motors	Switch Or Fuseholder Size	Minimum NEMA Starter Size	Minimum Copper Conductor Size†	Minimum Conduit Size‡
115VAC (120V), 1 ph. (LPN-RK-SP or LPJ-SP)	1/6	4.4	5	5	5.6	5.6	30	00	14	1/2
	1/4	5.8	7	6.25	7.5	7	30	00	14	1/2
	1/3	7.2	9	8	9	9	30	0	14	1/2
	1/2	9.8	12	10	15	12	30	0	14	1/2
	3/4	13.8	15	15	17.5	17.5	30	0	14	1/2
	1	16	20	17.5	20	20	30	0	14	1/2
	1½	20	25	20	25	25	30	1	12	1/2
	2	24	30	25	30	30	30	1	10	1/2
200V (208V), 3 ph. (LPN-RK-SP or LPJ-SP)	1/2	2.3	2.6	2.5	3	2.8	30	00	14	1/2
	3/4	3.22	4	3.5	4.5	4	30	00	14	1/2
	1	4.14	5	4.5	5.6	5	30	00	14	1/2
	1½	5.98	7	6.25	7.5	7	30	00	14	1/2
	2	7.82	9	8	10	9	30	0	14	1/2
	3	11	12	12	15	15	30	0	14	1/2
	5	17.5	20	20	25	25	30	1	12	1/2
	7½	25.3	30*	25*	35	30*	60	1	8	1/2
	10	32.2	40	35	45	40	60	2	6	3/4
	15	48.3	60	50	70**	60	60	3	4	1
	20	62.1	75	70	80	75	100	3	3	1
	25	78.2	90	80	100	90	100	3	1	1¼
	30	92	100	100*	125	110	200	4	1/0	1¼
	40	120	150	125	150	150	200	4	1/0	1¼
	50	150	175	150	200	175	200	5	3/0	1½
	60	177	200*	200*	225	225	400	5	4/0	2
	75	221	250	250	300	300	400	5	300	2
	100	285	350	300	400	350	400	6	500	3
	125	359	400*	400*	450	450	600	6	2-4/0	2-2
	150	414	500	450	600	500	600	6	2-300	2-2
230V (240V), 1 ph. (LPS-RK-SP or LPJ-SP)	1/6	2.2	2.5	2.5	2.8	2.8	30	00	14	1/2
	1/4	2.9	3.5	3.2	4	3.5	30	00	14	1/2
	1/3	3.6	4.5	4	4.5	4.5	30	00	14	1/2
	1/2	4.9	5.6	5.6	6.25	6	30	00	14	1/2
	3/4	6.9	8	7.5	9	8	30	00	14	1/2
	1	8	10	9	10	10	30	00	14	1/2
	1½	10	12	10	15	12	30	0	14	1/2
	2	12	15	12	15	15	30	0	14	1/2
	3	17	20	17.5	25	20	30	1	12	1/2
	5	28	35	30*	35	35	60	2	8	1/2
	7½	40	50	45	50	50	60	2	6	3/4
	10	50	60	50	70**	60	60	3	4	3/4
460V (480V), 3 ph. (LPS-RK-SP or LPJ-SP)	1/2	1	1.25	1⅛	1.25	1.25	30	00	14	1/2
	3/4	1.4	1.6	1.6	1.8	1.8	30	00	14	1/2
	1	1.8	2.25	2	2.25	2.25	30	00	14	1/2
	1½	2.6	3.2	2.6	3.5	3	30	00	14	1/2
	2	3.4	4	3.5	4.5	4	30	00	14	1/2
	3	4.8	5.6	5	6	5.6	30	0	14	1/2
	5	7.6	9	8	10	9	30	0	14	1/2
	7½	11	12	12	15	15	30	1	14	1/2
	10	14	17.5	15	17.5	17.5	30	1	14	1/2
	15	21	25	20	30	25	30	2	10	1/2
	20	27	30*	30*	35	35	60	2	8	1/2
	25	34	40	35	45	40	60	2	6	3/4
	30	40	50	45	50	50	60	3	8	3/4
	40	52	60*	60*	70	60*	100	3	4	1
	50	65	80	70	90	75	100	3	3	1
	60	77	90	80	100	90	100	4	1	1¼
	75	96	110	110	125	125	200	4	1/0	1¼
	100	124	150	125	175	150	200	4	2/0	1½
	125	156	175	175	200	200	200	5	3/0	1½
	150	180	225	200*	225	225	400	5	4/0	2
	200	240	300	250	300	300	400	5	350	2½

* Fuse reducers required. ** 100A switch required.
† THWN connected to 60°C terminations up to #1. AWG to 75°C terminations for 1/0 and larger. Consult equipment manufacturer for listed termination temperature rating. Higher equipment termination temperature ratings may allow smaller conductor and conduit size.
‡ Based on 3 conductors for 3 ph. circuits and 2 conductors for 1 ph. circuits.
MPG

FIGURE 10.5b Bussman® Motor Protection Guide.

branch circuit should be equipped with a combination of a 100 A fusible disconnect switch (see fourth column from the right) and 100 A short-circuit protection fuses.

Examination of Branch Circuit D4

As apparent from the load identifying symbol, LC-5, this branch circuit supplies power to lighting load of 5 hp. The 60 A disconnect switches and the 60 A fuses are based on NEC® Article 430. Note that the Bus® chart cannot be used in the case of lighting loads.

Wiring Diagram

Unlike a one-line schematic, a wiring diagram displays comprehensive information on all three phases of a three-phase AC system, along with the associated grounds. See Figure 10.6. The wiring diagram in Figure 10.6 pertains to three motor loads: 40-hp supply air fan motor, 75-hp return air fan motor, and a 15-hp pump motor.

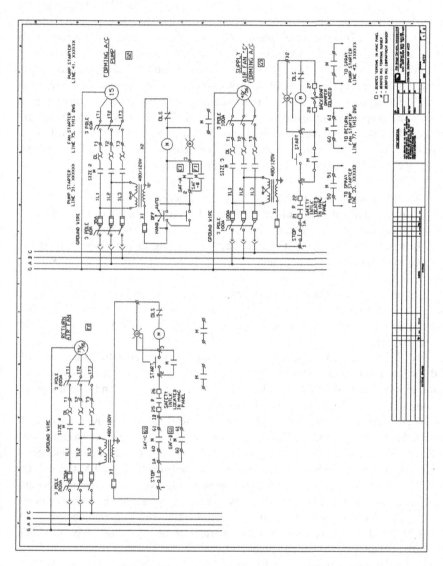

FIGURE 10.6 Wiring diagram for power distribution system.

We will limit our focus to the 75-hp return air motor circuit on the left. Notice the vertical lines, spanning from top to bottom, on the left side of the wiring diagram. These lines, as annotated on the drawing, represent the voltage bearing phase busses A, B, and C, as well as the ground bus (G).

As we work our way from the left side of this 75-hp motor circuit to the right, we notice the three energized phase busses are "tapped" by the 75-hp branch circuit via the bus stabs, represented by the "chevron" symbols. The MCC ground is shown routed from the MCC to the three-phase AC 75-hp return air motor. All three phases are shown bridged from the MCC bus stabs to the 200 A disconnect switch, labeled as "3 Pole, 200 A." Each of the three phases is then routed to respective motor starter contactor contacts via the short-circuit protection fuses. Note the strict adherence of unique labeling assigned to each continuous conductor between consecutive current interrupting points. For instance, the wire label "1L1" is reserved for the conductor between the load side of phase A 150 A fuse and the motor starter contact, labeled "Size 4 M," where "M" represents the motor contactor for this circuit. Unique numbering of conductors is crucial for the following few reasons:

a. Troubleshooting and tracing of circuits, using a drawing.
b. Ensuring that wires are not "cross-wired." Crossed wires can result in phase to phase and phase to ground faults, and control circuit malfunctions.
c. Unique wiring numbers facilitate accurate wiring or assembly of power distribution and control circuits.

Application of wiring numbers is not feasible when wire-type conductors are not used to connect one current interrupting point to another, such as the connection between a motor starter contact and an overload contact, as evident on Figure 10.6 wiring diagram.

Wires T1, T2, and T3 connect phases A, B, and C to respective line-side terminals of the safety disconnect switch poles at the motor in the field. Note that the identification of wires, completing the final connection of phases A, B, and C to the motor terminals changes to 1T1, 1T2, and 1T3, respectively.

The pilot device and control power for this 75-hp motor is shown "tapped" or branched off phases A and B. This phase to phase power, as expected, would be 480 V. However, the control circuit must be 120 V or less. Therefore, the tapped 480 V power, as shown on the wiring diagram, is "stepped down" to 120 V through the 480/120 V control transformer. Note that the secondary of this control transformer is grounded on one leg. This leg, serves as a neutral, ground, or low potential point. The other terminal, then, serves as the higher potential, 120 V, point, labeled as X1. This 120 V and neutral pair sustains and operates all of the pilot devices, switches, interlock contacts, and motor contactor coil.

The 120 V control power is protected through a fuse and arrives at the "STOP" switch's normally open contacts via conductor numbered as "1." The spring loaded, momentary, STOP push button contact is normally closed, as shown in the wiring diagram. So, the 120 V potential crosses over to point 1 A and three safety interlock contacts. These contacts are normally open. Such interlock contacts are often associated with system components such as doors – which must be closed before the motor is allowed to be energized. Let's assume that all safety conditions are met in this case. This would allow us to assume that all three of these interlock contacts are closed. Then 120 control voltage would bridge over to point 26 on wiring diagram. Point 26 and the terminal labeled "2" are at the same potential; there is an electrical short between these two points. Therefore, terminal 2 retains 120 V potential as long as the STOP button is **not** depressed. Then, if the START switch is depressed, its normally open contacts close allowing the 120 control voltage to be applied to the coil of the motor starter contactor "M." Since neutral or ground side of the coil is always grounded through the normally closed solid-state overload contacts (unless the overload protection device has tripped under an overload condition), the motor starter coil energizes when the START switch is depressed. The three motor starter contacts, shown in the motor branch circuit, close, thus releasing three-phase voltage and power to the 75-hp motor. Since the **pilot light** for the motor is connected in parallel with the motor starter coil, the motor "ON" light turns on as well. One of the motor starter contacts, referred to as the "latching contact "M," is connected in parallel with the START switch contacts. This contact seals the coil in energized mode, such that when the START button is released, the motor starter coil stays latched and the motor continues to operate. The motor continues to operate as long as all of the safety interlocks stay closed, the overload protection device does not sense an overload, and the STOP switch is not depressed.

When the motor STOP switch is depressed, its normally closed contacts open and the continuity of the 120 V circuit is broken. This results in de-energization of the motor starter coil, thus **unlatching** the sealing contacts of the motor starter coil. The motor ceases to operate.

The control sequence and logic described above represents the essence of approaches applied in most motor starter circuits, with some application specific variations.

CONTROL DIAGRAM/DRAWING

Our discussion in this section will be based on the control diagrams depicted in Figures 10.7 and 10.8. Both of these drawings are based on a control system premised on a PLC. In other words, the control algorithm is being implemented through a control architecture where the PLC CPU, Central Processing Unit, makes all control decisions in accordance with the PLC control program. Of course, the CPU makes the control decisions in **response** to the current state or change of state of all pertinent inputs.

The control drawing, or control schematic, in Figure 10.7 shows some of the inputs interfaced to, or monitored by, the PLC. The inputs are, essentially, signals coming from sensors and switches in the field. These signals can be "discrete" or "digital;" meaning, they are in the form of "ON's," "OFF's," "1's," "0's," switch **closed** or **open**, etc. The column to the right of the drawing, labeled with input numbers, i.e. Input 1, Input 2, and so on, represents a PLC input module. Note that this module is labeled as 115 VAC Input, Modicon, Cat. No.: B805-016. The role of this input module is to

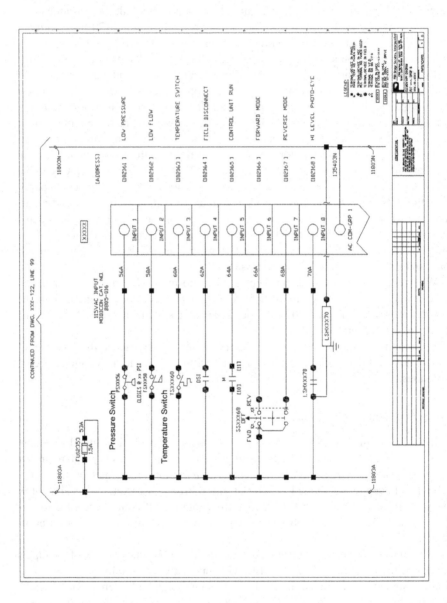

FIGURE 10.7 Input control drawing.

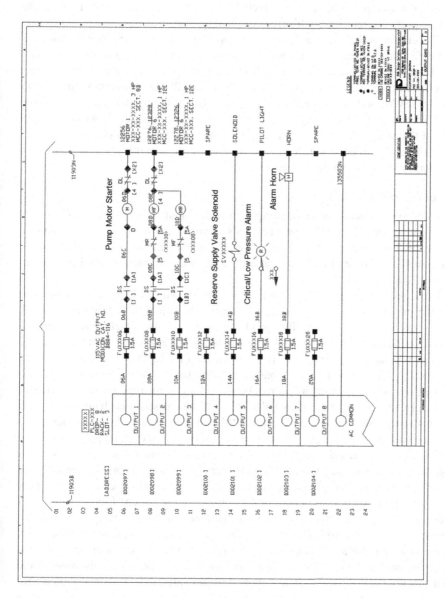

FIGURE 10.8 PLC output module-based control drawing.

receive 115 V AC signals from various sensors and switches in the field and convert them into low-voltage DC signals. A 115 or 120 VAC signal is transduced by the input module circuitry into a 5 V_{DC} or 10 V_{DC} signal. An absence of AC voltage at a given input is converted into 0 V_{DC} and interpreted as a "0." When a 115 V signal is received at an input point, it is interpreted by the PLC as a "**1**." In other words, the PLC expects discrete **1's** and **0's** from its input modules.

The output module shown in Figure 10.8 serves as an interface between the PLC and the control devices/equipment in the field in a manner similar to the input module. However, there are a few distinct differences. A discrete 115 VAC output module **receives** discrete signals, 1's and 0's, **from** the PLC CPU and **converts them** into 115 VAC or 0 VAC outputs or commands – that are conveyed to the control devices in the field, such as motor starter coils, solenoids, lamps, and horns – as shown on the right side of the control drawing in Figure 10.8. Notice that contiguous electrical connections or wires are identified by unique numbers or alphanumeric identifications on the input and the output control drawings.

Input Control Diagram

In order to get a better understanding of how the inputs from various field sensors and switches interface with the PLC, let's focus on the input control drawing in Figure 10.7 and follow a hypothetical, yet plausible, scenario. We will then extend the possible outcome of this hypothetical scenario to the output control diagram in Figure 10.8.

Let's assume that pressure switch, PSxxx56, shown in Figure 10.7, is monitoring the pressure in a vessel that must be maintained above a certain critical level. This pressure switch is **normally open** but **closed** when the vessel is pressurized, as shown in the control drawing. In this safe state, with the pressure switch closed, 120 V control voltage is passed on to Input #1 on the PLC input module. As described earlier, the 120 V potential at Input #1 gets transduced into a logic level "1" for interpretation by the CPU (microprocessor). As long as the pressure switch remains closed, Input #1 maintains a logic value of "1." If the pressure switch opens, 120 V at Input #1 would disappear and logic level "1" would be replaced by a logic level "0." Often switches – such as the pressure switch in this circuit – are applied such that they **open** under "**unsafe**" conditions. When electrical control devices are applied in this manner, they are said to be in "**fail safe**" mode. If the vessel in this scenario were to lose pressure to a level below a preset critical level, the pressure switch will open. The set point in such a case could be set at the pressure switch. The temperature of a downstream process in this scenario is being monitored by a temperature switch TSxxx60. See Figure 10.7. The process, downstream, relies on pressure in the vessel being maintained at an adequate level. This temperature switch is open under normal circumstances and would close only if safe operating temperature is exceeded. Note that this temperature switch is **not** applied in **fail safe** mode. Under normal operation, the temperature switch is open, and the 120 V control voltage stops at the left side of the temperature switch. This results in 0 V at Input #3. No voltage at terminal #3 of the PLC input module translates into logic level "0" at Input #3. The PLC CPU, or microprocessor, completes execution of one cycle of the control program, on average, in approximately 30 ms. So, as the CPU examines Input #3, every 30 ms, it sees a "0" and interprets that as a normal situation, requiring no specific action by the PLC system.

Now, let's assume that pressurized vessel develops a leak and loses pressure to the extent that the pressure switch opens. As the pressure switch opens, the 120 V control voltage at Input #1 disappears, resulting in the change of Input #1 logic state from a "1" to a "0." At the same time, as the pressure in the vessel drops below critical level, the temperature of the process rises to critical level and the temperature switch closes. The 120 V control voltage is bridged across to PLC input module Input #3, changing its logic state from a "0" to a "1." At the very next scan, the PLC senses the change of state at Inputs #1 and #3. The PLC logic and algorithm recognizes these two changes of state as an anomaly and develops a logical solution that it must execute through its outputs. This juncture serves as an appropriate segue into the output section of the control system.

Output Control Diagram

As described earlier, the 115 VAC Modicon output module, shown in Figure 10.8, serves as an interface between the PLC and the control devices in the field. Let's continue our analysis of the hypothetical scenario surrounding the loss of pressure in a vessel, with the focus, this time, on specific outputs associated with the proper response to the anomaly described above. Assume that the PLC's response, to the anomaly described above, involves some actions that yield instantaneous results and some that will require a finite amount of time to implement. Actions that the PLC-based control system must initiate and execute, immediately, are as follows:

1. Energize the reserve tank solenoid valve to boost the vessel pressure to a certain level.
2. Turn ON the alarm horn.
3. Turn ON the Critical/Low Pressure Alarm light.
4. Turn ON the pump motor starter.

In order to energize the reserve tank solenoid valve to boost the vessel pressure, the PLC CPU must change the logic level at Output #5 from a "0" to a "1." The output module transduces the "1" to a 120 V output at the terminal for Output #1. This 120 V control voltage propagates to the solenoid through the 1.5 A short-circuit protection fuse FUxxx14 and turns on the solenoid. The solenoid opens the reserve tank valve and begins to pressurize the vessel.

In order to implement full restoration of pressure in the vessel, the PLC CPU must change the logic level at Output #1 from a "0" to a "1." As a result, the 120 V control voltage propagates to the pump motor starter coil through the 1.5 A short-circuit protection fuse FUxxx06 and a normally open safety interlock contact. The safety interlock is assumed to be met, so, this safety interlock contact should be closed. The 120 VAC, therefore, arrives at the pump motor starter, "M," energizes it, and starts the 3-hp motor, shown in Figure 10.8.

At the same instant, the CPU changes the state of Output #7 from "0" to a "1," thus sending 120 V to the alarm horn through another 1.5 A fuse. The horn continues to annunciate the critical alarm condition, audibly, until the CPU has received verification of re-pressurization of the vessel via Input #1 or the alarm is acknowledged and silenced through a horn silence circuit on the input side of the control system. Note that the horn acknowledge and silence button is not shown on input control drawing in Figure 10.7.

In addition to turning ON Output #7, the CPU changes the logic state of Output #6 from a "0" to a "1," thus sending 120 V to an alarm light via 1.5 A fuse FUxxx16. The alarm light turns on and stays on until the CPU has received verification of re-pressurization of the vessel via Input #1.

RELAY LADDER LOGIC

An introduction to control circuits and control systems would be incomplete without some explanation on illustration of PLC programming method. Even though the brain of a PLC, like a PC (Personal Computer), consists of single or multiple microprocessors, unlike a PC it does not operate off a "Windows®" or Microsoft Office-based software systems. PLC operation is based on proprietary software provided by the PLC manufacturer. On the other hand, the PLC **programming** software, nowadays, is "Windows®" based. The programming software allows control engineers and technicians to program the operation of PLCs in the form of a diagrammatic programming language called **Relay Ladder Logic**. This programming language emulates the "pre-PLC" era, electromechanical relay and counter- and timer-based control logic, in symbolic form. Of course, the symbols and associated functions are represented as "virtual" symbols and logical functions that are used to implement functional specifications of the overall control system. In the pre-PLC era, these electromechanical relay, timer, and counter based functions had limited reliability, lacked speed, resolution, accuracy, flexibility, versatility, and occupied a great deal of control cabinet or control room space. PLC manufacturers, such as Rockwell®, provide training classes for control engineers and technicians, on PLC programming and installation, at local or regional offices.

A basic, relay ladder logic-based program example is shown in Figure 10.9. Brief excerpts of a comprehensive relay ladder logic program for a control system are shown in Figures 10.9 and 10.10. These excerpts are, actually, screen captures or "print screens" of the PLC program as viewed by a PLC programmer for program creation, program modification, or troubleshooting purposes. These excerpts represent a specific example of an RSLogix5000 program for an Allen Bradley Control Logix Programmable Controller.

Fan Start and Stop Program Segment: Note that, at first glance, **this** page of the overall program for the control system resembles a typical Windows®-based application program. However, a closer inspection reveals that – unlike computer programing languages like COBOL, Basic, C, C+, etc. – this is a diagrammatic logic code, wherein, program or algorithm is premised on symbols that appear as coils and contacts of relays, timers, counters, etc. As visible on the left side of the relay ladder program, there

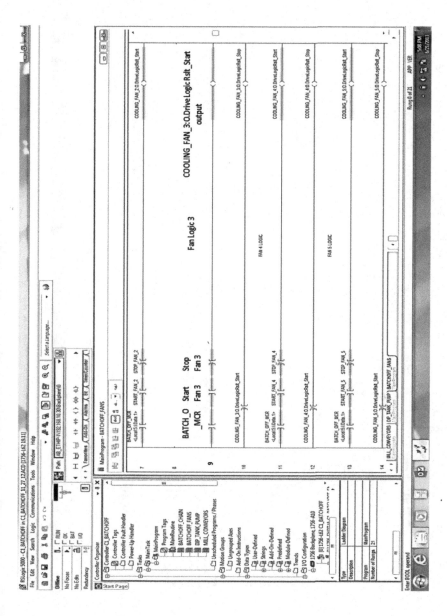

FIGURE 10.9   PLC relay ladder logic programming example – cooling fan control logic.

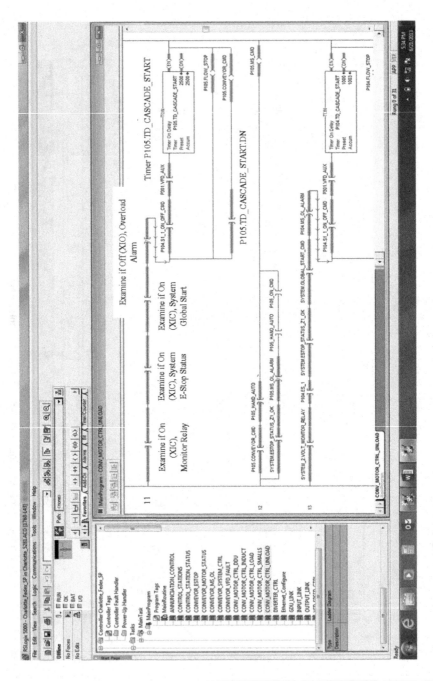

FIGURE 10.10 PLC relay ladder logic programming example – cascade start timer.

are seven tiers, lines, or "rungs" depicted on this page of the program, ranging from rung #7 to rung #14. The logic symbols in each rung represent **inputs** and **outputs**. See Chapter 8 for descriptions and illustrations of typical PLC inputs and outputs. These inputs and outputs, in a typical PLC program, can be **"internal"** or **"external"** inputs. External inputs (**discrete**) are inputs that represent "tangible" or physical switches and contacts in the field. In other words, these inputs emanate from devices and equipment that are "peripheral" to the PLC. On program segment depicted in Figure 10.9, the logic elements, input, and output symbols can be categorized as follows:

Batch MCR: Batch control system master control relay (MCR) contact. This contact is "normally open" but shown closed – at the time the program was viewed – through highlighting (darker shading) of the contact.

Start Fan 3: This symbol or element represents the normally open contact of the START switch for Fan #3. Even though this is a "normally open" contact "--| |--," it is shown **closed** – at the time the program was viewed – through highlighting (darker shading) of the contact.

Stop Fan 3: This symbol or element represents the normally closed contact belonging to the STOP switch for Fan #3. This is a "normally closed" contact "--|/|--"(look for the "/" within the parallel lines) and it is shown **closed** (True) – at the time the program was viewed – through highlighting (darker shading) of the contact. In other words, the STOP switch is not depressed at this time.

Cooling Fan #3 Dr Start: The coil symbol "—( )—," typically, represents an output. When this output is energized or **"True,"** it actually turns ON an output, such as those outputs we considered in the review of PLC Output Controls Drawing in Figure 10.8. In this case, this output is connected to a VFD, variable frequency drive.

Note that all elements in Rung #9 are highlighted, meaning, they were conducting current, ON, or True, at the time the snapshot of the program was taken. Note that when elements are **closed** or passing current, their status is said to be **"True."** When elements are not energized, not activated, their status is said to be **"False**." The current status of Rung #9 can be interpreted as follows:

At the time this snapshot of the logic was taken, **BATCH_OFF_MCR** input was **True,** closed, or High, the **START_FAN_3** bit (logical representation of input emanating from **Fan #3** start switch) is **True**, and the **STOP_FAN_3** bit (logical representation of input emanating from Fan #3 stop switch) is **false** then the **COOLING_FAN_3:O. DriveLogicRslt Start** output (same as Cooling Fan # 3 Dr Start output) will go high or **True**. This output is directly connected via the software to the Allen Bradley® (PowerFlex 70) VFD. This is accomplished through an Ethernet connection and will cause the motor connected to the VFD to start. In this case, there are seven fans that are used to cool freshly made rubber so that it can be used in the next process.

Example 10.2

Consider the PLC relay ladder logic program in Figure 10.9 and answer the following questions:

- a. What would happen to the "run" signal going to the VFD if MCR drops out or de-energizes?
- b. How does Rung #10 respond to the energization (becoming True) of **COOLING_FAN_3:O. DriveLogicRslt Start** output?

- a. **Answer**: If the MCR for the system drops out or de-energizes, the continuity of Rung #9 would break. This will result in de-energization of **COOLING_FAN_3:O.DriveLogicRslt Start** output to the VFD, and Cooling Fan #3 will **stop**.
- b. **Answer**: Upon inspection of Rung #10, we see that a normally closed contact belonging to output "**COOLING_FAN_3:O. DriveLogicRslt Start**" is in series in Rung #10. This normally closed contact opens when **COOLING_FAN_3:O. DriveLogicRslt Start** output is energized. That is the reason why both elements in this rung are shown de-energized (no shading or highlighting). Note that the output of this rung is labeled as **COOLING_FAN_3:O. DriveLogicRslt Stop.** This implies that when COOLING_FAN_3:O. DriveLogicRslt **Start** output is turned ON, the **STOP** signal to the Cooling Fan VFD is disabled through the normally closed contact of **COOLING_FAN_3:O.DriveLogicRslt Start** "coil" in Rung #10. In essence, Rung #10 serves as an interlock to ensure that the STOP command to Cooling Fan #3 is disabled when Cooling Fan #3 START bit is ON.

Timer-Based Starting Control Logic: The PLC control program examined in this section is shown in Figure 10.10. The logic elements, input and output symbols, employed in this section of the overall PLC program are shown on Rung 11, in Figure 10.10. The logic represented in this rung is more complex than the logic associated with Rung #9 in Figure 10.8. The essence of this program rung is as follows:

If all of the requirements, conditions, or bits for the Examine if ON (labeled as XIC, Examine if Closed) and Examine if Off (XIO), in Rung 11, are **True** then Timer P105.TD_CASCADE_START will start timing. Notice the rectangular block on the right side of this rung. This block represents the timer. The timer setting is annotated as 2,500. The set point of 2,500, in this case, implies 2,500 ms. In other words, the timer will time for 2.5 seconds before energizing the output. When 2.5 seconds expire, timer output labeled as P105.TD_ CASCADE_ START.DN becomes **True,** and P105_ CONVEYOR_CMD will turn ON starting P105 conveyor in this simple material handling conveyor system application.

Example 10.3

Consider the logic associated with Timer P105.TD in Rung # 11 of the PLC relay ladder logic program shown in Figure 10.10 and answer the following questions:

    a. Which "register" of field on Rung #11 represents the **time elapsed** while the timer is timing?

    b. What is the status of the timer? How many seconds remain on the clock before the timer times out?

    c. What is the function of Rung #11 branch consisting of the "P105.TD_CASCADE_START.DN" bit and the "P105 Conveyor _CMD" coil?

    a. **Answer:** As we examine the right side of Timer P105.TD in Rung # 11, we see the register or field labeled "Preset 2500." This register holds the timer preset time of 2,500 ms. The register directly below the Preset register is labeled "Accum 2500." This field can be interpreted as: Accumulated or elapsed time = 2,500 ms or 2.5 seconds.

    b. **Answer:** As observed in part (a), the accumulated or elapsed time = 2,500 ms. Since the accumulated or elapsed time is equal to the Preset time, the timer has timed out, and "0" ms remain.

    c. **Answer:** As we examine the right side of Timer P105.TD in Rung # 11, and observe the branch with two elements, the "P105.TD_CASCADE_START. DN" bit and the "P105 Conveyor _CMD" coil, we note the fact that P105 Conveyor is commanded to turn ON when the P105. TD_CASCADE_ START. DN contact or bit is closed or True. In other words, after the accumulated time reaches the preset time of 2,500 ms, the "**Done**" coil labeled as —**(DN)**— turns ON or becomes **True**. This results in the closure (or conversion to **True** state) of the timer **Done** bit: "--| |--" 105.TD_ CASCADE_ START. DN, and with the three XIC and one XIO conditions satisfied (**True**), the P105 Conveyor _CMD" coil is energized to turn on P105 Conveyor.

SELF-ASSESSMENT PROBLEMS AND QUESTIONS – CHAPTER 10

1. Answer the following questions pertaining to the branch circuit shown in the schematic diagram below:

    a. What is the maximum voltage the power distribution system for this branch circuit rated for?

    b. How many wires and phases is the power distribution system for this motor branch circuit rated for?

    c. What would be the proper rating for the branch circuit disconnect switch?

    d. What should the solid-state overload device be set for at commissioning of this branch circuit?

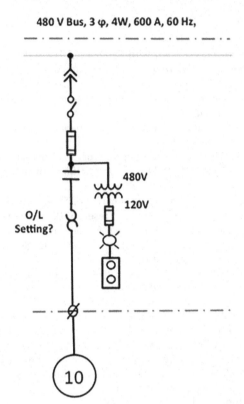

480 V Bus, 3 φ, 4W, 600 A, 60 Hz,

2. Determine the sizes/specifications of the following components in the branch circuit shown below using the Bus® table and information included in the questions:
   a. Conductor size.
   b. Conduit size.
   c. Overload setting based on the 115% NEC® stipulation.
   d. Disconnect switch size, safety and fusible.
3. Consider the wiring diagram for the 75-hp motor shown in Figure 10.6 and answer the following questions based on the control logic explained in this chapter:
   a. What would be the likely outcome if the START switch is depressed while the motor is operating?
   b. What would be the outcome if START and STOP switches are depressed simultaneously?
   c. What would be the likely outcome if the main disconnect switch for the MCC is opened?
   d. Can the motor be stopped if the motor starter latching contact "welds" shut due to overheating?

480 V Bus, 3 φ, 4W, 600 A, 60 Hz,

4. Consider the logic associated with Timer P105.TD in Rung # 11 of the PLC relay ladder logic program shown in Figure 10.10 and answer the following questions:
   a. What would be status of the timer **Done** bit: "--| |--" 105.TD_ CASCADE_ START. DN, if the XIC System E-Stop bit turns "**False**" 1 second after Timer P105.TD in Rung # 11 is turned on?
   b. Would the "P105 Conveyor _CMD" bit – commanding P105 Conveyor to turn ON be **True** if and when the system E-Stop switch is engaged/ pressed?

11 Electrical Power Bill Calculation and Electrical Energy Cost Reduction

INTRODUCTION

This chapter introduces the reader to four-faceted approach associated with comprehensive electricity cost reduction strategy and planning. This chapter allows readers an opportunity to enhance their knowledge about various electrical power rate schedules and options. Awareness of electrical energy cost structure allows an energy engineer, plant engineer, or a facilities manager an opportunity to make calculated and informed decisions associated with electrical energy cost minimization. We will also explore some electrical cost reduction ideas and will discuss the concepts of Energy Performance Contracting (EPC) and Energy Service Company (ESCO) as a means for pursuing electrical energy intensity improvement projects with little or no capital investment. In this chapter, we will leave the reader with some food for thought on the "source" aspect of electrical energy consideration through information related to proven energy sources that can be harnessed to produce electricity. We will also touch on energy storage methods and their role in electrical energy cost reduction.

FOUR-FACETED APPROACH TO ELECTRICAL ENERGY COST REDUCTION[1]

Most energy programs – adopted for residential, commercial, or industrial settings – focus, mainly, on **energy conservation**. However, in order to develop a more comprehensive and complete energy program, one should view an electrical energy intensity improvement program as a four-faceted approach. Those facets are as follows:

I. **Energy Conservation:** This facet consists of the application of high-efficiency electrical equipment, i.e. premium efficiency electric motors and energy-efficient lighting systems.
II. **The Supply Side:** Exploration, evaluation, and application of local electrical power generating systems that are cost competitive with the grid.
III. **Energy Cost Rate:** This approach focuses, primarily, on exploration and financial evaluation of electrical power schedules and contracts that offer the lowest overall cost rates for commercial and industrial applications.

---

[1] *Finance and Accounting for Energy Engineers* by S. Bobby Rauf.

IV. **Energy Storage Systems:** This approach focuses on application of economically and technically suitable energy storage systems. The underlying premise of this approach being storage of electrical energy during off-peak periods and consumption of that stored energy during peak periods.

ELECTRIC UTILITY RATE SCHEDULES[2]

Measurement, verification, and computation of utility bills are key elements in understanding, planning for, and managing the consumption of various utilities. Conservation of utilities is an essential and strategic component of overall utility cost reduction effort. Its primary impact is through enhancement of utility productivity or, in the strict energy realm, through maximization of energy productivity. The other vital determinants of the overall cost of utilities are contract and rate schedule. In a given billing month, once the utility usage has been recorded, it must be processed using the rate schedule or structure the contract calls for.

The thrust of our discussion, in this chapter, will entail introduction of various rate schedules and the impact of major rate schedules on the composition of the electric bill in the industrial and residential realm. The residential rate structures and computation methods are relatively straightforward. Consequently, our discussion on residential electric bills will be brief. Some of the contract categories and rate schedules are as follows:

- Industrial
- Commercial
- Residential
- Municipal or Co-ops
- OPT, or Time of the Day
- HP, or Hourly Pricing
- Interruptible

Most utility companies, around the United States, offer rate schedules similar to the ones stated above. However, in many instances, the names of the rate schedules may vary from one utility to another. Despite the apparent semantic differences among rate schedules offered by various utilities, the cost components, cost tiers, and rate application mechanisms are similar. The essence and format of electrical rate schedules are being introduced in this text using Duke Energy® Company as an example.

Duke Energy® is one of the largest electric power holding companies in the United States, providing electricity to 7.7 million retail customers in six states. Duke has approximately 51,000 MW of electric generating capacity in the Carolinas, the Midwest, and Florida – and natural gas distribution services serving more than 1.6 million customers in Ohio, Kentucky, Tennessee, and the Carolinas. Duke owns and operates diverse power generation assets in North America, including a portfolio of renewable energy assets.

---

[2] *Finance and Accounting for Energy Engineers* by S. Bobby Rauf.

Details on Duke® Energy electric power rates are available at the following Duke Energy® website:

https://www.duke-energy.com/our-company/about-us

Some of the rates schedules offered by Duke® are listed in Table 11.1, with the more common ones highlighted.

Three of the most common schedules are explained in more detail below through excerpts from the Duke Energy® website. These schedules are also more pertinent

TABLE 11.1
Some Rate Schedules Offered by Duke® Energy

Rate Code	Description
RS	**Residential Service**
RE	Residential Service, Electric Water Heating, and Space Conditioning
ES	Residential Service, Energy Star
RT	**Residential Service, Time of Use**
WC	Residential Service, Water Heating, Controlled/Sub-metered
SGS	Small General Service
OPT-G	**Optional Power Service, Time-of-Use General Service**
BC	General Service, Building Construction Service
LGS	**Large General Service**
FL	General Service, Floodlighting Service
OL	General Service, Outdoor Lighting Service
GL	Government Lighting Service
PL	General Service, Street and Public Lighting Service
NL	Non-standard Lighting Service
TS	General Service, Traffic Signal Service
I	**Industrial Service**
OPT-E	Optional Power Service, Time of Use, Energy Only (Pilot)
OPT-H	**Optional Power Service, Time of Use, General Service, High Load Factor**
OPT-I	**Optional Power Service, Time of Use, Industrial Services**
HP	**Hourly Pricing for Incremental Load**
PG	Parallel Generation
REPS	Renewable Energy Portfolio Standard Rider
PM	Power Manager Load Control Service Rider
NM	Net Metering Rider (NM)
SCG	Small Customer Generator Rider (SCG)
IS	**Interruptible Power Service Rider**
SG	**Standby Generator Control Rider**
EC	Economic Development Rider
ER	Economic Redevelopment Rider
PSC	Power Share Call Option
PP-N	Purchased Power Non-hydroelectric
PP-H	Purchased Power Hydroelectric

to the topics introduced in this text. The reader is advised to regard the following excerpted rate schedule information as **reference information** intended to **introduce** the reader to typical electrical billing terminology and general format of the electrical power bills in the residential and industrial sectors. The rates quantified in the discussion that follows represent a "snapshot" at the time this text was authored, and only the current prevailing rates should be used in the practice of engineering, energy, or facilities management.

Schedule RS (NC)

The RS stands for Residential Service. The specifics of this schedule, as published by Duke®, apply to the State of North Carolina only.

This schedule and associated rates are available only to residential customers in residences, condominiums, mobile homes, or individually metered apartments which provide independent and permanent facilities complete for living, sleeping, eating, cooking, and sanitation.

This schedule is applicable to all electric service of the same available type supplied to customer's premises at one point of delivery through 1 kWh meter.

The types of service to which this schedule is applicable are alternating current, 60 Hz, either single-phase two or three wires, or three-phase four wires, at Company's standard voltages of 240 V or less. Note that while the cost ($) figures used in this chapter, and most of this text, are representative in order of magnitude, they are used, herein, for concept or principle illustration purposes only. Actual cost rates are subject to change and are, indeed, updated frequently.

Monthly Rate

I. For Single-Phase Service:
   Bills Rendered During July–October
   A. Basic Customer Charge: $14.00 per month
   B. Kilowatt-Hour Charge: 10.556¢ per kWh
   Bills Rendered During November–June
   A. Basic Customer Charge: $14.00 per month
   B. Kilowatt-Hour Charge:10.083¢ per kWh
II. For Three-Phase Service:
   The bill computed for single-phase service plus $7.00.
III. Renewable Energy Portfolio Standard (REPS) Adjustment:
   The monthly bill shall include a REPS Adjustment based upon the revenue classification: Residential Classification – $1.45/month.

Schedule OPT-I (NC)

This rate schedule represents an **optional power industrial service** that is based on time of use. Therefore, it is often referred to as **Time of Use Industrial Service.**

This rate schedule, offered by Duke Energy®, as specified in this text, is available in North Carolina only. Furthermore, this rate schedule is available only to establishments classified as "Manufacturing Industries" by the Standard Industrial

Classification Manual published by the United States Government, and where more than 50% of the electric energy consumption of such establishment is used for its **manufacturing processes**.

The specifications of this rate schedule or service, as stipulated by Duke Energy®, are as follows:

- The Company (Duke Energy®) will furnish 60 Hz service through one meter, at one delivery point, at one of the following approximate voltages where available:
  - Single-phase, 120/240 V, 120/208 V, 240/480 V, or other available single-phase voltages at the company's option; or
  - Three-phase, 208Y/120 V, 460Y/265 V, 480Y/277 V; or
  - Three-phase, three-wire, 240, 460, 480, 575, or 2,300 V; or
  - Three-phase, 4,160Y/2,400, 12,470Y/7,200, or 24,940Y/14,400 V.
- Motors of less than 5 hP may be single phase. All motors of more than 5 hP must be equipped with starting compensators. The starting compensator methods, as discussed in earlier chapters, include "Star–Delta Starting," application of "Soft-Start" systems, utilization of VFDs (variable frequency drives), and selection of high slip motors. The Company (Duke Energy®) reserves the right, when in its opinion the installation would not be detrimental to the service of the Company, to permit other types of motors.

RATES UNDER OPT-I SCHEDULE

As evident below, due to the fact that this rate schedule is a function of time and seasons, rate structure under the OPT-I schedule is multi-dimensional and more complex. The rate (or charge) components of an OPT-I based electric power bill, for illustration purposes only, are shown in Table 11.2.

On-Peak and Off-Peak Hours, for schedule OPT-I, are classified as follows:

On-Peak Hours:

Summer Months

June 1–September 30, Monday–Friday

On-Peak Period Hours

1:00 PM–9:00 PM

Winter Months

October 1–May 31, Monday–Friday

On-Peak Period Hours

6:00 AM– 1:00 PM

Off-Peak Hours:

- All weekday hours not included under On-Peak hours and all Saturday and Sunday hours.
- All hours for the following holidays:
  - New Year's Day, Memorial Day, Good Friday, Independence Day, Labor Day, Thanksgiving Day, Day after Thanksgiving, and Christmas Day.

TABLE 11.2
Time of Use, OPT-I Rate Schedule Offered by Duke® Energy

I. RATE:	Basic Facilities Charge per Month	$39.79
II. Demand Charge	Summer Months	Winter Months
A. On-Peak Demand Charge	June 1–September 30	October 1–May 31
For the first 2,000 kW of Billing Demand per month, per kW	$14.0767	$8.2981
For the next 3,000 kW of Billing Demand per month, per kW	$12.8972	$7.1075
For all over 5,000 kW of Billing Demand per month, per kW	$11.7067	$5.9064
B. Economy Demand Charge	$1.1448	$1.1448
III. Energy Charges	All Months	
A. All On-Peak Energy, per Month, per kWh	5.7847¢	
B. All Off-Peak Energy, per Month, per kWh	3.4734¢	

DETERMINATION OF BILLING DEMAND

A. The **On-Peak Billing Demand** each month shall be the largest of the following:
   1. The maximum integrated 30-minute demand during the applicable summer or winter on-peak period during the month for which the bill is rendered.
   2. Fifty percent (50%) of the Contract Demand (or 50% of the On-Peak Contract Demand if such is specified in the contract)
   3. Fifteen kilowatts (kW)
B. **Economy Demand:** To determine the Economy Demand, the larger of
   1. The maximum integrated 30-minute demand during the month for which the bill is rendered; or
   2. Fifty percent of the Contract Demand shall be compared to the On-Peak Billing Demand as determined in A, above.

If the demand determined by the larger of B.1 and B.2 above exceeds the On-Peak Billing Demand, the difference shall be the Economy Demand. In addition, if the peak set during the Off-Peak period exceeds the peak set during the On-Peak period, the difference between the two demands would be construed as the Economy Demand.

POWER FACTOR CORRECTION

When the average monthly power factor of the Customer's power requirements is **less than 85%**, the Company (Power Company) may correct the integrated demand in kilowatts for that month by multiplying by **85%** and dividing by the average power factor in percent for that month.

HOURLY PRICING OPTION/SCHEDULE

HP option is, essentially, a DSM, Demand Side Management, instrument. If your business is expecting incremental loads and has the ability to manage consumption patterns on a day ahead, hourly basis, Duke Energy® can offer electric energy cost savings through rate schedule HP, also referred to as **Hourly Pricing**. This opportunity is available to customers who have the ability to be responsive to fluctuations in hourly prices (use more energy when prices are low, use less energy when prices are high).

The HP schedule is available to non-residential establishments with a minimum contract demand of 1,000 kW who qualify for service under the Duke Energy® rate schedules LGS, I, OPT-G, OPT-H, OPT-I, or PG, at the Company's (Duke Energy®) option on a voluntary basis. The maximum number of customers on the system to be served under this schedule is 150. Service under this schedule is available to contracting customer in a single enterprise, located entirely on a single, contiguous, premise.

AREAS OF OPPORTUNITY FOR ELECTRICAL ENERGY COST SAVINGS

Net energy cost reduction can be achieved through focus on one or more of the following approaches:

A. Actual reduction in energy consumption through concerted **energy conservation measures.**
B. Partial or complete switch to **renewable energy sources**. Partial reliance on renewable energy sources would require less capital and still empower the energy consumer with the ability to exercise demand control and to perform peak shaving. See information on renewable energy later in this chapter.
C. Identification of, and transition to, the most **suitable energy contract or billing rate schedule**. This approach may be, strictly, administrative and may not yield any reduction in energy consumption, but it could reduce energy costs substantially. This is often referred to as the **"pure cost" approach**.
D. **Energy storage.** Peak shaving and peak demand cost reduction can be achieved through storage of energy procured, or energy generated internally, during "off-peak" periods for consumption during higher cost peak-demand periods.

There are several areas of potential cost savings available to electrical power consumers, beyond the ones discussed in this text, so far. The reader is encouraged to see organized and comprehensive approach to energy cost reduction, supported by illustrative case studies, in **Finance and Accounting for Energy Engineers by S. Bobby Rauf**. Some areas of opportunity, related specifically to electrical rate schedules, are listed below:

Schedule or Rider SG, On-Site Generation

This program is designed to offer large industrial or commercial consumers an opportunity to take advantage of on-site emergency power generation assets they already own. Most emergency power generators have to be tested, under load, periodically, for predictive maintenance, preventive maintenance, and personnel training purposes. Why not get compensated or credited for this routine, but necessary, exercise by the power company? That is where Schedule or Rider SG offers the larger consumers an opportunity to do so.

Therefore, if a large consumer owns emergency standby generators and can make their capacity available for use by Duke Energy® during times of system emergencies, monthly credits are offered through rate rider SG.

Schedule or Rider IS, Interruptible Power Service

The interruptible power service schedule, IS, is available for non-residential customers receiving concurrent service from the Company (Duke Energy®) on schedules LGS, I, HP, OPT-G, OPT-H, or OPT-I served under continually effectively agreements for this Rider made prior February 26, 2009. Under this schedule, the customer agrees, at the Company's (Duke Energy's®) request, to reduce and maintain his load at a level specified in the individual contract. The Company's request to interrupt service may be initiated at any time the Company has capacity problems.

Contracts for interruptible power service are accepted by Duke Energy® on the basis of successive contracts, and each contract must specify an interruptible, integrated demand of not more than 50,000 KW to be subject to these provisions. The Company can limit the acceptance of contracts to a total of 1,100,000 KW of Interruptible Contract Demand on all non-residential schedules on the total system.

Duke Energy® reserves the right to test the provisions of this Rider once per year if there has not been an occasion during the previous 12 months when the Company requested an interruption. Duke Energy® gives advance notice of any test to customers served under this Rider. In return for participation in this schedule, Duke issues credits to the participating customers based on a specific formula that can be reviewed at Duke Energy® website.

Tips on Utility Rate Schedules and Contracts

It's, ultimately, the consumer's responsibility to maintain vigilance and awareness of opportunities and incentives available through the power company that serves the region. The following tips can prove to be useful to the consumer for minimizing the **cost** of electrical energy:

1. Electric Utility Company is not obligated to notify facilities about availability of more favorable rates or "schedules." Some utility companies, for example, Duke Energy®, assign Account Managers to certain segments of their market. In order to maximize the utility of the consumer's relationship with the assigned Account Manager, the consumer's representative, namely, the energy or utilities engineer, must note the following:
   - Utility company Account Managers will, at times, advise their larger accounts of favorable rate schedules as the customers' demand and usage changes. However, it is good practice to explore suitable contract alternatives when some of the following changes are experienced in load characteristics and profile:
     - **Addition or removal of loads** that constitute a substantial percentage of facility's overall load.
     - **Addition of highly reactive loads**, i.e. large motors and transformers. This might impact the facility's overall power factor and energy consumption.
2. Look for newfound **flexibility in load schedules**. For example, "Off-Peak" unloading of rail cars in industries where mass transport of raw materials is required.
3. Look for addition of **onsite generation**; Standby or Cogeneration assets.
4. Be vigilant of any change in facility's operation constraints. For instance:
   - Could the facility tolerate power interruption, with some advance notice?
   - Could facility participate in, online, diesel generator testing?
5. **Renewable Energy Portfolio Standard (REPS) Adjustment**:
   The monthly bill shall include a REPS Adjustment based upon the revenue classification:
   Commercial/Governmental Classification – **$3.22/month**
   Industrial/Public Authority Classification – **$32.20/month**
   Upon written request, only one REPS Adjustment shall apply to each premise serving the same customer for all accounts of the same revenue classification. If a customer has accounts which serve in an auxiliary role to a main account on the same premise, no REPS charge should apply to the auxiliary accounts regardless of their revenue classification (see Annual Billing Adjustments Rider BA). Additional information on renewable electrical energy is recounted later in this chapter.
6. **Transformation Discounts:** When Customer owns the step-down transformation and all other facilities beyond the transformation which Company (Duke Energy®) would normally own, except Company's metering equipment, the charge per kW of On-Peak Billing Demand will be reduced as follows:
   Transmission Service: $0.75/kW
   Distribution Service: $0.5/kW
   For customer to qualify for the Transmission Service Transformation Discount, the customer must own the step-down transformation and all

Electrical Engineering Fundamentals

other facilities beyond the transformation which Company (Duke Energy®) would normally own, except Company's metering equipment, necessary to take service at the voltage of the 69 kV, 115 kV, or 230 kV transmission line from which Customer received service.

ENERGY PERFORMANCE CONTRACTING AND ESCO OPPORTUNITIES[3]

EPC is a service vehicle for provision of energy conservation or energy productivity enhancement services. In most cases, EPC-type contracts and projects involve turnkey service. Turn-key service is defined as a comprehensive service provided by a vendor or contractor that begins with definition and scope of an energy project and ends with project start-up, commissioning, and subsequent verification of energy savings. Projects covered by EPC range from simple energy conservation efforts such as replacement of inefficient lighting systems to more complex projects that address the supply side of the energy equation through renewable energy systems.

While EPC and ESCO alternatives are often adopted for substantial energy conservation or renewable energy projects, initiatives that leverage lower cost power contracts can be included in the overall energy project portfolio. EPC projects and contracts often include guarantees that the savings produced by a project will be sufficient to finance the full cost of the project.

The guarantee to fund the project through the savings generated by the project is what distinguishes **EPC** projects from **non-EPC** or owner-funded energy projects. Even though it appears counterintuitive, often EPC projects are not limited to energy conservation or energy capacity enhancement projects, instead, the breadth of their scope includes water conservation, sustainable materials, and operations.

ESCO stands for Energy Service Company. One way to understand the relationship and distinction between EPC and ESCOs is to think of EPC as a **process** and the ESCOs as **entities that implement the EPC process**. ESCOs can provide a full range of services required to complete an energy project. Such services, often, include the following or a combination, thereof:

- Energy audit
- Design engineering
- Construction management and supervision
- Facilitation or provision of project financing
- Start-up and commissioning
- Operations and maintenance
- Monitoring and verification of energy savings

Historically, the inception of EPC could be traced as far back as the early 1980s. In the pre-1985 era, ESCOs were established as a part of the DSM efforts to provide personnel and equipment resources to the utilities as they strived to meet the energy conservation mandates imposed by federal and state governments.

---

[3] *Finance and Accounting for Energy Engineers* by S. Bobby Rauf.

From the mid-1980s through 2003, ESCOs, and EPC industry as a whole, have seen substantial ebb and flow in growth, acceptability, and revenue. Over this period, some of the ESCOs have transformed and some have grown either organically or accretively through consolidation. This evolution within the EPC domain was influenced, favorably, by the state and federal governments. The successes of the EPC industry, in the 1994–2002 period could, to a certain extent, be attributed to studies by LBNL, Lawrence Berkley National Laboratory, and NAESCO, National Association of Energy Service Companies, that highlighted the EPC successes and encouraged the state and federal governments to promote EPC. Another important event that could be credited for EPC growth and successes in the 1994–2002 period was the formulation and implementation of the IPMVP, or International Performance Measurement and Verification Protocol. The IPMVP provided standard methods for documenting project savings and provide commercial lenders the confidence to finance EPC projects.

As plausible, the EPC industry was impacted unfavorably by the ENRON debacle, as ENRON was a significant player in the ESCO market. The ENRON collapse coupled with the uncertainty about the deregulation of the electric utility industry impeded the growth of EPC in the 2002–2004 period.

The 2004–2006 period showed 20% growth in the EPC industry with comparable projected growth trend. The 20% growth and subsequent upward trend can be attributed to volatility in the energy market and the increasing energy prices. Other contributors to the heightened interest in the EPC are state and federal mandates, inadequate capital and maintenance budgets for federal and state facilities. Growing awareness of the greenhouse gas emissions and realization of the fact that large-scale, sustained, remediation is needed in this regard has made EPC more attractive at local, state, and federal levels and in the private sector.

While ESCOs, in response to customer requests, are constantly adding new measures and services to their project portfolios, electrical and non-electrical, they are not to be construed as stewards for technological research, development, and marketing in the energy domain. ESCOs and their clients tend to be fairly conservative and risk averse in selection of technologies for projects. Due to the fact that the cost of most ESCO projects are paid from energy savings, often secured with financial guarantees, ESCOs and their clients tend to lean in favor of proven technologies.

EPC involves distinct skills and areas of expertise in the following two areas:

- Energy procurement
- Commercial law.

Both of these disciplines involve risk management, risk allocation, isolating benefits, and option analysis. The following list could serve as a checklist for energy managers, utilities engineers, and facilities managers as they consider EPC and ESCOs in formulation of the energy program strategy:

- EPC is **one** way to finance and implement energy conservation projects.
- Remuneration sought by the EPCs is typically included in the overall cost of the project.

- Initial investment, maintenance cost, energy cost (over the life of the project) monitoring, and training cost are – and should be – included in the overall cost of the project.
- The energy and cost savings produced by the project need to be sufficient to cover all project costs over the term of the contract.
- EPC project contracts, typically, span a period of 10 years or more.

Some of the benefits associated with the EPC and ESCO approach for implementation Energy Programs and Strategies are as follows:

- EPCs save company capital for projects that lack financial justification.
- EPCs fund energy conservation projects from savings generated by the project.
- EPCs reduce repair and maintenance costs caused by inadequate, aging, or obsolete equipment, electrical or non-electrical.
- EPCs, and energy conservation projects in general, provide secondary benefits, i.e. increased employee productivity, safe and more comfortable working environment. For example, energy-efficient lighting projects/programs, computer/PLC-based automated HVAC systems and computer/PLC-based EMS, Energy Management Systems.
- Improve the environment and conserve energy resources.

ELECTRICAL ENERGY-RELATED MEASURES TYPICALLY INCLUDED IN ESCO OR EPC ENDEAVORS

Additional details on some of the mainstream technical measures, typically included in ESCO or EPC driven projects, are recounted for reference below:
- **Lighting**: Replacement of inefficient lighting systems with energy-efficient lamps, energy-efficient ballasts, and optimally designed light fixtures. Examples of such measures include:
  - Replacement of mercury lamps with higher efficacy sodium vapor lamps.
  - Replacement of incandescent and, in some cases, florescent lamps with high-efficacy LED, light emitting diode, lamps. An LED is constructed similar to a regular diode with the exception of the fact that when an LED is forward biased, holes and the electrons combine at the p–n junction releasing light energy or photons. Symbol for a light emitting diode is shown below:

Anode _____ Cathode
LED

  - Replacement of older florescent lamps and fixtures with high-efficiency T-8 or T-5 florescent light systems.
  - Replacement of lighting systems, that are inadequately designed by today's standards, with light systems that are designed with emphasis

on important factors such as lighting efficacy (lumens/watt), CU, coefficient of utilization (Lumens Reaching the Work Plane/Total Lumens Generated). See Chapter 12.

- **HVAC, Heating, Air-Conditioning, and Ventilation:** Optimization and improvement of chilled water systems, replacement of lower efficiency, and high-maintenance HVAC systems with HVAC systems that carry high Energy Star Rating, utilize high-efficiency chillers, use green technologies, i.e. geothermal, solar, thermal storage, etc. Convert manual HVAC control systems to BMCS, Building Management Control Systems, or Direct Digital Control Systems.[4]

- **Control Systems:** Control systems incorporating effectively designed and optimally applied control architecture in energy usage and energy generating systems. These control systems employ cutting edge – yet proven – sensors, transducers, and other control devices for field application. Furthermore, these control systems are driven by CPUs, Central Processing Units, or computers and PLCs, Programmable Logic Controllers that offer the latest improvements in hardware, firmware, application software, and HMI, Human Machine Interface, options.

- **Building Envelope Improvements:** Measures in this category include infrastructure improvements related to the building envelope or exterior.

- **Cogeneration and CHP:** Cogeneration and CHP, Combined Heat and Power, measures address the **supply side** of the energy equation through production of electricity while catering to other needs such as steam required for production processes. Measures in this category can include the following:

  - **Topping Cycle Cogeneration System:** In topping cycle cogenerating systems, electrical power is generated at the top, or beginning, of the cogeneration cycle.

  - **Bottoming Cycle Cogeneration System:** In bottoming cycle cogenerating systems, electrical power is generated at the bottom, or tail end, of the cogeneration cycle.

  - **Combined Cycle Cogeneration System:** In the combined cycle cogenerating systems, electrical power is generated at the top **and** bottom segment of the cogeneration cycle.

    Since the combined cycle systems employ both the topping cycle feature as well as the bottoming cycle feature, they offer higher efficiency in the production of electricity.

- **Demand Response Measures:** Demand response measures are projects or actions undertaken to avert the need for **electrical power generating** capacity expansion. Demand response measures are also referred to as DSM, or Demand Side Management, measures. DSM is an important tool to help balance supply and demand in electricity markets, to reduce price volatility, to increase system reliability and security. This enables the utility industry to rationalize investment in electricity supply infrastructure and to

---

[4] *Thermodynamics Made Simple for Energy Engineers*, by S. Bobby Rauf.

reduce greenhouse gas emissions. Examples of these measures include the following:

- Energy efficiency enhancement technologies, management practices, or other strategies in residential, commercial, institutional, or governmental arena that reduce electricity consumption.
- Demand response or load management technologies, management practices, or other strategies in residential, commercial, industrial, institutional, and governmental arena that shift electric load from periods of peak demand to periods of off-peak demand, including pump storage technologies.
- Industrial by-product technologies consisting of the use of a by-product from an industrial process, including the reuse of energy from exhaust gases, steam, or other manufacturing by-products that are used in the direct production of electricity.

Figure 11.1 is Quantity (Q)–Price (P) graph. This graph shows the effect of **demand response** on **demand elasticity**. The inelastic demand in the electrical power market place is represented by curve **D1**. The high price **P1** associated with the **inelastic** demand **D1** is extrapolated off the point of intersection of the supply curve **S** and the demand curve **D1**. When demand response measures are introduced, demand becomes **elastic**. The elastic demand is represented by curve **D2**. The point of intersection of elastic demand curve **D2** and the supply curve **S** precipitates in a substantially **reduced market price P2**.

It is estimated that a 5% lowering of the demand would result in a 50% reduction in price during peak hours, as demonstrated in the California Electricity Crisis of 2000–2001.[5]

Other studies, such as the two studies sponsored by Carnegie Mellon in 2006,[6] examined the impact of demand response measures on demand elasticity and price.

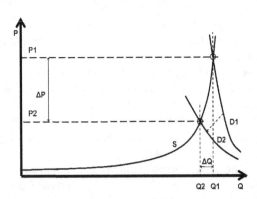

FIGURE 11.1   Demand elasticity and the effect of demand response.

[5] The Power to Choose – Enhancing Demand Response in Liberalized Electricity Markets Findings of IEA Demand Response Project, Presentation 2003.
[6] CEIC-07-01 "Demand Response and Electricity Market Efficiency," Kathleen Spees and Lester Lave. CEIC-07-02 "Impacts of Responsive Load in PJM: Load Shifting and Real Time Pricing" Kathleen Spees and Lester Lave.

The price reduction can be explained by the fact that operators generally plan to use the least expensive, or the lowest marginal cost, generating capacity first, and use additional capacity from more expensive plants as demand increases.

• **Renewable Electrical Energy:** Renewable energy is, mostly, derived from natural resources such as sunlight, wind, rain, tides, and geothermal heat. These are energy sources, or forms of energy, that are renewable and replenished naturally. Of course, the energy harnessed from these renewable natural sources is often in the form of **electrical energy**.

   Some of the more proven renewable energy technologies are as follows:
   • **Hydroelectric Power**
   • **Biomass Energy**
      – Biomass heat
      – Biomass power
   • **Solar Energy**
      – Solar heat energy
      – Solar photovoltaic electrical energy
   • **Wind Energy**
   • **Geothermal Energy**
      – Geothermal heat
      – Geothermal power
   • **Ocean or Tidal Energy**
   Published data shows that approximately 17% of the total US electrical energy is derived from renewable sources. See Figure 11.2. This 17% of the

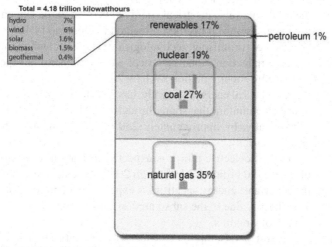

Note: Electricity generation from utility-scale facilities.

Source: U.S. Energy Information Administration, *Electric Power Monthly*, February 2019, preliminary data

FIGURE 11.2 Sources of US electricity generation. Based on 2019 data published by EIA, US. Energy Information Administration

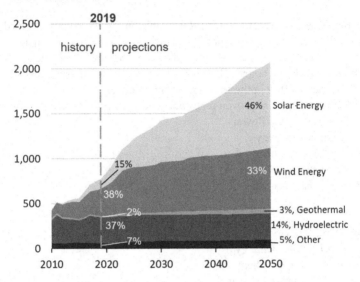

FIGURE 11.3   Renewable electricity generation breakdown – United States.

overall electricity generated in the United States can be segmented as shown
in Figure 11.3. The segmentation in Figure 11.3 shows that:

- While solar electrical energy, currently (2020), represents 15% of
  the total renewable electrical energy generated in the United States,
  it is on an upward trajectory, projected to climb up to 46% by 2050
  because of promising technological evolution and manufacturing cost reduction. Solar electrical energy capacity is expected to
  increase by about 1 trillion kWh by 2050.
- Wind electrical energy, currently, has 38% footprint and is expected
  to largely maintain that share through 2050 but is expected to grow
  substantially, by approximately 250 billion kWh, in incremental
  energy capacity.
- While hydroelectric power is expected to maintain a contribution
  of about 400 billion kWh through 2050, its relative contribution to
  the renewable energy spectrum is expected to contract from 37% to
  14%, largely due to the substantial expansion in the solar and wind
  technology realms.
- **Water and Sewer**: Consumption of water and discharge into sewage systems, at face value, may not appear to have a direct relationship with the
  electrical energy. However, as one examines the process behind the recla-
  mation and distribution of potable water, the vital and indispensable role
  that energy plays becomes palpable. It takes energy, and therefore dollars,

to drive the pumps to collect untreated water. The filtration and sanitization phases of water purification require hydraulic head to move water, which amounts to conversion of electrical energy to hydraulic head. In addition, significant and continuous amount of electrical energy is needed to store and pump water to consumers at a standard head. Hence, it's obvious that if the consumption of water is reduced, energy is conserved and the demand for electrical energy abates.

• **Sustainable Materials and Associated Operations:** Mining, transportation, refinement, treatment, sizing, packaging, marketing, sale, distribution, and application of all types of materials – pure metals, alloys, nonmetallic substances, polymers, and myriad synthetic substances – require energy in all phases of production. Even recycling of materials, as "green" as it is, requires energy. As we assume this perspective, it becomes easy to see how conservation in the use of materials can have an impact on energy demand.

SELF-ASSESSMENT PROBLEMS AND QUESTIONS – CHAPTER 11

1. HP program is a standard feature in all OPT, or Time of Use, schedules.
   A. True
   B. False

2. The energy charge rate structure with Electrical OPT schedule is:
   A. Flat, year round
   B. Tiered
   C. Exponential
   D. A function of time and season.

3. A large industrial electricity consumer set the peak demand in July's billing month at 40 MW. The demand rate structure is same as that included in Duke Energy® OPT-I, Time of Use, rate schedule, as shown in Table 11.2. Determine the demand cost for the month.
   A. $367,000
   B. $505,000
   C. $407,000
   D. $476,579

4. Calculate the **energy charge** for the month of July considered in Problem 3 if all of the energy is consumed during On-Peak hours, for 10 hours per day. Assume that there are 30 days in the billing month and that the load factor is 1, or 100%.
   A. $902,000
   B. $416,808
   C. $2,064,187
   D. None of the above

5. Which of the following statements describe the role of EPC and ESCO most accurately?

A. The terms EPC and ESCO are synonymous.

B. EPC is a method for implementing energy projects and ESCOs are entities that offer this alternative.

C. EPC is required by Department of Energy, ESCOs are not.

D. None of the above

12 Batteries and Capacitors as Energy Storage Devices

INTRODUCTION

Batteries and capacitors are similar in some respects. Both are charge and energy storage devices and consist of positive and negative terminals. Batteries and capacitors can be constructed from flat or cylindrical electrode surfaces; where electrodes constitute the negatively and positively charged surfaces in batteries and capacitors. While there are number of similarities between batteries and capacitors, for application purposes, it is important to note the following stark differences:

1. In batteries, the positively charged terminal, referred to as "anode," and the negatively charged terminal called "cathode" are separated by electrolytic, ionized, conductive liquid or paste. While, in capacitors, the anode and the cathode are separated by non-conductive dielectric medium.
2. Capacitors, typically, charge and discharge faster than batteries.
3. In general, batteries tend to hold more charge and energy than capacitors.
4. Almost all capacitors are rechargeable, while, as explained later in this chapter, not all batteries are rechargeable.
5. Capacitors can be applied in DC as well as AC systems, while batteries cannot. This is evident from the fact that capacitors can be applied to improve power factor in AC systems, but batteries cannot be applied directly to AC systems under any circumstance.
6. Because of the substantial differences in the construction, composition, and components of batteries and capacitors, the capacitors, typically, require little or no maintenance and tend to exhibit substantially longer life spans.
7. Also, because of the type of materials typically used in the construction of batteries, they tend to be notably heavier than capacitors.
8. As shown in Figure 12.1, the electrical or electronic symbols used to represent batteries and capacitors are distinctly different.

History of batteries can be traced as far back as the 18th century. Invention of battery is credited to Alessandro Volta, dating back to 1800. Volta constructed the first battery from stack of copper and zinc plates, separated by brine-soaked paper disks, which could produce a steady current for a considerable length of time. See Figure 12.2. Volta named his invention "Voltaic Pile." At the time, however, Volta did not fully recognize the fact that the voltage was due to chemical reactions, with a

Battery   Capacitor

FIGURE 12.1   The symbol (a) represents a battery. The symbol (b) represents a capacitor.

By: Luigi Chiaesa,
September 28, 2018

Alessandro Volta demonstrating his
pile to French emperor Napoleon
Bonaparte, Circa 1801

FIGURE 12.2   "Voltaic Pile," the first battery, by Alessandro Volta. Alessandro pictured demonstrating his invention to Napoleon Bonaparte.

finite life. He thought that his cells were an inexhaustible source of energy, and that the corrosion effects he noted at the electrodes were a mere nuisance. It was Michael Faraday who, later in 1834, proved that charge and energy held by the battery was indeed limited and exhaustible.

Ever since the invention of the first battery, scientists and engineers have relied on its unique and valuable, portability and energy storage, characteristics. Over the last two centuries, since the invention of the first battery, scientists and engineers have focused on the size, portability, mass, capacity, safe charging, and discharging characteristics of batteries. Today, in the 21st century, we find batteries to be vital, integral, and indispensable part of our daily lives. Batteries play a crucial role in our phones, computers, watches, light sources, kitchen appurtenances, cordless tools, health care equipment, our automobiles, aircraft, and spacecrafts. Batteries are a vital link in making renewable energy sources more practical, viable, and versatile. As the 21st-century all-electric automobiles – also known as EVs, or electric vehicles – have already proven, battery-powered electric motors cannot just replace the environmentally unfriendly, depletable fuel-dependent, internal combustion engine, they can, in many ways, *outperform* it. The key component – and sometimes a significant constraint – in EVs, and many other types of portable or transportable electric equipment, is the battery.

Therefore, to complete the introduction of engineers and non-engineers to electrical engineering, in this chapter, we will explore basic yet important facets of battery and its sustainable and safe operation. We will introduce the reader to the different types of commercially available batteries. We will examine the construction of common batteries, the charging and discharging characteristics of secondary batteries, influence of temperature on the performance of batteries, and more. While a significant portion of this chapter is devoted to fundamentals of battery technology, in this chapter, the reader will get an opportunity to learn about the not so obvious charging and discharging performance characteristics of batteries that can be determining factors in their selection, application, and optimal performance. In this text, in-depth discussion of the chemistry, charging, and discharging characteristics of batteries will focus, mainly, on the common lead–acid (SLI) battery.

As is the case with all of the electrical engineering topics introduced in the earlier chapters of this text, battery technology-related engineering and physics concepts, principles, and quantitative analytical methods will be explained and illustrated with the help of pictures, diagrams, and sample analytical problems. As before, the reader will have the opportunity to reinforce their learning by attempting the self-assessment problems and questions at the end of this chapter. The solutions to the self-assessment problems are made available in Appendix A.

BATTERY CLASSIFICATIONS

Overall, contemporary batteries can be classified into two major categories:

- Primary batteries
- Secondary batteries

Within these two major classifications listed above, batteries can be specified or classified into subcategories. The subcategories differ based on the following attributes:

- Construction, chemistry or chemical composition, maintenance-free or non-maintenance-free characteristics, etc.
- Physical size, weight, volume, and derived ratios, i.e. specific energy, specific power, and energy density
- Voltage, Amp-hour (energy) capacity, C-rating
- Discharging characteristics
- Charging characteristics.

A few commercially available primary and secondary batteries are depicted in Figure 12.3. Their voltages, energy (Amp-hour) ratings at specific discharge current levels are listed in latter sections of this chapter.

PRIMARY BATTERIES

Most **primary** batteries are of dry cell construction. With the exception of some especially designed rechargeable alkaline batteries, most primary batteries are not rechargeable. Most primary batteries are typically discarded when discharged.

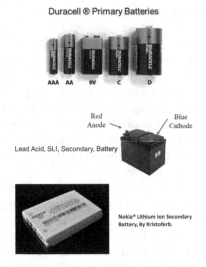

FIGURE 12.3   Example of common primary batteries.

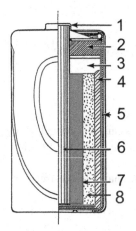

1. Brass cap or the (+) terminal, connected to the carbon rod

2. Plastic seal

3. Expansion space

4. Porous cardboard

5. Zinc can, or the (-) electrode, with high concentration of electrons

6. Inert center carbon rod and black carbon paste constitute the (+) electrode system

7. Manganese dioxide

8. Chemical mixture, zinc-chloride or ammonium chloride

FIGURE 12.4   Typical dry cell primary battery construction.

Therefore, primary batteries are sometimes referred to as "One-cycle" batteries. Primary batteries can be subdivided into two major categories:

- Zinc–carbon or zinc–manganese dry cell batteries
- Alkaline batteries or alkaline–manganese dry cell batteries

Cross-section of a typical zinc–carbon or zinc–manganese dry cell primary battery is shown in Figure 12.4.

Alkaline batteries are referred to as "alkaline" batteries because of the fact that the chemically active substance – the electrolyte – in these batteries is an alkaline, or a base, as compared to the acidic electrolyte used in zinc–carbon or zinc–manganese primary dry cell batteries. This chemically active substance, or electrolyte, in alkaline batteries is typically potassium hydroxide instead of the acidic ammonium chloride, or zinc chloride electrolyte, used in zinc–carbon primary batteries. Some alkaline primary batteries **are** rechargeable.

SECONDARY BATTERIES

Secondary batteries are batteries that generally **can** be recharged. Albeit, the number of discharge–recharge cycles – before the battery is retired – are finite and do vary based on battery types and depth of discharge (DOD). The concept of DOD is explained later in this chapter.

Some of the more commonly used, commercially available, secondary batteries are as follows:

- Li–ion, or lithium ion
- Ni–Cad, or nickle cadmium
- Lead–acid

Li–Ion Battery

The lithium–ion battery is sometimes abbreviated as LIB. Li–ion is a general term for several different lithium-battery chemistries that are used for different applications, i.e. LCO, LMO, NMC, LFP, LTO, and LNA. In lithium–ion batteries, lithium ions (Li^+) move through the electrolyte from the negative electrode to the positive electrode while the battery is being discharged or in use. However, when a Li–ion battery is being charged, the Li^+ ions move from the positive electrode to the negative electrode. This migration of Li^+ ions to the negative electrode, during the charging phase, eventually results in saturation of the negative electrode with positive charge and positive potential, which, when connected to load, can drive holes (electrons in the opposite direction) through the load, thus comprising the flow of electric current.

Lithium–ion batteries are commonly used as rechargeable batteries in consumer electronics and other common electric applications. Some of the strengths of Li–ion batteries are as follows:

- They offer high energy density (Wh/L, or Watt-hours per liter of volume). The concept of energy density is explained later in this chapter. For now, note that higher energy intensities are more desirable, in that, high energy density batteries deliver more energy while occupying less space. Note that a battery that contains more energy can perform more work before it needs recharging. In other words, a high energy density will operate those consumer electronics longer between charges while occupying less space, thus allowing those consumer electronics to be more compact in size.
- The memory effect is insignificant. In other words, the recharging performance of Li–ion battery is relatively independent of the magnitude and profile of discharging.

- They exhibit low self-discharge. Self-discharge is the non-work producing energy discharge that occurs while the battery is not in use.

Because of the strengths cited above – in addition to their popularity among consumer electronics – Li–ion batteries show a strong demand in the following applications:

- Military sector
- All-electric and hybrid automobiles
- Aerospace applications
- Energy storage, for base load needs, in renewable energy systems, i.e. wind turbines, PV farms, etc.
- Cordless tools
- Consumer electronics, i.e. cell phones, cameras, laptop PCs
- As replacements for the lead–acid batteries in general applications that have, traditionally, been dominated by lead–acid batteries, i.e. golf carts, scooters, ATVs (All Terrain Vehicles), mass transit equipment, like buses, etc.

Lithium–ion batteries can be dangerous and can pose a safety hazard under certain circumstances due to the fact that they contain flammable electrolyte under pressure. Therefore, the testing standards for Li–ion batteries are more stringent than those for acid–electrolyte, or lead–acid batteries, tests that span over a broad range of operating conditions, specific to the type and construction of batteries. This is in response to reported accidents and failures which have resulted in battery-related recalls by some battery manufacturers.

Major Li–ion batteries, in use today, are listed below:

- **LiFePO$_4$**: Also known as lithium phosphate battery, or LFP battery, with "LFP" standing for "lithium ferro-phosphate". LiFePO$_4$ batteries have a nearly constant discharge voltage. In other words, the voltage stays close to 3.2 V during discharge until the cell is exhausted. This permits the LFP cell to deliver virtually full power until it is discharged. This attribute greatly simplifies or even eliminates the need for voltage regulation circuitry. The nominal 3.2 V output of an LFP cell allows four cells to be placed in series for a nominal voltage of 12.8 V, which is close to the nominal voltage of a six-cell lead–acid battery. This advantage, coupled with good safety characteristics of LFP batteries, makes them a good potential replacement for lead–acid batteries. On the other hand, LFP batteries have a somewhat lower energy density than the more common LiCoO$_2$ design found in consumer electronics, but the LFP does offer longer lifetimes, better power density, expressed in W/L, or Watts per liter of overall battery volume. LiFePO$_4$ is a natural mineral of the olivine family (triphylite) and is used as a cathode material. Because of LFP batteries' low cost, non-toxicity, the natural abundance of iron, its excellent thermal stability, its safety characteristics, electrochemical performance, and specific energy capacity (170 mA·h/g), it has a respectable market share and acceptance. LFP's key barrier to initial commercialization was its intrinsically low electrical conductivity. This problem has, to a great extent, been overcome by reducing the particle size and the coating of the $LiFePO_4$ particles with conductive materials such as carbon. LFP batteries are now being used in industrial products by major corporations including Black and Decker's DeWalt brand, the Fisker Karma, Daimler, Cessna, and BAE Systems.

- **Li–Po**: Also known as lithium–polymer battery. The difference between Li–Po batteries and other Li–ion batteries is in the type of electrolyte used. In lithium–ion polymer batteries, a solid polymer electrolyte is used. This polymer may be a solid or a semi-solid gel. While, other Li–ion batteries consist of liquid electrolytes.

- **Li-Mn$_2$O$_4$**: Also known as lithium manganese oxide battery. In $Li\text{-}Mn_2O_4$ batteries, manganese oxide is used as cathode material. The $Li\text{-}Mn_2O_4$ architecture forms a three-dimensional "spinel" structure that improves ion flow on the electrode. This results in lower internal resistance and improved current flow. Another advantage of spinel is high thermal stability and enhanced safety. However, the cycle and operating life of $Li\text{-}Mn_2O_4$ batteries are relatively limited. The low internal cell resistance of lithium manganese oxide battery enables fast charging and discharging. It is also possible to apply 1-second load pulses of up to 50 A with $Li\text{-}Mn_2O_4$ battery. Continuous high load or current in manganese oxide battery can cause heat buildup. An operating – discharging or charging – temperature of 80°C (176°F) or higher can result in catastrophic failure of this battery. Li-manganese is used in power tools, medical instruments, as well as hybrid vehicles and EVs.

- **Li-CoO$_2$**: Also known as lithium cobalt battery. $Li\text{-}CoO_2$ battery's high specific energy makes it the popular choice for mobile phones, laptops, and digital cameras because of its high specific energy (Wh/kg). The $Li\text{-}CoO_2$ battery consists of a cobalt oxide cathode and a graphite carbon anode. The cathode has a layered structure and during discharge, lithium ions move from the anode to the cathode. The flow of lithium ions reverses on charge. The drawback of Li-cobalt is a relatively short life span, low thermal stability, and limited load capabilities. Nominal cell voltage of lithium cobalt battery is 3.7 V.

Lead–Acid or SLI Battery

Lead–acid battery is also known as the "SLI" battery. The name SLI stands for Starter, Light, and Ignition. The lead–acid or SLI battery is designed to deliver an instantaneous burst of energy to loads such as the motor starter of an automobile. Once the internal combustion engine of the automobile starts, the alternator – or the onboard DC generator of the automobile – assumes the task of recharging the lead–acid battery and also provides DC power and energy to all onboard electrical and electronic components.

The SLI battery is designed with a high density of thin electrode plates resulting in a large electrode surface area for fast electrochemical reaction to provide the needed instantaneous burst of energy. This distinctive design of the SLI battery is also conducive to faster charging characteristics. However, this distinctive property of the SLI battery does carry with it the vulnerability of this design to deep and permanent discharging.

Lead–acid batteries can be divided into two major categories based on their construction and application:

- Deep cycle battery
- Regular SLI battery

Deep Cycle Lead–Acid Battery versus Regular SLI Battery

The deep cycle battery is designed for endurance, longer and deeper discharge cycles without substantial depreciation of life. Since deep cycle batteries are designed to work reliably even when regular charging is not viable, they offer a better charge or energy storage alternative for renewable energy storage applications, as compared to the regular SLI batteries. From construction perspective, the key distinction between the deep cycle battery and the SLI battery is that the former consists of thicker cell or electrode plates. The thicker battery plates resist corrosion during extended charge and discharge cycles.

Deep cycle lead–acid batteries generally fall into two distinct categories:

- Valve-regulated lead acid (VRLA)
- Flooded (FLA)

VRLA Battery: The acronym VRLA stands for valve-regulated lead–acid battery The VRLA battery is also known as a sealed lead acid (SLA) battery or the maintenance-free battery. Even though the VRLA batteries do not require constant maintenance, their characterization as "Maintenance Free" is somewhat of a misnomer. The VRLA batteries still require cleaning and regular functional testing. The VRLA batteries are widely used in large portable electrical devices and off-grid power systems, where large amounts of storage are needed. The VRLA batteries, in such applications, in some cases, offer a lower cost alternative as compared to the Li–ion batteries. As implicit from its classification in the secondary battery category, the VRLA or SLA battery is rechargeable. The VRLA or SLA batteries can be sub-classified further as:

- Sealed VR wet cell batteries
- AGM batteries
- Gel batteries

AGM (absorbed glass mat) batteries feature fiberglass mesh between the battery plates which serves to contain the electrolyte. Gel cells add silica dust to the electrolyte, forming a thick putty-type substance with the consistency of a gel. The gel cells or gel batteries are sometimes referred to as "silicone batteries." Both designs offer advantages and disadvantages compared to conventional batteries, the sealed VR wet cells, as well as each other.

VRLA gel and AGM batteries offer several advantages compared with VRLA-flooded wet cell lead acid and conventional standard lead–acid batteries. Because the gel and AGM batteries don't contain low viscosity liquids, they can be mounted in any position, since the valves only operate in response to over-pressure conditions. Gel and AGM battery chemical system is designed to be recombinant with little or no emission of gases in case of overcharge. Therefore, when AGM and gel batteries are applied, room ventilation requirements are reduced as no acid fume is emitted during normal operation. In reality, even with flooded cell applications, gas emissions are of little consequence in all but the smallest confined areas, and these wet cell batteries pose little threat to residential users, who can benefit from the durability and the lower cost, in $s per kWh. Other advantages of gel cells or gel batteries are as follows:

- In case of damage, with AGMs or gel cells, the volume of free electrolyte that could be released would be significantly less than the potential release from a damaged flooded (wet) lead–acid battery.
- The AGM or gel cell batteries require little or no maintenance. While with flooded lead–acid batteries, the level of electrolyte needs to checked and replenished with distilled water, periodically. However, note that the wet cell batteries can be maintained by automatic or semi-automatic watering systems. The need for watering is dictated often by overcharging.

CONSTRUCTION, COMPOSITION, AND OPERATION OF A COMMON LEAD–ACID BATTERY

General construction of a common lead–acid battery is depicted in Figure 12.5. The term "flooded" is used in the flooded lead–acid batteries because in this type of battery the electrode plates are completely submerged in the electrolyte. The electrolyte level must be maintained above the tops of plates. The minimum electrolyte level is marked on the battery case. As shown by the charging phase reaction in Eq. 12.6, some of the electrolyte water is decomposed and depleted during the charging cycles. Therefore, regular maintenance of flooded batteries requires inspection of electrolyte level and addition of water. Major modes of failure of deep cycle batteries are loss of the active material due to the decomposition and shedding of the plates and the corrosion of internal grid that supports the active material. The limiting factor with capacity of a deep cycle battery is the decline in the concentration of the electrolyte. Because the deep cycle battery electrode plate mass is substantial, electrode plate attrition is seldom the cause of its failure.

As shown in Figure 12.5, the components actively participating in the electrochemical reaction of a lead acid battery are:

- Cathode constructed from solid lead peroxide or PbO_2.
- Anode constructed from relatively porous or "spongy" lead or Pb.
- Aqueous solution of sulfuric acid, H_2SO_4.

The nominal voltage in an individual lead–acid battery cell is 2.1 V.

A charged battery represents a state of chemical "imbalance." In the charged state, the cathode consists of lead peroxide, PbO_2, and the anode consists of "spongy" lead, Pb. These two electrodes – cathode and anode – are immersed in an aqueous solution

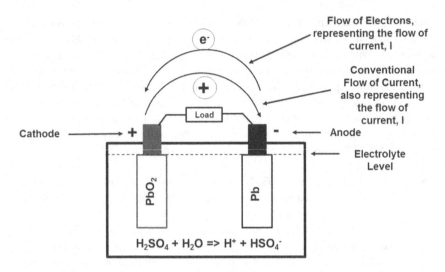

FIGURE 12.5 Lead–acid battery and the flow of current during battery discharging phase.

of sulfuric acid or H_2SO_4. The aqueous solution of sulfuric acid disperses into positively charged H^+ "cations" and negatively charged HSO_4^- anions.

During the discharging mode of a lead–acid battery, the electrochemical reaction that occurs at the lead (negative) terminal or anode is as follows:

$$HSO_4^- + Pb \rightarrow PbSO_4 + H^+ + e^- \tag{12.1}$$

Furthermore, during the discharge mode, the electrons released through the electrochemical reaction at the anode travel through the closed electrical circuit, as shown in Figure 12.4, thus establishing the flow of current through the load. These electrons arrive at the cathode and engage in the electrochemical reaction stated in Eq. 12.3. The flow of current through the load delivers energy to the load, which can be converted to work or simply heat, in accordance with Eq. 12.2.

$$\text{Energy} = \text{Work} = W = I^2 \times R \times t \tag{12.2}$$

where
  W = Work or energy can be measured in N-m, or Joules, and can be converted
    into ft-lbf or BTUs.
  R = Resistance offered by the load.
  t = Time, measured in seconds, representing the time span over which the battery is used or duration of discharge phase of the battery.

The electrochemical reaction that occurs during the discharging mode of a lead–acid battery, at the lead (negative) terminal or anode, is shown in Eq. 12.3.

$$HSO_4^- + PbO_2 + 3H^+ + e^- \rightarrow PbSO_4 + 2H_2O \tag{12.3}$$

Overall, combined, reaction at the two electrodes:

$$2HSO_4^- + Pb + PbO_2 + 2H^+ \rightarrow 2PbSO_4 + 2H_2O + Energy \qquad (12.4)$$

During the discharging mode of the battery, and as a result of the electrochemical reactions shown in Eqs. 12.1 and 12.3, the amount of H_2SO_4 is gradually depleted and replaced by H_2O; the lead peroxide, PbO_2, cathode, and the lead, Pb, anode transform progressively into lead sulfate, $PbSO_4$. In other words, the chemical differential between the anode and the cathode diminishes or neutralizes. This reduction of chemical differential is characteristic of the discharging of the battery and loss of energy.

Because the concentration of H_2SO_4 is depleted and replaced by H_2O during the discharging process of a battery, one method for assessing the level of charge or energy left in the battery consists of testing the concentration and specific gravity of the electrolyte in the battery fluid by using a hydrometer. In other words, the state of charge, or SOC, of a lead–acid battery can be measured by testing the specific gravity of the battery fluid. The specific gravity of a fully charged battery – with high concentration of H_2SO_4 in the electrolyte – is substantially higher than 1.0. The specific gravity of the electrolyte would be expressed in the form of Eq. 12.5.

$$\text{Specific Gravity of the Electrolyte} = \frac{\text{Density of the Electrolyte}}{\text{Density of Water}} \qquad (12.5)$$

The density of electrolyte with high concentration of H_2SO_4 is substantially higher than the density of pure H_2O, or approximately, 1.3. As substantial concentration of H_2SO_4 is replaced by H_2O during the discharging phase, the specific gravity of the electrolyte approaches 1.1.

> It is important to note that since such hydrometer-based testing for the level of charge left in the battery requires opening of each cell, and the hazard of exposure to a strong acid, the voltage measurement method – especially under load – is preferred.

Since a lead–acid battery is a secondary battery and it, of course, can be recharged, recharging of a lead acid battery, essentially, involves a reaction that is the reverse of the discharging reaction. The charging electrochemical reaction is shown in Eq. 12.6. The charging phase is also illustrated in Figure 12.6.

$$2PbSO_4 + 2H_2O + Energy \rightarrow 2HSO_4^- + Pb + PbO_2 + 2H^+ \qquad (12.6)$$

During the charging phase, electrons are driven by the battery charger into the terminal labeled "Anode." In order to avoid confusion, it is best to ignore the polarity of the battery terminals at this point in time and to focus attention on the dynamics of the charge flow, the current flow, and the electrochemical reaction shown in Eq. 12.6. As apparent from the chemical reaction in Eq. 12.6, during the charging phase, the battery charger pumps electrons – and, therefore, energy – into the battery via the "anode." This injection of electrons at the "anode," the positive holes at the "cathode," and energy from the battery charging system result in the restoration of the anode to lead (Pb) and the cathode to lead per oxide (PbO_2). So, in essence, the

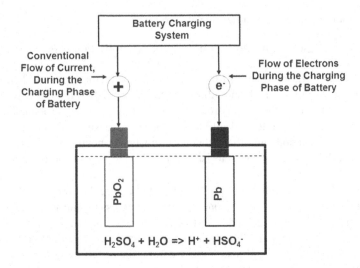

FIGURE 12.6   Lead–acid battery and the flow of current during battery charging phase.

charging process reconstitutes the electrochemical differential between the cathode and the anode as the battery is fully charged.

From battery sustainability point of view, it is important to note that when lead–acid batteries are fully discharged and kept in the discharged state for prolonged periods of time, the lead sulfate ($PbSO_4$) compound formed on the electrodes (anode and cathode) crystallizes and hardens. An attempt to recharge a battery in such a state is often unsuccessful.

Distinctive discharging and charging characteristics of batteries lead to two major subcategories within the lead–acid secondary class of batteries, namely, the SLI battery and the deep cycle battery.

Having introduced some of the major families of batteries, and their general characteristics, let's examine some of the key parameters that are used to compare the performance of various batteries, quantitatively.

SPECIFIC ENERGY

Specific energy is energy expressed in per unit mass basis. Therefore, the units for assessment of specific energy of batteries would be a conventional unit of energy per unit mass. Common unit for quantification of electrical energy is watt-hours, or Wh, often expanded to kilowatt-hours, or kWh, Megawatt-hours, or MWh, etc. Common unit for specification of battery-specific energy is Wh/kg. Specific energy is also referred to as the "gravimetric energy density," not to be confused with "energy density" – which is measured in Wh/L, as discussed later.

Examples of specific energy:

- **Li-Po** or lithium polymer cell battery's specific energy is approximately 180 Wh/kg. This specification can, conversely, be interpreted as 5.6 kg mass of battery per 1 kWh of energy capacity of the battery.

- **LiFePO$_4$** or lithium phosphate cell battery's specific energy is approximately 100 Wh/kg. Conversely, it could be said that a lithium phosphate cell battery with an energy capacity of 1 kWh would have a mass of 10 kg.
- **Pb-H$_2$SO$_4$** or lead–acid secondary battery's specific energy is approximately 40 Wh/kg. Conversely, it could be stated that a lead–acid battery with an energy capacity of 1 kWh would have a mass of 25 kg.

Because energy delivered by a battery, in Wh (or kWh, MWh, etc.) – as explained later in the discussion of the "C" ratings of batteries – can vary based on the rate of discharge, the specific energy of a battery is not a constant or permanent characteristic.

ENERGY DENSITY

Energy density is energy expressed on per unit volume basis. Therefore, the units for assessment of energy density of batteries would be a conventional unit of electrical energy per unit volume. Therefore, the units for energy density would be Watt-hours per Liter, or Wh/L, or kWh/L, etc.

Examples of energy density:

- **Li-Po** or lithium polymer cell battery's energy density is, approximately, 440 Wh/L. Conversely, a Li-Po would occupy 2.3 L of volumetric space for each kWh of energy capacity.
- **LiFePO$_4$** or lithium phosphate cell battery's energy density is, approximately, 220 Wh/L. Conversely, a lithium polymer cell battery capable of holding and delivering 1 kWh of electrical energy would occupy a volume of approximately 4.6 L.
- **Pb-H$_2$SO$_4$** or lead–acid secondary battery's energy density is, approximately, 85 Wh/L Conversely, a lithium polymer cell battery capable of holding and delivering 1 kWh of electrical energy would occupy a volume of approximately 11.8 L.

It is important to note that just as specific energy of a battery can vary with non-nominal, or abnormal demand, so can energy density, due to that fact that energy delivered by a battery, in Watt-hours, can vary based on the rate of discharge. Therefore, the energy density of a battery is not a constant or permanent fixed characteristic. In addition to specific energy and energy density, another important performance parameter of batteries is specific power.

SPECIFIC POWER

Specific power is power (in Watts) supplied by the battery, expressed on per unit mass (kg) basis. Therefore, the units for specific power would be Watts per kilogram, or W/kg.

Examples of specific power:

- **Li-Po** or lithium polymer cell battery's specific power is, approximately, 300 W/kg. Conversely, a lithium polymer battery capable of delivering 1 kW of power would have a mass of 3.33 kg.

- **LiFePO$_4$** or lithium iron phosphate cell battery's specific power is, approximately, 240 W/kg. Conversely, a lithium iron phosphate battery capable of delivering 1 kW of power would have a mass of 4.17 kg.
- **Pb-H$_2$SO$_4$** or lead–acid secondary battery's specific power is, approximately, 180 W/kg. Conversely, a lead–acid battery capable of delivering 1 kW of power would have a mass of 5.55 kg.

COMPARISON OF ELECTRICAL ENERGY AND CHARGE STORAGE SYSTEMS

Electrical charge or energy storage devices are not limited to batteries. Other practical and comparable electrical charge and energy storage devices include fuel cells and capacitors. Comprehensive discussion on the fuel cells is not included in this text. However, to facilitate a brief comparison between various, mainstream, energy storage systems, fuels cells can be defined as systems that, similar to batteries, convert energy from chemical form to electrical form. The chemical substance being hydrocarbon fuel, such as methane, or CH$_4$ and oxygen or air. In a methane (or alkane)-based fuel cell, the fuel is "reacted" with oxygen in the air and steam to produce ions that move between two electrodes, the anode and the cathode. As we learned earlier in this text, movement of charge or charged particles constitutes current. The current in the case of fuel cells is DC or direct current. The DC is then converted to AC through an inverter. Of course, unlike the electrolyte in batteries, in fuel cells the key source of energy is the hydrocarbon fuel. In addition, unlike the electrolyte in batteries, the hydrocarbon fuel in the fuel cells is consumed and must be replenished. Like the recharging phenomenon in secondary batteries, the replenishment of hydrocarbon fuel serves as a supply of energy, storage of energy, and subsequent conversion and availability of energy in electrical form.

Among capacitors, as energy storage systems, there are double-layered capacitors, ultra-capacitors, and aluminum-electrolytic capacitors. The electrical energy storage in a capacitor, as introduced earlier in this text, can be quantified as $\frac{1}{2}CV^2$, where "C" is the capacitance of the capacitor, measured in Farads, and "V' is the voltage of the capacitor, in volts. Having introduced the mainstream contemporary batteries, with prior introduction to capacitors, and the brief introduction to fuel cells above, let's examine Figure 12.7 to delineate the differences, strengths, and weaknesses between these various charge storage devices. The chart depicted in Figure 12.7 contrasts the charge or energy storage devices discussed in this chapter on the basis of their specific energy (in Wh/kg) and specific power (in W/kg).

ENERGY OR CHARGE STORAGE DEVICE COMPARISON – SPECIFIC ENERGY PERSPECTIVE

From **specific energy** (kWh/kg) perspective, the charge storage devices compared in Figure 12.7 can be ranked in descending specific energy capacity as follows:

1. Fuel cells
2. Batteries: lead–acid, Ni–Cad, and Li–ion

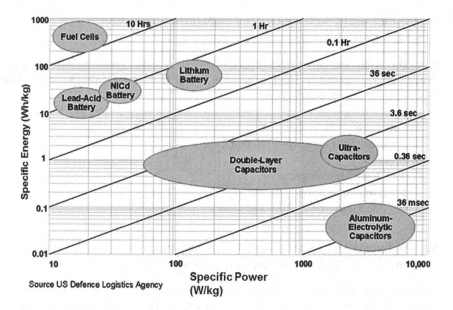

FIGURE 12.7   Specific power and specific energy comparison of common electrical energy storage devices.

3. Capacitors: double-layer capacitors, ultra-capacitors, and aluminum-electrolytic capacitors

ENERGY OR CHARGE STORAGE DEVICE COMPARISON – SPECIFIC POWER PERSPECTIVE

From specific power (W/kg) perspective, the charge storage devices compared in Figure 12.7 can be ranked in descending specific energy capacity as follows:

1. Capacitors:
    a. Aluminum-electrolytic capacitors
    b. Ultra-capacitors
    c. Double-layer capacitors
2. Batteries:
    a. Li–ion
    b. Ni–Cad
    c. Lead–acid
3. Fuel cells

Careful reflection on the information depicted in Figure 12.7 tells us that for **fast charging and discharging** applications, capacitors would be more suitable than batteries or fuel cells. On the other hand, Figure 12.7 also shows us that for **greater energy storage capacity** perspective, batteries would outperform capacitors – one would need to construct much larger and more massive capacitors to store the same amount of energy.

ENERGY OR CHARGE STORAGE DEVICE COMPARISON – DURATION OF ENERGY SUPPLY AND STORAGE PERSPECTIVE

From **Duration** of energy supply (hours, minutes, seconds, or milliseconds) perspective, the charge storage devices compared in Figure 12.7 can be ranked in descending energy supply duration as follows:

1. Fuel cells
2. Batteries: lead–acid, Ni–Cad, and Li–ion
3. Capacitors:
   a. Double-layer capacitors
   b. Ultra-capacitors
   c. Aluminum-electrolytic capacitors

SPECIFIC ENERGY AND ENERGY DENSITY COMPARISON BETWEEN LEAD–ACID AND OTHER CONTEMPORARY BATTERIES

Having compared the specific energy and specific power characteristics of major energy and charge storage devices, let's narrow or focus to the specific energy and energy density characteristics of just batteries. In order to compare the specific energy and energy density characteristics of batteries, let's examine Figure 12.8. The chart depicted in Figure 12.8 compares: lead–acid, Ni–CAD, or nickle cadmium, NMH, or nickle metal hydride, and Li–ion batteries. The specifications, such as "cylindrical," "can," or "prismatic," relate to the physical design and construction of the various batteries. As ostensible from Figure 12.8, Li–ion, Li–Po, and Li–phosphate batteries outperform all other batteries insofar as the specific energy and energy density attributes are concerned; of course, at a price. In other words, if cost is not a determining factor and the objective is to compact the maximum amount of energy in the least amount of volume, while adding the least amount of mass or weight, then the best alternative would be one

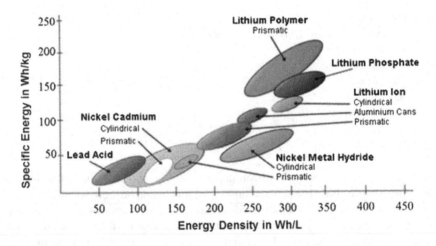

FIGURE 12.8   Specific energy and energy density of mainstream secondary batteries.

of the lithium secondary batteries. Note that, because of slight cost advantage in some cases, in some applications, and in some markets – despite the obvious disadvantages associated with low specific energy and energy density – the lead–acid batteries may be more suitable, particularly, in applications that require large amounts of energy storage. Examples of such applications include: some UPS (Uninterruptible Power Supply) systems and some renewable energy systems like photovoltaic farms, wind turbines, etc. However, even in these large battery applications, as shown in the Cost Comparison sections below, Li–ion batteries have gained ground over their lead–acid counterparts.

Example 12.1

A battery must be selected to operate onboard electronics on a NASA rover for an exploratory mission to Mars. If the energy required is 1,000 Wh (1 kWh) and the weight of the battery must not exceed 6 kg, based on specific energy considerations, which of the following battery systems would be suitable?

  a. Li-Po, or lithium polymer
  b. Lead acid
  c. $LiPO_4$

Solution/Answer: *As stated in the specific energy discussion:*

"Li-Po or lithium polymer cell battery's specific energy is approximately 180 Wh/kg. This specification, as shown below, can conversely be interpreted as 5.6 kg mass of battery per 1 kWh of energy capacity of the battery."

$$\frac{1}{180 \text{ Wh/kg}} \Rightarrow 0.00556 \text{ kg/Wh}; \therefore 1,000 \text{ Wh of energy would entail:}$$

$$\text{Mass} = (0.00556 \text{ kg/Wh}) \cdot (1,000 \text{ Wh}) \approx 5.6 \text{ kg}$$

Therefore, a Li-Po battery that offers a specific energy of 180 WH per kg of mass would be the best choice because an Li-Po battery capable of supplying 1,000 Wh, or 1 kWh, of energy and would at a mass less than 6 kg.
  Note: Based on charts in Figures 12.7 and 12.8, and other information provided in this chapter, a lead–acid battery system would have a mass of, approximately, 25 kg and a $LiPO_4$ battery would have a mass of, approximately, 10 kg. In addition, more important from the physical application point of view, because this problem entails battery application in a spacecraft – which is designed to undergo three-dimensional motion – the lead–acid battery would not offer a viable alternative.

COST COMPARISON OF MAJOR SECONDARY BATTERIES

Cost of major secondary, rechargeable, batteries has continued to trend down over the last two decades. Table 12.1 compares estimated cost of mainstream rechargeable batteries commonly used in consumer electronics, tools, and electric automobiles – also referred to as EVs. Car maker GM® revealed at their 2015 Annual Global Business Conference that they expected the price of EV (Li–ion cell) batteries to drop to

TABLE 12.1

Mainstream Battery Cost Estimate in $/kWh

Battery Type	Year	Cost ($/kWh)
Li–ion	2016	145
Li–ion	2014	200–300
Li–ion	2012	500–600
Nickel metal hydride	2004	750
Nickel metal hydride	2013	500–550
Lead–acid	2013	256.68

$145/kWh in 2016 – substantially lower than other analyst's earlier cost estimates. GM also expects a cost of US $100/kWh by the end of 2021. However, the reader is advised to bear in mind that cost of batteries – or energy storage devices, in general, especially, those that have not reached technological maturity – is volatile, dynamic, and in state of flux. Therefore, the only most current and verifiable costs should be used in formal work. The cost data provided herein is for relative comparison purposes only.

UNIVERSAL SYMBOL FOR BATTERY

Most batteries consist of a stack of cells connected in series. Therefore, as shown in Figure 12.9, the universal battery symbol depicts a stack of cells in a battery. From another perspective, the individual cells within the overall battery are the "building blocks" of the battery. The more the number of cells in a battery, the more the overall voltage of the battery and, generally, the more the energy storage capacity. As introduced in earlier chapters of this text, and later in this chapter, the symbol for battery depicted in Figure 12.9 is used to represent a battery in electrical circuits and circuit analysis.

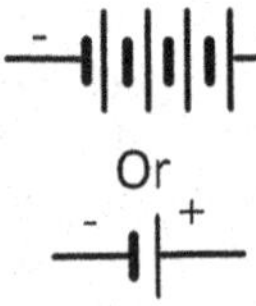

FIGURE 12.9   Standard symbol for a multi-cell battery.

As evident, application of batteries beyond the ordinary applications such as toys, remote controls watches, cell phones, and other myriad portable electronic devices is limitless and transformative, especially with the focus on renewable energy technologies and the need to cater to base load.

DISCHARGING CHARACTERISTICS OF BATTERIES

An **ideal** battery would always deliver the same amount of energy and Amp-hours regardless of the rate of discharge or the number of Amps, at a constant voltage, as depicted in Figure 12.10.

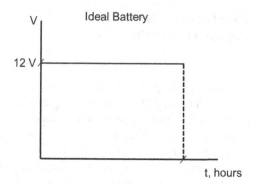

FIGURE 12.10   Discharge profile of an **ideal** battery.

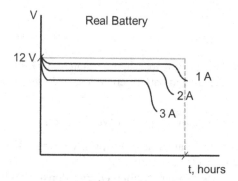

FIGURE 12.11   Discharge profile of real battery.

The performance of **real** batteries, however, could be explained more accurately through the voltage versus time graph shown in Figure 12.11. As shown in Figure 12.11, when the rate of discharge, or amperage (current), is higher, the operating voltage of the battery drops in conjunction with the decrease in the amount of energy the battery is capable of delivering before cutting off.

Depth of Discharge

DOD is an indication of the amount of energy or charge depleted. The SOC on the other hand is the amount of energy or charge left in the battery. In percent, the relationship between DOD and SOC can be expressed as:

$$DOD(\%) = 100\% - SOC(\%)$$

or,

$$SOC(\%) = 100\% - DOD(\%)$$

ADVANTAGES AND DISADVANTAGES, LITHIUM ION VERSUS LEAD ACID

CHARGING

- A lithium ion could be charged in around 1 hour versus, approximately, 8 hours for a lead–acid battery. However, charging of a lithium–ion battery would be difficult at sub-zero temperatures; its temperature would need to be raised through the use of an external heater before it can be charged.
- Lithium polymer cells, with a nominal voltage of 3.7 V, can experience catastrophic failure if charged to 4.7 V or greater. Therefore, robust protection design is required to avoid unsafe charging of lithium–ion batteries. Lead–acid batteries, in contrast, are more tolerant of overcharging. While sudden overcharging of lead–acid batteries can also be damaging, it is less likely to result in a catastrophe.

SELF-DISCHARGE RATES

- A lithium–ion battery can lose approximately 10% of its charge over a month. While, a lead–acid battery could lose as little as 5%.

SPECIFIC POWER, SPECIFIC ENERGY, AND POWER DENSITY

- A lithium–ion battery carries **higher specific energy** (Wh/kg) than a lead–acid battery; a ratio of greater than 4:1 in favor of lithium–ion battery.
- A lithium–ion battery offers **higher energy density** (Wh/L) than a lead–acid battery; a ratio of greater than 5:1, in favor of lithium–ion battery.
- A lithium–ion battery offers **higher specific power** (W/kg) than a lead–acid battery.

LIFE SPAN OF BATTERIES

Life span of a battery depends on the following factors:

- **Temperature:** Batteries operated within temperature ranges specified by the manufacturer tend to outlast those that are operated at extreme temperatures, outside the specified range.
- **Storage:** Batteries stored within temperature ranges specified by the manufacturer tend to outlast those that are operated at extreme temperatures.
- **Treatment:** Frequent and high DODs tend to shorten the life span of batteries.

C-RATING OF BATTERIES

C-rating of a battery relates the capacity of the battery to the rate of discharge. C-rating is also used to correlate the battery capacity to the charging rate. For example, C_{20} rating of a battery would be the energy capacity of the battery if it were to deliver a constant current for a period of 20 hours. The C-rating of a battery also marks the duration **after** which the battery operation must be ceased. In other words, if a C_{20} battery is not cut-off after continuous operation of 20 hours, it can be damaged permanently.

Additionally, a C_{20} battery capacity rating of 7 Ah would imply that this battery would deliver 350 mA of current over a period of 20 hours. In other words:

Current Delivered by a 7 Ah C_{20} Battery = 7 Ah / 20 h = 0.350 A or 350 mA

So, a C_{20} battery rated for nominal 7 Ah would **not** be capable of delivering 7 A for a period of 20 hours. Instead, as shown in Table 12.2, a nominal 7 Ah C_{20} battery would deliver 7 A for a period of only 1 hour. Therefore, a more complete and explicit rating of a battery should include specification of the discharge rate of the battery in addition to the "C" rating. So, as indicated in Table 12.2, a more complete specification of a 7 Ah, C_{20}, battery, being applied in a 1 A discharge mode would be: Nominal 7 Ah, 1 C_{20}. For additional clarification, let's examine the 7 Ah C_{20} battery in a 0.14 A discharge mode. A complete and proper C-rating specification of the battery, in this case, would be **7 Ah, 0.02 C_{20}**. Note that the coefficient of the C-rating represents the "Discharge Rate of the Battery," as stated in the second column of Table 12.2. From the second column in Table 12.2, we also see that when a nominal 7 Ah, C_{20} battery is discharged at only 0.14 A, it can sustain the demanded current and wattage for up to 50 hours – a period that is 50 times longer than the "20-hour" safe discharge specification of the battery. So, in this 0.14 A discharge case, the battery discharge rate – or the coefficient of C_{20} –would be 1/50 or 0.02. As apparent from further examination of Table 12.2, batteries, almost always, will last longer than the "C" rating when discharged at a rate lower than the specified "Discharge Rate" of the battery.

The C_{20} rating is more common among smaller batteries. The larger batteries, commonly used for "off-the-grid" applications, are rated C_{100}, or have C-rating of 100, and are rated based on 100-hour discharge.

TABLE 12.2

Voltage and Energy Capacity of Common Primary Batteries

Nominal Amp Rating, in Ah	Discharge Rate of a C_{20} Battery	C-Rating of the Battery, in hours	Current Draw, in A	Time to Complete Discharge or Cut-off, in hours
7	0.02	20	0.14	50
7	0.05	20	0.35	20
7	0.1	20	0.7	10
7	0.2	20	1.4	5
7	0.5	20	3.5	2
7	1	20	7	1
7	2	20	14	0.5

BATTERY DISCHARGE RATE AND NOMINAL CAPACITY

The performance of a battery, in terms of its nominal capacity, is a function of the discharge rate. Relationship of battery's nominal capacity and the rate at which it is discharged is illustrated in Figure 12.12. The graph in Figure 12.12 represents a 12 V, 7 Ah, C_{20}, battery. As apparent from the graph, the performance of the battery, or the nominal capacity of the battery, would be 100% of the rated value if it is discharged at 0.05 C_{20} rate. As the discharge rate is increased to 0.5 C_{20} and beyond, the nominal capacity and performance of the battery declines; delivering only 34.4% of rated capacity at 2 C_{20}.

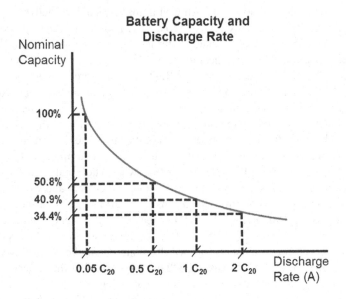

FIGURE 12.12   Battery discharge rate and its nominal capacity.

TEMPERATURE AND BATTERY DISCHARGE CHARACTERISTICS

The performance of a battery, when in use or being discharged, is influenced by temperature as shown in Figure 12.13. Figure 12.13 depicts the performance of a 12 V, 7 Ah, 0.5 C_{20} battery relative to temperature. A battery, typically, exhibits optimal 100% performance at standard temperature of approximately 20°C (68°F). The graph also shows that at a sub-zero temperature of 0°C (23°F), the capacity of the battery declines, substantially, delivering only 80% of the rated energy.

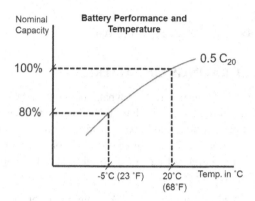

FIGURE 12.13 Discharge performance of a 12 V, 0.5 C_{20} 7 Ah battery relative to temperature.

POWER AND ENERGY PERFORMANCE ASSESSMENT OF BATTERY

The power and energy assessment of battery will be illustrated through Example 12.2, which is based on the battery performance versus temperature information stated in the previous section.

Example 12.2

The data pertaining to a 7 Ah 0.5 C_{20} battery is as follows:

Rated battery charge: 7 Ah (Amp-hours)
Rated discharge period: 20 hours (C_{20})
Discharge current, based on the 0.5 coefficient of the 0.5 C_{20}: 3.5 A
Battery voltage: 12 V

Determine the following:

a. Power and energy drawn or demanded from the battery at 20°C (68°F)
b. Power and energy drawn or demanded from the battery at −5°C (23°F)

Solution:

Based on the specifications of the 12 V, 7 Ah, 0.5 C_{20} battery stated above, and the graph depicted in Figure 12.12:

a. Power, at 20°C (68°F):
   In W: 12 V×3.5 A = 42 W

   Energy, at 20°C (68°F):

   In Ah: 3.5 A×2 h = 7 Ah
   In Wh: 12 V×3.5 A×2 h = 84 Wh

b. Power, at −5°C (68°F):
   In W: 0.8×12 V×3.5 A = 33.6 W

   Energy, at −5°C (68°F):

In Ah: 0.8×3.5 A×2 h = 5.6 Ah
In Wh: 0.8×12 V×3.5 A×2 h = 67.2 Wh

SAFE DISCHARGE RATINGS OF BATTERIES

Safe discharge ratings of batteries are somewhat analogous to the concept of "Service Factor" discussed earlier in this text, in that, if a battery is rated eight times the nominal "Ah" rate, it has been tested, under extreme circumstances, to discharge at eight times the nominal current, safely. So, in the example of nominal 7 Ah, C_{20} battery stated above, a safe discharge rating of "8" implies that the battery is capable of safely discharging at the rate of:

$$8 \times C_{20}\text{-Rating, or, } 8 \times 7 \text{ Ah nominal, or } 56 \text{ A}$$

Note that as a battery discharges, its voltage decreases slightly.

POWER AND ENERGY (AH) FORMULAS
AND NUMERICAL EXAMPLES

Example 12.3

How much does the energy capacity of a D-cell drop when the drain current is increased from 10 mA to 100 mA?

Solution:

From the Table 12.3, the capacity at 10 mA:

$$C = (10 \times 10^{-3} \text{ A}) \cdot (500 \text{ h}) = 5,000 \times 10^{-3} \text{ A-h}$$

TABLE 12.3
Voltage and Energy Capacity of Common Primary Batteries

Battery Type	Voltage Rating, in Volts	Drain Current, in mA	Service Capacity, in hours	Energy Capacity, in mAh
AAA	1.5	2	290	580
	1.5	10	45	450
	1.5	20	17	340
AA	1.5	3	350	1,050
	1.5	15	40	600
	1.5	30	15	450
C	1.5	5	430	2,150
	1.5	25	100	2,500
	1.5	50	40	2,000
D	1.5	10	500	5,000
	1.5	50	105	5,250
	1.5	100	45	4,500

The capacity at 100 mA is:

$$C = (100 \times 10^{-3} \text{ A}) \cdot (45 \text{ h}) = 4{,}500 \times 10^{-3} \text{ A-h}$$

The decrease in capacity is:

$$1 - \frac{4{,}500 \times 10^{-3} \text{ A-h}}{5{,}000 \times 10^{-3} \text{ A-h}} = 0.1$$

BATTERY EQUIVALENT CIRCUIT

Batteries possess and exhibit internal resistance which results in lower output voltage as current increases. The internal resistance of a battery is labeled as R_{int} in the battery equivalent circuit shown in Figure 12.14. The internal resistance of a battery increases with the age of the battery and the frequency and magnitudes of the DOD. This is the reason why the terminal voltage of a battery declines with the age of the battery.

Example 12.4

A 1.5 V battery with an internal resistance of 0.3 Ω is connected to a load of 1 Ω.

    a. What is the load current?
    b. What is the power dissipated in the battery resistance?
    c. What is the power delivered to the load?

Solution:

    a. Load current:

$$V = IR$$

$$I = \frac{V}{R} = \frac{V}{R_{int} + R_{load}}$$

$$= \frac{1.5 \text{ V}}{0.3\,\Omega + 1\,\Omega} = 1.15 \text{ A}$$

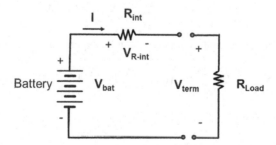

FIGURE 12.14 Battery equivalent circuit.

b. Power dissipated in the internal resistance of the battery:
   The power is determined by

$$P = I^2 R_{int}$$

$$= (1.15\ A)^2 (0.3\ \Omega) = 0.397\ W$$

c. Power delivered to the load:

$$P = I^2 \cdot R_{load} = (1.15\ A)^2 \cdot (1\ \Omega) = 1.32\ W$$

BATTERY VOLTAGE REGULATION

Voltage regulation of a battery is an indication of how much voltage drop is encountered in a battery application, or in a battery circuit, when a battery is supporting a load.

Voltage Regulation of a Battery : V_{Reg}

$$\%V_{Reg} = \frac{V_{NL} - V_{FL}}{V_{FL}} \cdot 100\% \tag{12.7}$$

In Figure 12.13, V_{NL}, the no-load voltage, would be the voltage (V_{term}) measured at the terminals before the load R_{load} is connected, with no current flowing in the circuit. The no-load voltage V_{NL} will also be the rated voltage specified on the battery, V_{bat}, i.e. 1.5 V for AAA, AA, C, and D cell primary batteries, as indicted in Table 12.3. Once the load, R_{load}, is connected into the circuit, the current "I" begins to flow and a voltage drop, V_{R-int}, develops across the internal resistor, R_{int}, of the battery. This results in a finite reduction in the terminal voltage. This lower terminal voltage with the current flowing in the circuit is V_{FL}.

Application of Kirchhoff's voltage law (KVL) in the battery circuit, underload, yields:

$$+V_{term} - V_{Bat} + V_{R-int} = 0$$

or,

$$V_{term} = V_{Bat} - V_{R-int} = V_{FL}$$

$$\text{since } V_{Bat} = V_{NL}, \quad V_{term} = V_{FL} = V_{NL} - V_{R-int}$$

Therefore, the voltage regulation equation can be expanded as follows:

$$\% V_{Reg} = \frac{V_{NL} - V_{FL}}{V_{FL}} \cdot 100\% = \frac{V_{NL} - (V_{NL} - V_{R-int})}{V_{Bat} - V_{R-int}} \cdot 100\%$$

or,

$$\% V_{Reg} = \frac{I \cdot R_{int}}{V_{Bat} - I \cdot R_{int}} \cdot 100\% \tag{12.8}$$

Example 12.5

What is the full-load voltage across the battery terminals in Example 12.4? What is the voltage regulation of the battery with the full load of 1 Ω in the circuit?

Strategy: Apply KVL in the circuit and then use Eq. 12.8 to assess the % voltage regulation.

Application of KVL in the battery circuit, underload, yields:

$+V_{term} - V_{Bat} + V_{R\text{-}int} = 0$

or,

$V_{term} = V_{Bat} - V_{R\text{-}int} = 1.5 - (I) \cdot (R_{int}) = 1.5 - (1.15\,A) \cdot (0.3)$

$\therefore V_{term} = 1.155\,V$

$\% V_{Reg} = \dfrac{I \cdot R_{int}}{V_{Bat} - I \cdot R_{int}} \cdot 100\% = \dfrac{(1.15\,A) \cdot (0.3\,\Omega)}{1.5\,V - (1.15\,A) \cdot (0.3\,\Omega)} \cdot 100\% = 29.87\%$

Note: The regulation in this example is relatively high due to the high current demand from the 1.5 V battery.

PEUKERT'S LAW

Although we have explored lead–acid battery discharging characteristics in foregoing sections of this chapter, we would be remiss if didn't discuss Peukert's law contributions toward understanding of lead–acid battery operation. Peukert's law describes the relationship between the discharge rate and the available capacity, SOC, of a lead–acid battery. According to Peukert's law, as the rate of discharge increases, the battery's available capacity decreases. Note: **Peukert's law holds true only if a battery is discharged at a constant discharge current and constant temperature.** If the exponent constant "k" was equal to "1," the delivered capacity would be independent of the current. For a lead–acid battery, the value of **k** is typically between **1.1** and **1.3**. The Peukert constant varies according to the age of the battery, generally **increasing** with age.

The approximate change in capacity of rechargeable lead–acid batteries for different **k** values, and at different rates of discharge, can be assessed using Eqs. 13.9 and 13.10.

$$C_p = I^k t, \tag{12.9}$$

$$t = H\left(\frac{C}{IH}\right)^k. \tag{12.10}$$

where
H = Rated discharge time (in hours)
C_p = The capacity at a 1 A discharge rate, expressed in Ah (ampere-hours)

C = Rated capacity at that discharge rate (in ampere-hours)
I = Actual discharge current (in amperes)
k = The Peukert constant (dimensionless)
t = Actual time to discharge the battery (in hours)

Example 12.6

A battery is rated 100 Ah, when discharged over 20 hours. The battery is connected to a circuit that draws 10 A. If Peukert's exponent of the battery is approximately 1.4, the amount of time the battery can sustain the current in the circuit is most nearly:

    A. 10 hours
    B. 20 hours
    C. 7.5 hours
    D. 25.5 hours

Solution:

Apply the following Peukart's law Eq. 12.10:

$$t = H\left(\frac{C}{IH}\right)^k$$

where
    H = Rated discharge time (in hours) = 20 hours
    C = Rated capacity at that discharge rate (in ampere-hours) = 100 Ah
    I = Actual discharge current (in amperes) = 10 A
    k = The Peukert constant (dimensionless) = 1.4
    t = Actual time to discharge the battery (in hours) = ?

$$t = H\left(\frac{C}{IH}\right)^k = 20\left(\frac{100}{(10)\cdot(20)}\right)^{(1.4)} = 7.58 \text{ hours}$$

Answer: **(C)**

Peukert's exponent can be determined algebraically. Peukert's constants for some of the more common batteries are listed in Table 12.4.

$$\frac{Q}{Q_0} = \left(\frac{T}{T_0}\right)^{\left(\frac{k-1}{k}\right)} \tag{12.11}$$

where Q and Q_0 represent charging or discharging in times T and T_0, respectively.

TABLE 12.4

Voltage and Energy Capacity of Common Primary Batteries

α	k	Description/Comments
0	1	Full charge regardless of the applied current, i.e. an ideal battery
0.1	1.090909	VRSLAB AGM batteries
0.2	1.166667	VRSLAB AGM batteries; available with a different α
0.25	1.200000	Gelled
0.3	1.230769	Gelled; available with a different α
0.33	1.249998	Flooded lead–acid battery
0.8	1.444444	Flooded lead–acid battery; available with a different α
0.9	1.473684	Additional example of α and k for a flooded lead–acid battery
0.4	1.285714	Regular lithium–ion battery
0.5	1.333333	Diffusion control, Cottrell–Warburg
0.75	1.428571	Additional example of α and k for diffusion control, Cottrell–Warburg battery

Another commonly used the form of Peukert's law is:

$$\frac{Q}{Q_0} = \left(\frac{I}{I_0}\right)^{\alpha} \tag{12.12}$$

where exponent or constant "α" can be defined mathematically as:

$$\alpha = \left(\frac{k-1}{2-k}\right) \tag{12.13}$$

CHARGING OF SECONDARY BATTERIES

Charging of secondary batteries is a function of voltage, current, time, chemistry, and the design of the battery. In the ideal world, one would wish to charge a battery to its full capacity, in the least amount of time, safely, without negatively impacting the life of the battery. In the real world, however, one has to choose between capacity of recharge, the speed of recharge, physical constraints – i.e. size, weight, energy density, specific power, specific energy, power density, etc. – and economics.

GASSING VOLTAGE AND MAINTENANCE OF LEAD–ACID SECONDARY BATTERIES

Gassing voltage is the voltage at which electrolysis begins to electrically break down water, H_2O, into hydrogen, H_2, and Oxygen, O_2. The gassing voltage for a lead–acid battery is, approximately, 2.4 V per cell. A typical 12 V lead–acid SLI battery consists of six (6) cells, therefore, its gassing voltage is approximately 14.4 V. Because hydrogen and oxygen are highly flammable, they must be vented safely during the charging process. A certain amount of gassing is expected and it serves to circulate the electrolyte. However, excess gassing depletes the water and increases the concentration of electrolyte, H_2SO_4. Therefore, as a regular maintenance requirement, distilled water must be added to the battery, periodically, to sustain proper electrolyte concentration. This periodic maintenance requirement can be circumvented through "maintenance-free" battery design, wherein, by design, the vented hydrogen and oxygen gases are recombined to form water, thus maintaining the electrolyte concentration.

Charge rate of a battery is a function of the voltage applied and its capacity or SOC. Note that, with constant internal resistance of the battery, charging voltage determines the charging current. In other words, higher voltage would result in higher charging currents and accelerated rate of electrolysis, or gassing, at or greater than gassing voltage. Safe and optimal charging of a battery requires that the battery voltage not exceed gassing voltage for prolonged periods of time. If the battery capacity is substantially depleted, or its SOC is low, for instance 10%, it should be charged at lower voltage and low charging current to avoid prolonged gassing. If a battery is already at a high SOC, say 90%, aggressive charging would likely drive the voltage up to or beyond the 14.4 V electrolysis voltage level, resulting in higher current and subsequent gassing. It is for this reason that most battery charging systems are designed with voltage and current control systems.

A battery charging system can be designed to meet one of the following functional criteria, or a combination of both:

SINGLE-STEP CONSTANT CURRENT CHARGING

In a single-step **constant current** charging process, voltage is ramped up gradually as the battery is charged or the SOC increments from low SOC to high SOC. However, the rate of increase of the voltage, as depicted by the slope of the voltage line, in Figure 12.15 is controlled in such a fashion that the charging current remains constant at a low magnitude. While this type of charging approach is safe, it can prolong the charging or recharging time.

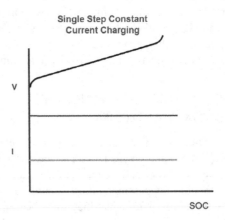

FIGURE 12.15 Voltage and current profile in single-step constant current charging process.

MULTI-STEP CONSTANT CURRENT CHARGING

The slow charging rate shortcoming of the single-step constant current charging approach can be overcome with the addition of more controls, such that a voltage is ramped up, initially, at a high rate. But, as the voltage approaches the gassing level, it is reduced or stepped down by a finite amount to lower the current level as shown in Figure 12.16. As shown in Figure 12.16, this closed-looped control approach is repeated until the battery has reached 100% SOC.

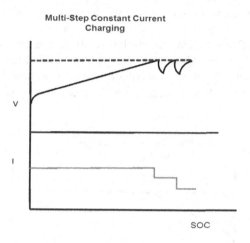

FIGURE 12.16 Voltage and current profile in multi-step constant current charging process.

CONSTANT VOLTAGE CHARGING

A constant voltage charging approach, as shown in Figure 12.17, is relatively simple and can be implemented at low cost through the incorporation of a voltage regulator. With the application of a constant voltage, at a magnitude just below the gassing voltage, the current would decline as the battery gets charged or as the SOC increases.

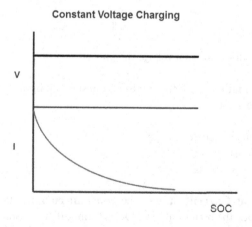

FIGURE 12.17 Voltage and current profile in multi-step constant current charging process.

However, this simplicity and cost advantage is coupled with the risk of battery drawing a large, unsafe, amount of current when it is deeply discharged or at a very low SOC level.

MODIFIED CONSTANT VOLTAGE CHARGING

Single-step constant current charging, multi-step constant current charging, and constant voltage charging approaches, in isolation, present disadvantages. However, a hybrid approach that leverages the desirable characteristics of all the three charging approaches discussed above is the approach employed by most contemporary secondary battery chargers. This approach is referred to as the "Modified Constant Voltage Charging" or "Fast Charging." The principle behind the modified constant voltage charging method is that it combines the benefits of constant voltage and constant current approaches to yield versatile and robust battery charging systems. See Figure 12.18.

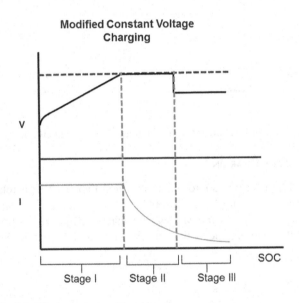

FIGURE 12.18 Modified constant voltage charging process.

As depicted in Figure 12.18, the modified constant charging process can be categorized into three stages:

- Stage I – Constant current stage
- Stage II – Constant voltage stage
- Stage III – Float voltage stage

Stage I – Constant Current Stage: The **constant current** stage pertains to the charging phase when the battery at low SOC. The modified constant voltage charging phase begins with a gradual increase in voltage, while keeping the current constant and below the gassing voltage level and at or below the maximum current allowable by the battery manufacturer. This, controlled, low level of current ensures that the battery does not overheat. This stage, typically, takes up almost **half** of the required battery charge time. This stage is also referred to the "bulk stage" because the battery receives bulk of its charge in this stage.

Stage II – Constant Voltage Stage: In this stage, the battery charger switches to **constant voltage** mode and allows the current to ramp down as the battery absorbs more and more charge. That is the reason why this stage is sometimes referred to as the "Absorption Stage," or the "Topping Charge Stage." Since the battery has already achieved a substantial magnitude in voltage when it reaches this stage, the voltage differential between the battery and the charger is much smaller and the current driven through the battery continues to decline. The battery voltage at this stage is usually just below the gassing voltage, or approximately, 14.4 V for a typical lead acid battery.

Stage III – Float Voltage Stage: The battery enters the float voltage stage when it is almost saturated with charge or energy, and it only needs a small amount of "trickle charge" to keep the battery in fully charged (100% SOC) state. In typical lead–acid batteries, at this stage, the charge switches to a constant voltage of, approximately, 13.8 V.

BATTERIES – THE FUTURE

Recent advances in nanotechnology hold a promising future for lithium ion. Egg-line nanoparticles of sulfur present an option, improving energy transfer and hugely increasing capacity. Use of silicon nanoparticles in the Li–ion battery anode is also expected to boost their performance.

Research into the overcoming current challenges with lithium–ion tech – such as reducing the tendency for lithium to gather around the battery electrodes – is underway.

Work is underway on lithium–air technology. This is an offshoot of lithium–ion batteries and one that could significantly increase energy density. In essence, at this vantage point in time, it appears that Li–ion technology will be around as viable and effective electrochemical energy storage technology for years to come.

SELF-ASSESSMENT PROBLEMS/QUESTIONS

1. What type of battery would be most suitable for the following applications?
   a. UPS, Uninterruptible Power Supply system
   b. Starting an automobile
   c. Electric camp light
   d. Remote wind turbine
   e. Remote solar PV farm
2. In Example 13.1, if the volume of space reserved for the battery is 3 L or less, based on energy density considerations, which of the given battery designs would be suitable?

    a. Li-Po, or lithium polymer

    b. Lead–acid

    c. $LiPO_4$

3. It has been disclosed that the battery being selected for the mission to Mars, as outlined in problem 2, must be capable of serving as an emergency source of power for 1 kW of ignition power without overheating. However, the mass of the battery must not exceed 3.5 kg. Based on the specific power data provided in this chapter, which of the three batteries would meet the specifications?

4. A $LiFePO_4$ cell battery is discharged at three times the nominal discharge rate. Will the energy density of the battery:

    a. Remain 220 Wh/L?

    b. Be greater than 220 Wh/L?

    c. Be less than 220 Wh/L?

    d. None of the above?

5. A Li-Po, or lithium–polymer cell battery, is discharged at half the nominal discharge rate. Will the specific energy of the battery be:

    a. 180 Wh/kg?

    b. Greater than 180 Wh/kg?

    c. Less than 180 Wh/kg?

    d. None of the above?

6. A double-layered capacitor is being considered as an energy storage device in an electrical system. The amount of initial energy needed by the system is 10 Wh. Based on the specific energy and specific power chart in Figure 12.7, the approximate weight of the double-layered capacitor would be:

    a. 1 kg

    b. 10 kg

    c. 100 kg

    d. 1,000 kg

7. If, in problem 8, the maximum weight allowed for the charge storage system is 1 kg, based on the specific energy and specific power chart in Figure 12.7, which charge storage system would be the most suitable?

    a. Lead–acid battery

    b. Double-layered capacitor

    c. Ultra-capacitor

    d. None of the above

Appendix A
Solutions for End-of-Chapter Self-Assessment Problems

This appendix includes the solutions and answers to end-of-chapter self-assessment problems and questions.

CHAPTER 1 – SOLUTIONS

1. In an AC system, a voltage source $V(t) = 120Sin(377t + 0°)$ V rms, sets up a current of $I(t) = 5Sin(377t + 45°)$ A rms. Calculate the maximum values of voltage and current in this case.

   Solution:
   According to Eq. 1.2:

   $$V_p = V_m = \sqrt{2}V_{RMS} = \sqrt{2}V_{EFF}$$

   $$\therefore \quad V_p = \sqrt{2}V_{RMS} = \sqrt{2}(115) = 163 \text{ V}$$

   According to Eq. 1.5:

   $$I_p = I_m = \sqrt{2}I_{RMS} = \sqrt{2}I_{EFF}$$

   $$\therefore \quad I_p = \sqrt{2}I_{RMS} = \sqrt{2}(5) = 7.1 \text{ A}$$

2. A phase conductor of a transmission line is 1 mile long and has a diameter of 1.5 inch. The conductor is composed of aluminum. Calculate the electrical resistance of this conductor.

   Solution:
   Solution strategy: Since the resistivity value of aluminum, as stated in Chapter 1, is in metric or SI unit system, the length and diameter specifications stated in this problem must be streamlined in metric units before application of Eq. 1.9 for determination of resistance in ohms (Ω's).
   L = 1 mile = 1,609 m
   Diameter = 1.5 inch = 0.0381 m; $\therefore$**R** = Radius = D/2 = 0.019 m
   A = Area of cross-section = $\pi \cdot R^2$ = (3.14) $(0.019)^2$ = 0.00113 m^2
   $\rho_{aluminum}$ = 28.2 n Ωm = 28.2 $\times$ 10^{-9} Ωm; given in Chapter 1

$$R = \rho \cdot \frac{L}{A} \qquad (1.9)$$

$$\therefore \quad R = \rho \cdot \frac{L}{A} = 2.8 \times 10^{-9} \left(\frac{1,609}{0.00113} \right) = 0.039 \, \Omega$$

3. What is the resistance of the following circuit as seen from the battery?

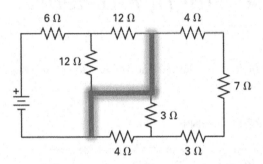

Solution:
No current will flow through the two 4 Ω resistors, the two 3 Ω resistors, or the 7 Ω resistor. The current finds the path of least resistance through the high-lighted short-circuit segment of the circuit. Therefore, the circuit reduces to one 6 Ω in series with two 12 Ω in parallel.

$$R = 6\Omega + \frac{(12\,\Omega) \cdot (12\,\Omega)}{(12\,\Omega + 12\,\Omega)}$$

$$R = 6\,\Omega + 6\,\Omega = 12\,\Omega$$

4. Consider the RC circuit shown in the diagram below. The source voltage is 12 V. The capacitor voltage is 2 V before the switch is closed. The switch is closed at t = 0. What would the capacitor voltage be at t = 5 seconds?

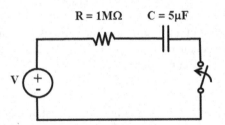

Solution:
This particular case represents a capacitor charging scenario. Given the value of **R, C, v$_c$(0)**, and the source voltage **V**, Eq. 1.18 allows us to calculate the volt-age after and elapsed time "**t**," during the capacitor charging phase.

$$v_c(t) = v_c(0)\, e^{-\frac{t}{RC}} + V \left(1 - e^{-\frac{t}{RC}} \right) \qquad (1.18)$$

Given:

$R = 1\,M\Omega = 1{,}000{,}000\,\Omega$

$C = 5\mu F = 5 \times 10^{-6}\,F$

$v_c(0) = 2\,V =$ Voltage across the capacitor at $t = 0$

$v_c(t) =$ Voltage across the capacitor at a given time $t = ?$

$V =$ Voltage of the power source $= 12\,V$

$t = 5$ seconds

$\tau = RC$ circuit time constant $= \mathbf{RC}$

$\quad = (1{,}000{,}000)(5 \times 10^{-6}) = 5$ seconds

Substitute the given and derived values in Eq. 1.18:

$$v_c(t) = v_c(0) \cdot e^{-\frac{t}{RC}} + 12 \cdot \left(1 - e^{-\frac{t}{RC}}\right)$$

$$= (2\,V) \cdot \left(e^{-\frac{5}{5}}\right) + (12) \cdot \left(1 - e^{-\frac{5}{5}}\right)$$

$$= (2\,V) \cdot (0.368) + (12) \cdot (0.632)$$

$$= 8.32\,V$$

5. Determine the equivalent capacitance for the DC circuit shown in the circuit diagram below if $C_1 = 5\,\mu F$ and $C_2 = 10\,\mu F$.

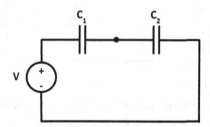

Solution:

Application of Eq. 1.21 to the two-capacitor series circuit shown in the given circuit diagram yields:

$$C_{EQ} = \frac{C_1 C_2}{C_1 + C_2}$$

$$C_{EQ} = \frac{(5 \times 10^{-6})(10 \times 10^{-6})}{(5 \times 10^{-6}) + (10 \times 10^{-6})}$$

$$= 3.33 \times 10^{-6} = 3.33\,\mu F$$

6. Determine the equivalent capacitance for the DC circuit shown below if this circuit consists of twenty 100 μF capacitors in series.

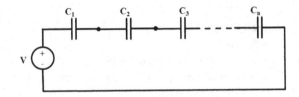

Solution:
Apply Eq. 1.22 to "n" series capacitor circuit shown in diagram:

$$C_{EQ} = C_{EQ\text{-}n} = \frac{C}{n}$$

where $n = 20$ and $C = 100\ \mu F$

$$C_{EQ} = \frac{100\ \mu F}{20}$$

$$= 5\ \mu F$$

7. Determine the equivalent capacitance in series and parallel combination circuit shown below. The capacitance values are: $C_1 = 10\ \mu F$, $C_2 = 10\ \mu F$, $C_3 = 20\ \mu F$, and $C_4 = 20\ \mu F$.

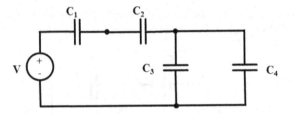

Solution:
The capacitors in this circuit that lend themselves to linear combination are C_3 and C_4. Therefore, the combined capacitance, C_{34}, would be as follows:

$$C_{34} = C_3 + C_4 = 20\ \mu F + 20\ \mu F = 40\ \mu F$$

Then, by applying Eq.1.24 to this special hybrid capacitor combination case:

$$C_{EQ} = \frac{C_1 C_2 C_{34}}{C_1 C_2 + C_2 C_{34} + C_1 C_{34}}$$

$$C_{EQ} = \frac{(10 \times 10^{-6})(10 \times 10^{-6})(40 \times 10^{-6})}{(10 \times 10^{-6})(10 \times 10^{-6}) + (10 \times 10^{-6})(40 \times 10^{-6}) + (10 \times 10^{-6})(40 \times 10^{-6})}$$

$$C_{EQ} = 4.44\ \mu F$$

8. Assume that the circuit in Problem 6 is powered by a 60 Hz AC source instead of the DC source. Determine the total capacitive reactance, X_c, seen by the AC source.

Solution:
If the DC source is replaced by an AC source, the circuit would appear as follows:

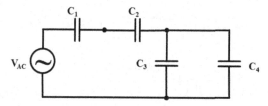

As computed in Problem 6, the combined or net capacitance contributed to the circuit by the parallel and series network of capacitors is $C_{EQ} = 4.44\ \mu F$. Then, by applying Eq.1.26:

$$X_c = \frac{1}{\omega C} = \frac{1}{2\pi f C}$$

where
f = Frequency = 60 Hz
$C_{EQ} = 4.44\ \mu F$

$$X_c = \frac{1}{\omega C} = \frac{1}{2\pi f C} = \frac{1}{2(3.14)(60)(4.44 \times 10^{-6})} = 597\,\Omega$$

9. Consider the series RL circuit shown in the diagram below. The source voltage is 12 V, $R = 10\ \Omega$, and $L = 10$ mH. The switch is closed at $t = 0$. What would be the magnitude of current flowing through this circuit at $t = 2$ ms?

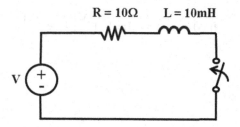

Solution:
In most series RL cases, the current value at a certain time "t" can be predicted through Eq. 1.31.

$$i_L(t) = i_R(t) = i(0)e^{-\frac{R}{L}t} + \frac{V}{R}\left(1 - e^{-\frac{R}{L}t}\right)$$

Given:
$t = 2 \times 10^{-3}$ s
$L = 10 \times 10^{-3}$ H

$$R = 10\,\Omega$$
$$V = 12\,V$$
$$i(0) = 0$$

$$i_L(t) = (0)e^{-\frac{10}{0.01}(0.002)} + \frac{12}{10}\left(1 - e^{-\frac{10}{0.01}(0.002)}\right)$$

$$i_L(t) = \frac{12}{10}\left(1 - e^{-\frac{10}{0.01}(0.002)}\right)$$

$$i_L(t) = (1.2)(1 - 0.135) = 1.04\,A$$

10. Consider the series RL circuit given in Problem 9, in discharge mode, with voltage source removed. Parameters such as $R = 10\,\Omega$ and $L = 10$ mH are the same. The switch has been closed for long period of time, such that the current has developed to the maximum or steady-state level 1.04 A. How much time would need to elapse for the current to drop to 0.5 A after the switch is opened.

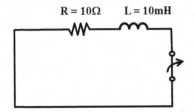

$$R = 10\Omega \qquad L = 10mH$$

Solution:
Apply series RL current equation, Eq. 1.31.

$$i_L(t) = i_R(t) = i(0)e^{-\frac{R}{L}t} + \frac{V}{R}\left(1 - e^{-\frac{R}{L}t}\right)$$

Given:
  t = ?
  $L = 10 \times 10^{-3}\,H$
  $R = 10\,\Omega$
  $V = 0$
  $i(0) = 1.04\,A$
  $i_L(t) = 0.5\,A$

$$i_L(t) = (0.5) = (1.04)e^{-\frac{10}{0.01}t} + (0)\left(1 - e^{-\frac{10}{0.01}t}\right)$$

$$0.5 = (1.04)e^{-\frac{10}{0.01}t}$$

$$0.481 = e^{-\frac{10}{0.01}t}$$

$$\ln(0.481) = \ln\left(e^{-\frac{10}{0.01}t} \right)$$

$$-0.7324 = -1,000t$$

$$t = 0.00073 \text{ s or } 0.73 \text{ ms}$$

11. Determine the equivalent inductance L_{EQ} for three parallel inductor DC circuit shown in the diagram below if $L_1 = 2$ mH and $L_2 = 5$ mH, and $L_3 = 20$ mH.

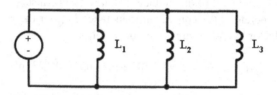

Solution:
Apply Eq. 1.36 to compute L_{EQ} for the three parallel inductor circuit.

$$L_{EQ} = \frac{L_1 L_2 L_3}{L_1 L_2 + L_2 L_3 + L_1 L_3}$$

$$= \frac{(2 \text{ mH})(5 \text{ mH})(20 \text{ mH})}{(2 \text{ mH})(5 \text{ mH}) + (5 \text{ mH})(20 \text{ mH}) + (2 \text{ mH})(20 \text{ mH})}$$

$$= 1.33 \text{ mH}$$

12. Calculate the net or total inductance as seen from the 24 V source vantage point in the circuit shown below.

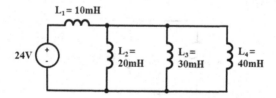

Solution:
Focus on the parallel combination of L_2, L_3, and L_4, first. Apply Eq. 1.36 to calculate the equivalent inductance L_{234} for the three parallel inductors:

$$L_{234} = \frac{L_2 L_3 L_4}{L_2 L_3 + L_3 L_4 + L_1 L_4}$$

$$= \frac{(20 \text{ mH})(30 \text{ mH})(40 \text{ mH})}{(20 \text{ mH})(30 \text{ mH}) + (30 \text{ mH})(40 \text{ mH}) + (20 \text{ mH})(40 \text{ mH})}$$

$$= 9.23 \text{ mH}$$

This reduces the circuit as shown below:

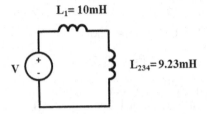

$$L_1 = 10mH$$

V

$$L_{234} = 9.23mH$$

Inductors L_1 and L_{234}, in this reduced circuit, lend themselves to a linear combination. Therefore, the equivalent inductance $\mathbf{L_{EQ}}$ for the entire parallel and series inductor hybrid circuit would be as follows:

$$\mathbf{L_{EQ}} = \mathbf{L_1} + \mathbf{L_{234}} = 10 \text{ mH} + 9.23 \text{ mH} = 19.23 \text{ mH}$$

13. Assume that the circuit in Problem 12 is powered by a 60 Hz AC source. Calculate the inductive reactance, X_L, as seen by the AC voltage source.

Solution:
If the DC source is replaced by an AC source, the circuit would appear as follows:

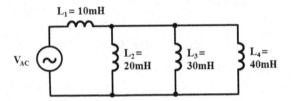

$L_1 = 10mH$

V_{AC}

$L_2 = 20mH$

$L_3 = 30mH$

$L_4 = 40mH$

L_{EQ}, as seen by the AC voltage source, is shown in the simplified equivalent circuit below:

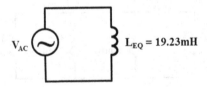

V_{AC}

$L_{EQ} = 19.23mH$

As computed in Problem 12, the combined or net inductance contributed to the circuit by the parallel and series network of inductors is $L_{EQ} = 19.23 \text{ mH}$. Then, by applying Eq. 1.37, the inductive reactance, $X_{L\text{-EQ}}$, as seen by the AC voltage source V_{AC}, would be as follows:

$$X_{L\text{-EQ}} = \omega \cdot L = (2\pi f) \cdot L_{EQ}$$

$$= 2(3.14)(60 \text{ Hz})(19.23 \text{ mH})$$

$$= 7.25 \ \Omega$$

CHAPTER 2 – SOLUTIONS

1. Determine the following for the DC circuit shown below if $R_1 = 5\ \Omega$, $R_2 = R_3 = 10\ \Omega$, and $R_4 = R_5 = 20\ \Omega$:
   a. Current flowing through resistor R_1
   b. Voltage across resistor R_5

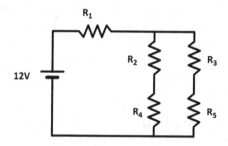

Solution:

a. R_{eq} was derived in Example 1.2 as follows:
   Combination of R_2 and $R_4 = R_{2,4} = R_2 + R_4 = 10\ \Omega + 20\ \Omega = 30\ \Omega$
   Combination of R_3 and $R_5 = R_{3,5} = R_3 + R_5 = 10\ \Omega + 20\ \Omega = 30\ \Omega$

   Combination of $R_{2,4}$ and $R_{3,5} = R_{2-5} = \dfrac{(30\ \Omega) \cdot (30\Omega)}{(30\Omega + 30\Omega)} = \dfrac{900}{60} = 15\ \Omega$

$$R_{eq} = R_1 + R_{2-5} = 5\ \Omega + 15\ \Omega = 20\ \Omega$$

   Current through R_1 would be the same as the current through the 12 V supply:

$$I = \frac{V}{R_{eq}} = \frac{12\ V}{20\ \Omega} = 0.6\ A$$

b. One method for determining V_{R5}, voltage across R_5, is to first calculate V_{R2-5}, the voltage across the combined resistance of resistances R_2, R_3, R_4, and R_5. Then, by applying voltage division, calculate V_{R5}:
   According to Ohm's law:

$$V_{R2-5} = I \cdot (R_{2-5}) = (0.6\ A) \cdot (15\ \Omega) = 9\ V$$

   Then, by applying the voltage division rule:

$$V_{R5} = (9\ V) \cdot \left(\frac{R_5}{R_5 + R_3} \right)$$

$$= (9\ V) \cdot \left(\frac{20\ \Omega}{20\ \Omega + 10\ \Omega} \right)$$

$$= (9\ V) \cdot (0.67) = 6\ V$$

2. What is the current through the 6 Ω resistor?

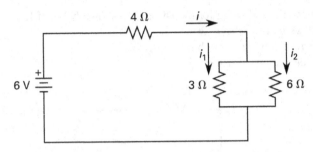

Solution:
Simplify the circuit.

3 Ω in parallel with 6 Ω = 2 Ω
2 Ω in series with 4 Ω = 6 Ω

$$i = \frac{6\text{ V}}{6\,\Omega} = 1\text{ A}$$

$R_{parallel} = 3\,\Omega$
$R_{total} = 3\,\Omega + 6\,\Omega = 9\,\Omega$

$$i = (1\text{ A})\left(\frac{3\,\Omega}{3\,\Omega + 6\,\Omega}\right) = 1/3\text{ A}$$

3. Find the current through the 0.5 Ω resistor.

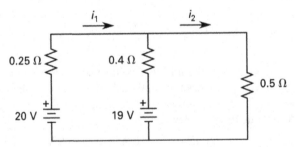

Solution:
The voltage sources around the left loop are equal to the voltage drops across
the resistances.

$$20\text{ V} - 19\text{ V} = 0.25\,\Omega i_1 + 0.4\,\Omega(i_1 - i_2)$$

The same is true for the right loop.

$$19\text{ V} = 0.4\,\Omega(i_2 - i_1) + 0.5\,\Omega i_2$$

Solve two equations and for two unknowns, using the simultaneous equation
method:

$$0.65\,\Omega i_1 - 0.4\,\Omega i_2 = 1\,\text{V}$$

$$-0.4\,\Omega i_1 + 0.9\,\Omega i_2 = 19\,\text{V}$$

$$i_1 = 20\,\text{A}$$

$$i_2 = 30\,\text{A}$$

The current through the 0.5 Ω resistor is 30 A.

4. Determine the value of currents I_1, I_2, and I_3 in the circuit shown below if the voltage source V_3 fails in short-circuit mode. The specifications of all components are listed in the table below:

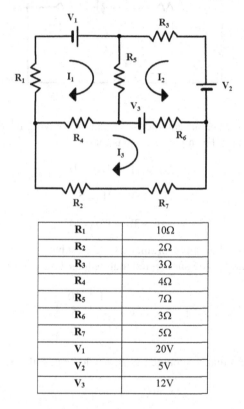

R_1	10Ω
R_2	2Ω
R_3	3Ω
R_4	4Ω
R_5	7Ω
R_6	3Ω
R_7	5Ω
V_1	20V
V_2	5V
V_3	12V

Solution:

Two noteworthy observations are in order before formulation of the three equations necessary for the derivation of the three unknown currents:

1. Even though the stated value of voltage for source V_3 is 12 V, the value used in the formulation of second and third loop equations would be 0 V because, in this scenario, voltage source V_3 is assumed to have failed in short-circuit mode.

2. As a matter of simplification, series resistors R_2 and R_7 are combined into one resistor R_{2-7}

$$R_{2-7} = R_2 + R_7 = 2\,\Omega + 5\,\Omega = 7\,\Omega$$

The revised, simplified, would then be as follows:

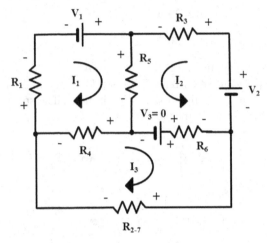

The three simultaneous equations derived by applying KVL to loops 1, 2, and 3, as described in Example 2.5, are as follows:

$$21I_1 + 7I_2 - 4I_3 = 20$$

$$7I_1 + 13I_2 + 3I_3 = 5$$

$$-4I_1 + 3I_2 + 14I_3 = 0$$

As in Example 2.5, apply Cramer's Rule to solve for the three unknown currents I_1, I_2, and I_3. The augmented matrix thus developed would be as follows:

$$\begin{vmatrix} 21 & 7 & -4 & \vdots & 20 \\ 7 & 13 & 3 & \vdots & 5 \\ -4 & 3 & 14 & \vdots & 0 \end{vmatrix}$$

The coefficient matrix, denoted as $\mathbf{A}$, would be as follows:

$$\begin{vmatrix} 21 & 7 & -4 \\ 7 & 13 & 3 \\ -4 & 3 & 14 \end{vmatrix}$$

The determinant of the coefficient matrix, denoted as $|\mathbf{A}|$, would be as follows:

$$|\mathbf{A}| = 21\{(13 \times 14) - (3 \times 3)\} - 7\{(7 \times 14) - (-4 \times 3)\}$$
$$-4\{(7 \times 3) - (-4 \times 13)\} = 2{,}571$$

The determinant of the substitutional matrix, $\mathbf{A}_1$, for determining the value of I_1 is denoted as $|\mathbf{A}_1|$, and

$$\mathbf{A}_1 = \begin{vmatrix} 20 & 7 & -4 \\ 5 & 13 & 3 \\ 0 & 3 & 14 \end{vmatrix}$$

$$|\mathbf{A}_1| = 20\{(13\times14)-(3\times3)\}-7\{(5\times14)-(0\times3)\} \\ -4\{(5\times3)-(0\times13)\}=2,910$$

The determinant of the substitutional matrix, $\mathbf{A}_2$, for determining the value of I_2 is denoted as $|\mathbf{A}_2|$, and

$$\mathbf{A}_2 = \begin{vmatrix} 21 & 20 & -4 \\ 7 & 5 & 3 \\ -4 & 0 & 14 \end{vmatrix}$$

$$|\mathbf{A}_2| = 21\{(5\times14)-(0\times3)\}-20\{(7\times14)-(-4\times3)\} \\ -4\{(7\times0)-(-4\times5)\}=-810$$

The determinant of the substitutional matrix, $\mathbf{A}_3$, for determining the value of I_3 is denoted as $|\mathbf{A}_3|$, and

$$\mathbf{A}_3 = \begin{vmatrix} 21 & 7 & 20 \\ 7 & 13 & 5 \\ -4 & 3 & 0 \end{vmatrix}$$

$$|\mathbf{A}_3| = 21\{(13\times0)-(3\times5)\}-7\{(7\times0)-(5\times-4)\} \\ +20\{(7\times3)-(-4\times13)\}=1,005$$

Applying Cramer's Rule, the unknown variables, currents I_1, I_2, and I_3, can be calculated by dividing the determinants of substitutional matrices $\mathbf{A}_1$, $\mathbf{A}_2$, and $\mathbf{A}_3$, respectively, by the determinant of the coefficient matrix $\mathbf{A}$. Therefore,

$$I_1 = \frac{|\mathbf{A}_1|}{|\mathbf{A}|} = 1.132 \text{ A}$$

$$I_2 = \frac{|\mathbf{A}_2|}{|\mathbf{A}|} = -0.315 \text{ A}$$

$$I_3 = \frac{|\mathbf{A}_3|}{|\mathbf{A}|} = 0.391 \text{ A}$$

Note: The negative sign for I_2 indicates that the counterclockwise direction assumed for this current is incorrect and that the correct direction of the flow of current in loop 2 is clockwise.

5. Use current division to determine the value of current I_1 in the circuit below:

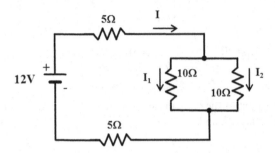

Solution:
We must determine the value of source current **I**, first. In order to determine the value of current **I** flowing through the source and the two 5 Ω resistors, we must consolidate all resistors into an equivalent resistance R_{EQ} and then apply Ohm's law.

$$R_{EQ} = 5\,\Omega + \left(\frac{(10\,\Omega)\cdot(10\,\Omega)}{(10\,\Omega)+(10\,\Omega)} \right) + 5\,\Omega = 5\,\Omega + 5\,\Omega + 5\,\Omega = 15\,\Omega$$

$$I = \frac{V}{R_{EQ}} = \frac{12\text{ V}}{15\,\Omega} = 0.8\text{ A}$$

Apply current division formula in the form of Eq. 2.7:

$$I_{10\,\Omega} = I_1 = \left(\frac{R_{parallel}}{R_{total}} \right) \cdot I = \left(\frac{10\,\Omega}{10\,\Omega+10\,\Omega} \right) \cdot (0.8\text{ A}) = 0.4\text{ A}$$

6. Using Kirchhoff's voltage law, calculate the current circulating in the series resistor network below:

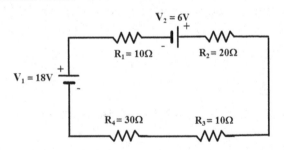

Solution:
This problem is similar to the Example 2.2, with following exceptions:

1. The two voltage sources are driving the current in the same direction, i.e. clockwise.
2. There are four resistors in series instead of three.

Using the strategy described in Example 2.2 and preparing the circuit for KVL application, the circuit would appear as follows:

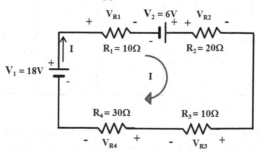

Apply Ohm's law to define the voltages, or voltage drops, across the four resistors. **Note**: since all four of the resistors are in series, we **could** combine them into a single REQ before applying KVL. However, in this case, we will keep resistors separate just to maintain consistency with the approach adopted in Example 2.2.

$$\therefore \; V_{R1} = IR_1 = 10(I), \quad V_{R2} = IR_2 = 20(I),$$

$$V_{R3} = IR_3 = 10(I), \text{ and } V_{R4} = IR_4 = 30(I)$$

With all voltages – voltage source, voltage load, and voltage drops across the resistors – identified and their polarities noted, apply KVL by "walking" the annotated circuit beginning at the cathode or negative electrode of the voltage source, V_{1s}. Add all voltages, with respective polarities, as you make a complete circle around the circuit, in the clockwise direction.

$$\Sigma V = 0$$

$$-18 \text{ V} + 10(I) + -6 \text{ V} + 20(I) + 10(I) + 30(I) = 0$$

$$70(I) = 24 \text{ V}$$

or,

$$I = \frac{24}{70} = 0.343 \text{ A}$$

Ancillary exercise: Verify the derived value of current through the alternative, R_{EQ} and Ohm's law method, as illustrated in Approach 1 of the solution for Example 2.2.

7. Determine the value of voltage source **current** in the parallel circuit below using KCL, Kirchhoff's current law.
     Ancillary question: If one of the 5 Ω resistors is removed (or replaced with an open circuit) and the other one is replaced with a short circuit, what would be the source current?

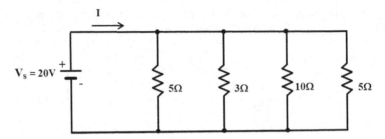

Solution:

KCL is applied to the given circuit after the node has been identified and circuit has been annotated with voltage designation, voltage polarity, branch currents, and current directions. See the circuit diagram below:

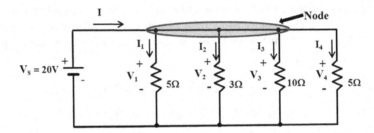

Subscribing to the definition of a node as a point where three or more conductors merge, the shaded segment in the diagram above is designated as the node for this circuit. Then, before applying KCL to determine the source current – using Ohm's law – define the individual currents through each of the resistors, in terms of the specific resistance values and the voltages around them:

$$I_1 = \frac{V_1}{R_1} \quad I_2 = \frac{V_2}{R_2} \quad I_3 = \frac{V_3}{R_3} \quad I_4 = \frac{V_4}{R_4}$$

Since all of the resistors are in parallel with the voltage source,

$$V_1 = V_2 = V_3 = V_4 = V_s = 20 \text{ V}$$

Therefore,

$$I_1 = \frac{20}{5} = 4 \text{ A}, \ I_2 = \frac{20}{3} = 6.67 \text{ A}, \ I_3 = \frac{20}{10} = 2 \text{ A}$$

$$I_4 = \frac{20}{5} = 4 \text{ A}$$

Then, application of KCL at the designated node yields the following equation:

$$I = I_1 + I_2 + I_3 + I_4$$

or

$$I = 4 \text{ A} + 6.67 \text{ A} + 2 \text{ A} + 4 \text{ A} = 16.67 \text{ A}$$

Ancillary Question: If one of the 5 Ω resistors is open-circuited and the other one short-circuited, the parallel resistor network would appear as follows:

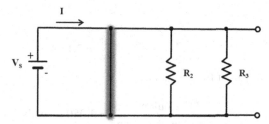

The highlighted segment in the circuit above represents the short circuit that replaces R1. Since R1 is replaced by a short circuit – regardless of the disposition of other circuit elements – it becomes a path of least resistance for the entire circuit. In other words, the voltage source is short-circuited. Interpreted in terms of Ohm's law, this would mean:

$$I = \frac{V_s}{0} = \infty$$

Since infinite current is not practical, this means that a very large amount of current would pass through the shunt or short circuit resulting in a catastrophic failure (burning or melting) of the short-circuiting conductor, the interconnecting wires or a fault in the voltage source.

CHAPTER 3 – SOLUTIONS

1. A plating tank with an effective resistance of 100 Ω is connected to the output of a full-wave rectifier. The AC supply voltage is 340 V_{peak}. Determine the amount of time, in hours, it would take to perform 0.075 Faradays worth of electroplating?

Solution:
Therefore, the amount of charge transfer, in Coulombs, in this electrodeposition case would be as follows:

$$q = \left(\frac{96,487 \text{ Coulombs}}{1 \text{ Faraday}} \right) \cdot (0.075 \text{ Faraday}) = 7,237 \text{ Coulombs}$$

As explained in Example 3.1, we are interested in the amount of time it takes to transfer a known amount of charge, so, rearrangement of Eq. 3.1 results in:

$$t = \frac{q}{I}$$

The next step entails determination of the DC voltage and current produced by the full-wave rectification of 340 V_{peak} AC, which is the same as 340 V_{max}.

$$V_{max} = \sqrt{2} \cdot V_{rms} = 340 \text{ V}$$

$$V_{DC} = 2 \cdot \left(\frac{V_{max}}{\pi}\right) = 2 \cdot \left(\frac{340}{3.14}\right) = 216 \text{ V} \qquad (3.2)$$

and

$$I_{DC} = \frac{V_{DC}}{R} = \frac{V_{Ave}}{R} = \frac{216}{100} = 2.16 \text{ A} \qquad (3.4)$$

Then, application of Eq. 3.1 yields:

$$t = \frac{7,237 \text{ Coulombs}}{2.16 \text{ Coulombs/second}} = 3,350 \text{ seconds}$$

or

$$t = \frac{3,350 \text{ seconds}}{3,600 \text{ seconds/hour}} = 0.93 \text{ hours}$$

2. Determine the source current I_{rms} in the AC circuit below.

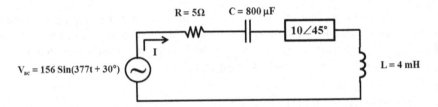

Solution:
Solution strategy: Convert the given inductance value of L = 4 mH into its equivalent inductive reactance X_L. Convert the given capacitance value C = 800 μF into its equivalent capacitive reactance X_C. Convert the given "black box" impedance of $Z = 10\angle45°$ Ω into the rectangular form. Combine **R, X_L, X_C,** and **Z**, linearly, to determine Z_{EQ}. Then apply Ohm's law to calculate the source current **I**. Note that the given polar version of the "black box" impedance $Z = 10\angle45°$ Ω can be converted into the equivalent rectangular form through Pythagorean Theorem, as explained in this chapter, or through a scientific calculator to:

$$Z = 10\angle45° = 7.07 + j7.07 \ \Omega$$

$$\therefore Z_{Eq} = R + Z_C + Z_L + Z$$

$$= R - jX_C + jX_C + Z$$

$$= R - j\frac{1}{\omega C} + j\omega \cdot L + Z$$

$$= R - j\frac{1}{(2) \cdot (\pi) \cdot (f) \cdot (C)} + j(2) \cdot (\pi) \cdot (f) \cdot (L) + Z$$

$$= 5 - j\frac{1}{(2)\cdot(3.14)\cdot(60)\cdot(800\times10^{-6})}$$
$$+j(2)\cdot(3.14)\cdot(60)\cdot(4\times10^{-3})+7.07+j7.07$$

$$= 5 - j\frac{1}{0.301}+j1.51+7.07+j7.07$$

$$= 5 - j3.32+j1.51+7.07+j7.07$$

$$= 12.07+j5.26 = 13.2\angle23.55° \ \Omega$$

Current "**I**" calculation:

V(t), in rms form $= 110\angle30°$ V$_{rms}$ as determined in the chapter examples.

Then, according to Ohm's Law, $I = \dfrac{V}{Z_{Eq}} = \dfrac{110\angle30°}{13.17\angle23.55}$

$\therefore$ The source rms current, $I = 8.35\angle6.5°$ A

3. Calculate the impedance Z_{EQ} as seen by the AC voltage source in the circuit below:

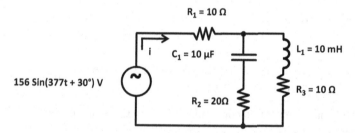

Solution:
Solution strategy: Convert the given inductance value of $L_1 = 10$ mH into its equivalent inductive reactance X_L. Convert the given capacitance value $C_1 = 10$ μF into its equivalent capacitive reactance X_C. Combine the load elements in each of the two parallel branch circuits into a single branch circuit impedance and then combine the resulting parallel branch circuits into a single impedance, using parallel load combination formula. The impedance thus derived would be combined, through series combination approach, with resistor $R_1 = 10$ Ω to arrive at the combined equivalent impedance representing all load elements driven by the 156 V$_p$ AC source.

The 10 mH inductor and the 10 Ω resistive branch circuit, series combination:

$$Z_{R3L} = R_3 + Z_L$$

$$= R_3 + jX_L$$

$$= R_3 + j\omega \cdot L$$

$$= R_3 + j(2) \cdot (\pi) \cdot (f) \cdot (L)$$

$$= 10 + j(2) \cdot (3.14) \cdot (60) \cdot (10 \times 10^{-3})$$

$$= 10 + j3.77 = 10.7\angle 20.7° \ \Omega$$

The 10 μF capacitor and the 20 Ω resistive branch circuit, series combination:

$$Z_{R_2C} = R_2 + Z_C$$

$$= R_2 - jX_C$$

$$= R_2 - j\frac{1}{\omega C}$$

$$= 20 - j\frac{1}{(2) \cdot (3.14) \cdot (60) \cdot (10 \times 10^{-6})}$$

$$= 20 - j265.4 = 266\angle -85.7° \ \Omega$$

The formula for parallel combination of the RC and RL parallel circuits:

$$\frac{1}{Z_{EqRLC}} = \frac{1}{Z_{R_2C}} + \frac{1}{Z_{R_3L}}$$

or,

$$Z_{EqRLC} = \frac{(266\angle -85.7) \cdot (10.7\angle 20.7)}{266\angle -85.7 \ + 10.7\angle 20.7} = \frac{2,848\angle -65°}{263.3\angle -83.5°} = 10.8\angle 18.5°$$

The final combination consisting of $R_1 = 10 \ \Omega$ and the newly derived impedance $Z_{EqRLC} = 10.8\angle 18.5 \ \Omega$:

$$Z_{Eq} = R_1 + Z_{EqRLC}$$

$$= 10 + 10.8\angle 18.5 = 10 + 10.26 + j3.43 = 20.26 + j3.43$$

$$= 20.55\angle 9.6 \ \Omega$$

4. A single-phase 1 kVA resistive load, designed to operate at 240 VAC, has to be powered by a 480 VAC source. A transformer is applied as shown in the diagram below. Answer the following questions associated with this scenario:

a. Would the transformer be connected in a "step-up" configuration or a "step-down" configuration?

b. When installing the transformer, what turns ratio should it be connected for?

c. What would be the secondary current, I_s, when the load is operating at full capacity?

d. What would be the primary current at full load?

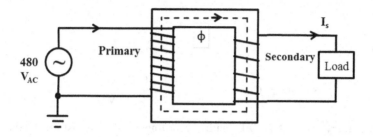

Solution:

The scenario described in this problem can be illustrated as follows:

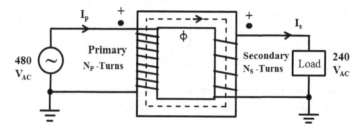

a. As discussed in this chapter, when voltage must be reduced to accommodate the load, the transformer must be connected in a step-down configuration. Therefore, the answer is: **step-down configuration.**

b. **Turns ratio:** According to Eq. 3.15, the relationship between the turns ratio, the primary voltage, and the secondary voltage is defined, mathematically, as follows:

$$V_s = \frac{V_p}{a} = \left(\frac{N_s}{N_p}\right) \cdot V_p \qquad (3.15)$$

or

$$a = \frac{V_p}{V_s}$$

Therefore,

$$\text{Turns Ratio} = a = \frac{480}{240} = 2:1$$

c. **Secondary current:** As explained in the single-phase transformer section of this chapter, the single-phase power is mathematically expressed in the form of Eq. 3.17:

Single Phase AC Apparent Power $= |S_{1-\phi}| = V_s \cdot I_s = V_p \cdot I_p$

Rearranging of this equation yields:

$$I_s = \frac{|S_{1-\phi}|}{V_s}$$

or

$$\text{Secondary Current} = I_s = \frac{|S_{1-\phi}|}{V_s} = \frac{1{,}000 \text{ VA}}{240 \text{ V}} = 4.2 \text{ A}$$

d. **Primary current at full load:** According to ideal transformer equation Eq. 3.4,

$$I_s = a \cdot I_p = \left(\frac{N_p}{N_s}\right) \cdot I_p$$

or

$$I_p = \frac{I_s}{a}$$

Since the turns ratio was determined to be 2:1 in part (b), and the secondary current I_s was determined to be 4.2 A in part (c),

$$I_p = \frac{I_s}{a} = \frac{4.2}{2} = 2.1 \text{ A}$$

5. Calculate the equivalent impedance as seen from the vantage point of the AC source V_{ac} in the circuit shown below. The transformer in the circuit is assumed to be ideal. The values of the primary and secondary circuit elements are: $X_{Lp} = 1\,\Omega$, $R_p = 4\,\Omega$, $R_s = 10\,\Omega$, $X_{Ls} = 5\,\Omega$, $X_{Cs} = 10\,\Omega$, $N_p = 100$, and $N_s = 200$.

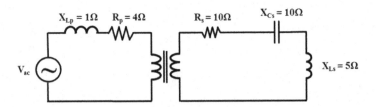

Solution:
As in Example 3.2, for simplicity, the primary and secondary inductors are given in the form of their reactances "X_L" instead of inductance values. In this problem, a capacitor is introduced on the secondary side. This capacitor is represented in the form its X_{Cs}. As explained earlier in this chapter, in the process of reducing the AC circuits to an equivalent impedance form, the individual

inductive and capacitive reactances are combined in the form of their respective impedance.

Therefore, on the primary side:

$$Z_{Lp} = jX_{Lp} = j1 \ \Omega$$
Hence, $Z_P = R_p + Z_{Lp} = 4 + j1 \ \Omega$

On the secondary side:

$Z_{Ls} = jX_{Ls} = j5 \ \Omega$, and
$Z_{Cs} = -jX_{Cs} = -j10 \ \Omega$; the reason for the negative sign, as explained in Chapter 3, lies in the fact that $Z_{Cs} = (1/j)X_{Cs}$ and that $1/j = -j$

Hence, the total secondary impedance $= Z_s = R_s + Z_{Ls} + Z_{Cs}$

$$= 10 + j5 - j10 \ \Omega$$

$a =$ Turns ratio $= N_p/N_s = 100/200 = 1/2$

$$Z_s' = a^2 \cdot Z_s$$

$$Z_s' = (1/2)^2 \cdot (10 + j5 - j10 \ \Omega) = 2.5 + j1.25 - j2.5 \ \Omega$$

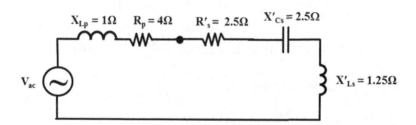

Then, combination of the primary impedance and the total reflected secondary impedance would result in the equivalent impedance Z_{eq}:

$$Z_{eq} = Z_p + Z_s' = (4 + j1 \ \Omega) + (2.5 + j1.25 \ \Omega - j2.5 \ \Omega)$$

$$= 6.5 - j0.25 \ \Omega$$

This final equivalent impedance Z_{eq} can be represented in rectangular complex form as follows:

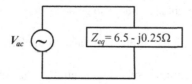

The equivalent impedance $\mathbf{Z}_{eq}$, derived in the rectangular form above, can be stated in polar or phasor form as follows:

$$\mathbf{Z}_{eq} = 6.5 - j0.25\,\Omega = 6.5\angle -2.2°$$

This conversion from rectangular to phasor form can be accomplished through a scientific calculator, with complex math feature or, as illustrated earlier in this chapter, through application of Pythagorean Theorem and trigonometry.

6. The no-load voltage at the main switch yard of a manufacturing facility is 13,400 V_{AC}. The voltage regulation of the main switch yard is 4%. What is the rated full-load voltage that is most likely to be measured on the load side of the main switch yard?

Solution:

$\mathbf{V_{NL}} = 13,400\,V$
Voltage Regulation = **4% = 0.04**
$\mathbf{V_{FL}} = \mathbf{V_{Rated}} = ?$

Apply Eq. 3.19:

$$\text{Voltage Regulation in \%} = \left[\frac{V_{NL} - V_{FL}}{V_{FL}}\right] \times 100$$

$$\text{Voltage Regulation in \%} = 4 = \left[\frac{13,400 - V_{FL}}{V_{FL}}\right] \times 100$$

$$0.04\,V_{FL} = 13,400 - V_{FL}$$

$$V_{FL} = \frac{13,400}{1.04} = 12,885\,V_{AC}$$

7. Consider the power distribution system shown in the schematic below. Determine the following unknown parameters on the Y load side of the transformer given that the turns ratio is 2:1:

a. $|I_{L\text{-}Sec}|$ = Magnitude of load or secondary line current
b. $|I_{P\text{-}Sec}|$ = Magnitude of secondary phase current or load phase current
c. $|V_{P\text{-}Pri}|$ = Magnitude of phase voltage on the source or primary side of the transformer
d. $|V_{L\text{-}Sec}|$ = Magnitude of line voltage on the load or secondary side of the transformer
e. $|V_{P\text{-}Sec}|$ = Magnitude of phase voltage on the load or secondary side of the transformer
f. $|V_{L\text{-}N,\,Sec}|$ = Magnitude of line to neutral voltage on the load or secondary side of the transformer

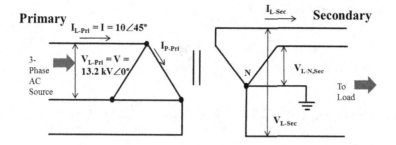

Solution:

$$V_{\text{P-Pri}} = V_{\text{L-Pri}} \tag{3.30}$$

$$V_{\text{L-Sec}} = V_{\text{P-Sec}} \cdot \sqrt{3} \tag{3.31}$$

$$I_{\text{P-Pri}} = \frac{I_{\text{L-Pri}}}{\sqrt{3}} = \frac{I}{\sqrt{3}} \tag{3.32}$$

$$I_{\text{P-Sec}} = I_{\text{L-Sec}} \tag{3.33}$$

where

$V_{\text{P-Pri}}$ = Primary phase voltage = **V**
$V_{\text{L-Pri}}$ = Primary line voltage = $V_{\text{Line-Delta}}$ = **V**
$V_{\text{P-Sec}}$ = Secondary phase voltage
$V_{\text{L-Sec}}$ = Secondary line voltage = $V_{\text{Line-Y}}$
$I_{\text{L-Pri}}$ = Primary line current = **I**
$I_{\text{P-Pri}}$ = Primary phase current
$I_{\text{P-Sec}}$ = Secondary phase current = $I_{\text{Line-Y}}$
$I_{\text{L-Sec}}$ = Secondary line current = $I_{\text{Line-Y}}$

For a Δ–Y three-phase transformer, as illustrated in Figure 3.26, the voltage and current transformations can be assessed using the following equations:

$$\frac{V_{\text{Line-Y}}}{V_{\text{Line-Delta}}} = \frac{\sqrt{3}}{a}$$

or   (3.34)

$$V_{\text{Line-Y}} = \frac{\sqrt{3}}{a} \cdot V_{\text{Line-Delta}} = \frac{\sqrt{3}}{a} \cdot V$$

$$\frac{I_{\text{Line-Y}}}{I_{\text{Line-Delta}}} = \frac{a}{\sqrt{3}}$$

or   or,   (3.35)

$$I_{\text{Line-Y}} = \frac{a}{\sqrt{3}} \cdot I_{\text{Line-Delta}} = \frac{a}{\sqrt{3}} \cdot I$$

a. According to Eq. 3.35:

$$I_{Line-Y} = \frac{a}{\sqrt{3}} \cdot I_{Line-Delta} = \frac{a}{\sqrt{3}} \cdot I$$

or,

$$\left| I_{L-Sec} \right| = \frac{a}{\sqrt{3}} \cdot |I| = \frac{2}{\sqrt{3}} \cdot \left| 10\angle 45° \right| = \frac{20}{\sqrt{3}} = 11.55 \text{ A}$$

b. As characteristic of **Y** three-phase AC circuits and in accordance with Eq. 3.33

$$I_{P-Sec} = I_{L-Sec}$$

or,

$$\left| I_{P-Sec} \right| = \left| I_{L-Sec} \right| = 11.55 \text{ A}$$

c. As characteristic of **Δ** three-phase AC circuits and in accordance with Eq. 3.30

$$V_{P-Pri} = V_{L-Pri}$$

and,

$$\left| V_{P-Pri} \right| = \left| V_{L-Pri} \right|$$

$$\therefore \ \left| V_{P-Pri} \right| = 13.2 \text{ kV}$$

d. According to Eq. 3.34:

$$V_{Line-Y} = \frac{\sqrt{3}}{a} \cdot V_{Line-Delta} = \frac{\sqrt{3}}{a} \cdot V$$

or,

$$\left| V_{L-Sec} \right| = \frac{\sqrt{3}}{a} \cdot |V| = \frac{\sqrt{3}}{2} \cdot \left| 13.2\angle 0° \right| = 11.432 \text{ kV} = 11,432 \text{ V}$$

e. As shown in Figure 3.28, on the secondary side of the transformer:

$$V_{P-Sec} = \frac{V_{Line-Delta}}{a} = \frac{V}{a}$$

and,

$$\left| V_{P-Sec} \right| = \frac{|V|}{a} = \frac{\left| 13.2\angle 0° \right|}{2} = 6.6 \text{ kV} = 6,600 \text{ V}$$

f. As apparent from Figure 3.25a:

$$V_{L\text{-}N, \text{Sec}} = V_{P\text{-}Sec}$$

$$\therefore \ |V_{L\text{-}N, \text{Sec}}| = |V_{P\text{-}Sec}| = 6.6 \text{ kV or } 6{,}600 \text{ V}$$

CHAPTER 4 – SOLUTIONS

1. Consider a hydroelectric reservoir where water is flowing through the turbine at the rate of 1,100 ft³/s. See the diagram below. The turbine exit point is 700 ft lower than the elevation of the reservoir water surface. The turbine efficiency is 90% and the total frictional head loss through the penstock shaft and turbine system is 52 ft.

   a. Calculate the power output of the turbine in MWs.
   b. If the efficiency of the electric power generator is 92%, what would the electric power output be for this hydroelectric power generating system?

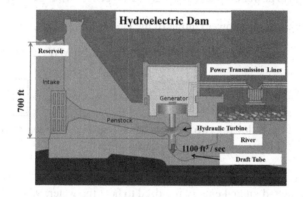

Solution:
Given:

$\dot{\gamma}$ = Specific weight = 62.4 lbf/ft³
h_f = Frictional head loss in the penstock and elsewhere in the system, upstream of the turbine = 52 ft
h_z = Total head available due to the height of the water level in the reservoir
   = 700 ft

$\therefore$ Net head delivered by the water to the turbine = Head added
   = $h_A = h_z - h_f = 648$ ft

$\dot{V}$ = Volumetric Flow Rate = 1,100 ft³/s

Turbine efficiency 90 %
Generator efficiency 92 %

a. Calculate the power output of the turbine in MWs.

$$P_p = WHP$$

$$= \text{Hydraulic Horsepower in hp} = \frac{h_A \cdot \gamma \cdot \dot{V}}{550} \qquad (4.39)$$

$$= \frac{(648)(62.4)(1,100)}{550} = 80,870 \text{ hp}$$

$$P_{out} = \text{Power Output of the Turbine}$$

$$= (0.90 \times 80,870 \text{ hp})$$

$$= 72,783 \text{ hp}$$

$$P_{out} = \text{Power Output of the Turbine in MW} = \frac{72,783 \text{ hp} \times 0.746 \text{ kW/hp}}{1,000 \text{ kW/MW}}$$

$$= 54 \text{ MW}$$

b. If the efficiency of the electric power generator is 92%, what would the electric power output be for this hydroelectric power generating system?

$$P_{Gen.\ Syst.-out} = \text{Power Output of the Generator in MW}$$

$$= (\text{Power Output of the Turbine in MW}) \cdot (\eta_{Turbine})$$

$$= (54 \text{ MW} \times 0.92)$$

$$= 50 \text{ MW}$$

2. Which of the following two water heaters would cost the least to operate, on annual cost basis, under the given assumptions?

A. Electric water heater:
   Estimated annual energy required to heat the water: 9,000 kWh
   Efficiency: **95%**
   Cost rate: **$0.10/kWh**
B. Natural gas water heater:
   Estimated annual energy required to heat the water: Same as the Electric water heater
   Efficiency: **98%**
   Cost rate: $10.87/DT

Solution:
Note the given 9,000 kWh worth of electrical energy is the **energy actually absorbed by the water.** Since the electric water heater efficiency is 95%, the electrical energy **pulled from the utility** would be as follows:

$$= \left(\frac{9,000 \text{ kWh}}{0.95} \right) = 9,474 \text{ kWh}$$

And the annual operating cost for the electric water heater would be as follows:

$$= (9,474 \text{ kWh}) \cdot (\$0.10/\text{kWh}) = \$947.40$$

The **annual energy absorbed by the gas water heater**, assumed to be the same as the electric water heater = 9,000 kWh. Since the gas water heater efficiency is 98%, the electrical energy pulled from the utility would be as follows:

$$= \left(\frac{9,000 \text{ kWh}}{0.98} \right) = 9,184 \text{ kWh}$$

Then, the annual energy consumption by the gas water heater, in DT or MMBtu, would be = (9,184 kWh)·(3,412 Btu/kWh)(1DT/1,000,000Btu) = **31.33 DT**. Since the natural gas cost rate is given as $10.87/DT, the annual operating cost for the gas water heater would be = (31.33 DT)·($10.87/DT) = **$340.61.** **Answer: The gas water heater would cost substantially less to operate** than the electric water heater.

3. A computer manufacturing company is testing a prototype for the amount of heat it dissipates as wasted energy over a 10-hour period of operation. The computer is powered by a 24 V DC power supply and is designed to draw 3 A of current. Determine the total energy dissipated in Btu.

Solution:
Apply Eq. 4.10:

$$\text{Energy} = V \cdot I \cdot t = (24 \text{ V}) \cdot (3 \text{ A}) \cdot (10 \text{ h}) = (72 \text{ W}) \cdot (36,000 \text{ s})$$

$$= \left(72 \frac{\text{J}}{\text{s}} \right)(36,000 \text{ s})$$

$$= 2,592,000 \text{ J} = 2,592 \text{ kJ}$$

Since 1.055 kJ = 1.0 BTU

$$\text{Energy Dissipated, in Btu} = (2,592 \text{ kJ}) \left(\frac{1 \text{ Btu}}{1.055 \text{ kJ}} \right) = 2,457 \text{ Btu}$$

4. In response to a significant near-miss incident and midair fire on a new commercial jet aircraft, a governmental agency is performing forensic analysis on the type of lithium–ion aircraft battery suspected to be the root cause. Estimate the amount of **current** involved in the suspected fault on the basis of the following forensic data:

- Total energy released in the catastrophic failure of the battery: **866 kJ**
- Estimated duration of fault: **2 seconds**
- Rated voltage of the battery: $3.7V_{DC}$

Solution:
Apply Eq. 4.10:

$$\text{Energy} = V \cdot I \cdot t$$

or,

$$I = \frac{\text{Energy}}{V \cdot t} = \frac{866,000 \text{ J}}{(3.7 \text{ V}) \cdot (2 \text{ s})} = \frac{866,000(V \cdot A \cdot s)}{(3.7 \text{ V}) \cdot (2 \text{ s})} = 117,027 \text{ A}$$

5. A **156Sin377t** sinusoidal voltage is connected across a load consisting of a parallel combination of a **20 Ω** resistor and a **10 Ω** capacitive reactance.

   a. Determine the real power dissipated by the resistor.
   b. Determine the reactive power stored in a **10 Ω** parallel capacitive reactance.
   c. Calculate the total apparent power delivered to this parallel **R** and **X** circuit by the AC voltage source.

Solution:
The circuit diagram for this scenario would be as depicted below:

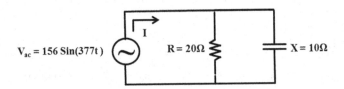

$V_{ac} = 156 \text{ Sin}(377t)$ $R = 20\Omega$ $X = 10\Omega$

   a. We can apply Eq. 4.19 to determine the power dissipated or consumed in the 20 Ω resistor. However, we must first derive the V_{RMS} from the given AC voltage of **156Sin377t**. This is due to the fact that coefficient of 156 stated in the give AC voltage function of **156Sin400t** is the peak or maximum voltage, V_m.
   As discussed in Chapter 3 and stipulated by Eq. 3.3:

$$V_m = \sqrt{2}V_{RMS} \text{ or } V_{RMS} = \frac{V_m}{\sqrt{2}}$$

$$\therefore V_{RMS} = \frac{156}{\sqrt{2}} = 110.3 \text{ V}$$

$$P = \frac{V_{RMS}^2}{R} = \frac{(110.3)^2}{20} = 608 \text{ W}$$

   b. Apply Eq. 4.20 to determine the reactive power sequestered in the 10 Ω parallel capacitive reactance.

$$Q = \frac{V_{RMS}^2}{X} = \frac{(110.3)^2}{10} = \frac{12,166}{10} = 1,217 \text{ VAR}$$

c. Apply Eq. 4.12 to calculate the total apparent power **S** delivered to this parallel R and X circuit by the AC voltage source.

$$\bar{S} = P + jQ = 608 - j1,217 = 1,361 \angle -63.4° \, \Omega$$

Note that jQ reactive power entity is entered into the apparent power calculation above as "−jQ" because of the fact that capacitance in an AC circuit results in negative impedance contribution or "−jX." Therefore, the reactive power Q due to a capacitor is applied as "−j1217," in overall **S** calculation.

We can also apply Eq. 4.22 to verify the magnitude of the total apparent power **S** delivered to this parallel R and X circuit:

$$|S| = \text{Magnitude of AC Apparent Power} = \sqrt{P^2 + Q^2}$$

$$= \sqrt{608^2 + 1,217^2} = 1,361 \text{ VA}$$

Ancillary: The reader is encouraged to verify the apparent power of 1,361 VA by applying Eq. 4.21. **Hint**: The Z, in this case, must be computed through parallel combination of R and Z_C as shown below:

$$|Z| = \left| \frac{R \cdot Z_C}{R + Z_C} \right|$$

6. A **156Sin400t** sinusoidal voltage is connected across an unknown resistive load. If the power dissipated in the resistor is 1,000 W, what is the resistance of the resistive load?

Solution:
We can apply Eq. 4.19 to determine the value of the resistor using the given power dissipation value of 1,000 W. However, we must first derive the V_{RMS} from the given AC voltage of **156Sin400t**. This is due to the fact that coefficient of 156 stated in the give AC voltage function of 156Sin400t is the peak or maximum voltage, V_m.

As discussed in Chapter 3 and stipulated by Eq. 3.3:

$$V_m = \sqrt{2} V_{RMS} \text{ or } V_{RMS} = \frac{V_m}{\sqrt{2}}$$

$$\therefore V_{RMS} = \frac{156}{\sqrt{2}} = 110.3 \text{ V}$$

$$P = \frac{V_{RMS}^2}{R} \qquad \text{Eq. 4.19}$$

$$1{,}000 = \frac{(110.3)^2}{R}$$

$$\therefore R = 12.17\ \Omega$$

7. The AC circuit shown below depicts a three-phase, one-line schematic of a hydroelectric power generating station. Assume that there is no voltage drop between the generator and the primary side of the transmission system transformer. The line current is indicated by an EMS system to be **10 kA**, RMS. Calculate the following if the power factor is known to be **0.95**:

  a.   Magnitude of the apparent power presented to the transmission lines.
  b.   Magnitude of the real power presented to the transmission lines.
  c.   The RMS line to neutral voltage at the source.

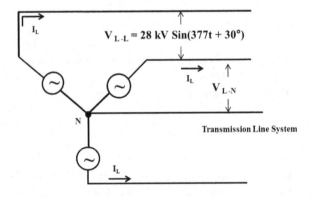

Solution:

  a.   Magnitude of the apparent power presented to the transmission lines:
      Note that the AC voltage function is NOT specified in RMS form; hence, by convention, it is "peak" or "maximum" voltage. Therefore, we need to derive the RMS voltage for subsequent computations. The line current is given in RMS form.

$$\left|V_{L\text{-}L,\,RMS}\right| = \left(\frac{V_{L\text{-}L,\,Peak}}{\sqrt{2}}\right) = \left(\frac{28{,}000\ V}{\sqrt{2}}\right) = 19{,}799\ V$$

According to Eq. 4.30:

$$\left|\bar{S}_{3\text{-}\phi}\right| = 3\left|\bar{S}_{1\text{-}\phi}\right| = 3\left(\frac{V_{L\text{-}L} \cdot I_L}{\sqrt{3}}\right) = \sqrt{3}\,V_{L\text{-}L} \cdot I_L,\ \text{for three phase Y and } \Delta \text{ circuits.}$$

Therefore,

$$|\bar{S}_{3\text{-}\phi}| = \sqrt{3}V_{\text{L-L}} \cdot I_L = \sqrt{3}(19,799 \text{ V}) \cdot (10,000 \text{ A}) = 342,928,564 \text{ VA}$$

$$= 342,929 \text{ kVA} = 342.93 \text{ MVA}$$

b. Magnitude of the real power presented to the transmission lines can be determined by rearranging and using Eq. 4.23:

$$PF = \text{Power Factor} = \frac{|P|}{|S|}$$

or,

$$|P| = PF \cdot |S| = (0.95) \cdot (342.929 \text{ MVA}) = 326 \text{ MW}$$

c. **Hint:** Use the $V_{\text{L-L}}$ computed in part (a). Line to neutral voltage, as introduced in Chapter 3, can be stated mathematically as follows:

$V_{\text{L-N, Y}}$ = Line to neutral voltage in a Y source or load

$$\text{configuration} = \frac{V_{\text{L-L}}}{\sqrt{3}} = \frac{19,799}{\sqrt{3}} = 11,431 \text{ V} = 11.43 \text{ kV}$$

8. A pump is to be installed on the ground floor of a commercial building to supply 200 ft³/s of water up to an elevation of 100 ft. Determine the minimum size of the motor for this application. Assume that the efficiency of the pump is 80%. The weight density of water $\gamma = 62.4$ lbf/ft³

Solution:
Solution strategy in this case would be to use Eq. 4.39 to compute the WHP. Then, the amount of real power "**P**" delivered by the motor would be computed based on the given efficiency of the pump.

Given:

$h_A = 100$ ft
$\dot{V}$ = Volumetric Flow Rate = 200 ft³/s
Pump efficiency = 80%
$\gamma = 62.4$ lbf/ft³

According to Eq. 4.39, water horsepower delivered by the pump to the water can be stated as follows:

WHP = P_p = Fluid horse power delivered by the pump to the water

$$= \frac{(h_A)(\gamma)(\dot{V})}{550} = \frac{(100 \text{ ft})(62.4 \text{ lbf/ft}^3)(200 \text{ ft}^3/s)}{550 \text{ ft-lbf/s/hp}}$$

$$= 2,269 \text{ hp}$$

Then, according to the wire to water (hydraulic pump) power flow diagram in Figure 4.4:

$$\text{BHP delivered by the motor to the} = \frac{\text{WHP}}{\eta_p} = \frac{2,269}{0.8}\text{hp} = 2,836 \text{ hp}$$

Therefore, a commercially available motor size above 2,836 hp should be selected.

CHAPTER 5 – SOLUTIONS

1. Determine the power factor of the circuit shown below, as seen by the AC source.

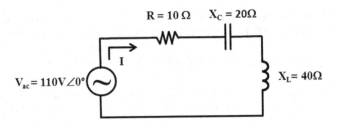

Solution:
This problem can be solved through multiple approaches. Two of those approaches are as follows:

i. Calculate the equivalent or combined impedance of the circuit as seen by the source. Then take the cosine of the angle of that impedance.
ii. Calculate the equivalent or combined impedance of the circuit as seen by the source. Apply Ohm's law to compute the AC. Then take the cosine of the angular difference between given AC voltage and the computed current.
    We will utilize the first approach:

$$Z_{Eq} = 10\,\Omega + j20\,\Omega - j40\,\Omega = 10\,\Omega - j20\,\Omega = 22.36\angle -63.43°$$

$$\left\{ \text{Note} |Z_{Eq}| = \sqrt{10^2 + 20^2} = 22.36\,\Omega, \text{ and } \theta_Z = -\text{Tan}^{-1}\left(\frac{20}{10}\right) = -63.43 \right\}$$

Therefore, $PF = Cos\theta_Z = Cos(-63.43°) = 0.45$ or 45%

The reader is encouraged to verify the result through approach ii.

2. Assume that the circuit depicted below represents one phase of a special power transmission line. Determine the power factor of the circuit shown below, as seen by the AC source.

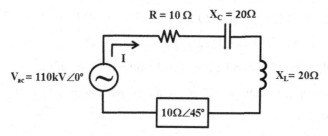

R = 10 Ω $X_C = 20$ Ω

I

$V_{ac} = 110kV\angle 0°$

$X_L = 20$ Ω

10 Ω$\angle 45°$

Solution:

As with Problem 1, this problem can be solved through multiple approaches. Two of those approaches are as follows:

i. Calculate the equivalent or combined impedance of the circuit as seen by the source. Then take the cosine of the angle of that impedance.

ii. Calculate the equivalent or combined impedance of the circuit as seen by the source. Apply Ohm's law to compute the AC. Then take the cosine of the angular difference between given AC voltage and the computed current.

In this case will illustrate the application of approach (ii):

$$Z_{Eq} = 10\,\Omega + j20\,\Omega - j20\,\Omega + 10\angle 45°\,\Omega = 10\,\Omega + 7.07\,\Omega + j7.07\,\Omega$$

$$= 17.07\,\Omega + j7.07\,\Omega = 18.48\,\Omega\angle 22.5°\,\Omega$$

$$\left\{ \text{Note} |Z_{Eq}| = \sqrt{17^2 + 7.07^2} = 18.48\,\Omega, \text{ and } \theta_Z = \text{Tan}^{-1}\left(\frac{7.07}{17.07}\right) = 22.5° \right\}$$

$$I = \frac{V}{Z_{Eq}} = \frac{110\text{ kV}\angle 0°}{18.48\,\Omega\angle 22.5°\,\Omega} = 5.95\text{ kA}\angle -22.5°, \text{ or } \phi_I = -22.5°$$

Therefore, $\text{PF} = \text{Cos}(\theta_V - \phi_I) = \text{Cos}(0 - (-22.5°)) = \text{Cos}(22.5°)$

$$= 0.923 \text{ or } 92.3\%$$

The reader is encouraged to verify the result through approach (i), which, actually, would require fewer steps.

3. If the power factor in Problem 2 is less than 1.0, how much capacitance or inductance must be added in series to raise the power factor to unity?

Solution:

This problem can be solved by simply focusing on the total or equivalent impedance Z_{Eq} and determining the amount of reactance needed to offset the reactance in the original impedance.

$$Z_{Eq} = 10\,\Omega + j20\,\Omega - j20\,\Omega + 10\angle 45°\,\Omega = 10\,\Omega + 7.07\,\Omega + j7.07\,\Omega$$

$$= 17.07\,\Omega + j7.07\,\Omega = 18.48\,\Omega\angle 22.5°\,\Omega$$

Now the question is: how much and what type of reactance must be added such that Z_{Eq} reduces to its resistance portion, or $17.07\,\Omega$.

or,

$17.07\,\Omega + j7.07\,\Omega + jX\,\Omega = 17.07\,\Omega$

$\therefore\ jX = -j7.07\,\Omega$, the negative sign indicates that the reactance to be added must result in negative impedance; it must be capacitive.

and,

$$|X| = |X_C| = 7.07\,\Omega$$

$X_C = 7.07\,\Omega$, and by definition, $X_C = \dfrac{1}{\omega C}$. Frequency, f, in the US, can be assumed to be 60 hz, and by definition, $\omega = 2\pi f$. Therefore,

$$\omega = 2\pi f = 2(3.14)(60\ Hz) = 377\ rad/s$$

and,

$$C = \frac{1}{\omega X_C} = \frac{1}{(377)\cdot(7.07)} = 0.000375\ F,\ or\ 375\ \mu F$$

4. The output of a variable frequency drive, as shown in the circuit below, is $157 Sin\omega t$. The VFD output is currently set at 50 Hz. This drive is connected to a resistive load, capacitive reactance, and an inductive reactance.

    a. What should be the new frequency setting to attain a power factor of 1, or 100%.

    b. What is the existing power factor, at 50 Hz?

    c. What would be the power factor if all circuit elements remain unchanged and the VFD frequency is lowered to 30 Hz.

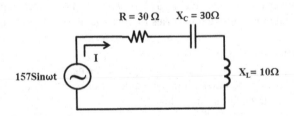

Solution

a. We must convert reactances X_C and X_B to corresponding capacitance, **C**, and inductance, **L**, values.

Since $X_L = 2\pi fL$, $L = \dfrac{X_L}{2\pi f}$, or, $L = \dfrac{10\,\Omega}{2\pi(50\text{ Hz})} = 0.03185\text{ H}$

and,

$X_C = \dfrac{1}{2\pi fC}$, $C = \dfrac{1}{2\pi fX_C}$, or, $C = \dfrac{1}{2\pi(50\text{ Hz})(30\,\Omega)} = 106 \times 10^{-6}\text{ F}$

$f = f_0 = \dfrac{1}{2\pi\sqrt{LC}} = \dfrac{1}{2(3.14)\sqrt{(0.03185)\cdot(106 \times 10^{-6})}}$ Eq. 5.8

$\therefore\ f = f_0 = 86\text{ Hz}$

b. The power factor can be calculated through the impedance angle, using Eq. 5.3. We must compute the circuit's total impedance first, **at 50 Hz**.

As derived in part (a), $L = 0.03185$ H and $C = 0.000106$ F

$\therefore\ X_L = 2\pi fL = 2(3.14)(50\text{ Hz})(0.03185\text{ H}) = 10\,\Omega$ {Given}

and,

$X_C = \dfrac{1}{2\pi fC} = \dfrac{1}{2(3.14)(50\text{ Hz})(0.000106\text{ F})} = 30\,\Omega$ {Given}

$X_t = X_Z =$

$\therefore\ Z = 30\,\Omega - j30\,\Omega + j10\,\Omega = 30\,\Omega - j20\,\Omega = 36\angle-33.7°\,\Omega$

and,

Power Factor $= \text{Cos}(\theta_Z) = \text{Cos}(-33.7°) = 0.83.3$, or, 83.3% leading.

c. The power factor at 30 Hz:

As derived in part (a), $L = 0.03185$ H and $C = 0.000106$ F

$\therefore\ X_L = 2\pi fL = 2(3.14)(30\text{ Hz})(0.03185\text{ H}) = 6\,\Omega$

and,

$X_C = \dfrac{1}{2\pi fC} = \dfrac{1}{2(3.14)(30\text{ Hz})(0.000106\text{ F})} = 50\,\Omega$

$\therefore\ Z = 30\,\Omega - j50\,\Omega + j6\,\Omega = 30\,\Omega - j44\,\Omega = 53.3\angle-55.7°\,\Omega$

and,

Power Factor $= \text{Cos}(\theta_Z) = \text{Cos}(-55.7°) = 0.5631$, or, 56.31% leading.

5. The HMI (Human Machine Interface) monitor of an automated HVAC system, monitoring an air washer supply fan motor is indicating a reactive power, Q_1, of 60 kVARs. This system is located in the United Kingdom, where the AC frequency is 50 Hz. Determine the amount of capacitance that must be added to improve the power factor of the motor branch circuit such that the reactive power is reduced to Q_2 of 20 kVARs. The branch circuit is operating at 240 V_{RMS}.

Solution:
Apply Eq. 5.4:

$$C = \frac{(Q_1 - Q_2)}{2\pi f V^2} = \frac{(60\ kVAR - 20\ kVAR)}{2\pi(50\ Hz)(240\ V)^2}$$

$$= \frac{(60,000\ VAR - 20,000\ VAR)}{2\pi(50\ Hz)(240\ V)^2} = 0.002211\ F = 2.211\ mF$$

6. **Power Factor Improvement and Cost Savings**: In conjunction with the local utility company DSM program, a manufacturing plant is being offered $2 per kVA for improvement in power factor from 0.75 to 0.85. The plant is operating at its contract level of 30 MW. Determine estimated annual pre-tax revenue if the plant accepts the offer.

Solution:
Since the objective is to assess the cost savings on the basis of apparent power (S) reduction, we must determine the apparent power S_1 being drawn by the air compressor motor at the existing power factor of 0.75 (75%) and the apparent power S_2 at the desired power factor of 0.85 (85%). Rearrange and apply Eq. 5.1.

$$\text{Magnitude of Apparent Power } |S| = \frac{|P|}{PF}$$

$$\text{Magnitude of existing apparent power } |S_1| = \frac{|P|}{PF_1} = \frac{30,000\ kW}{0.75}$$

$$= 40,000\ kVA$$

$$\text{Magnitude of improved apparent power } |S_2| = \frac{|P|}{PF_2} = \frac{30,000\ kW}{0.85}$$

$$= 35,294\ kVA$$

Annual DSM revenue expected from proposed power factor improvement

$$= (40,000\ kVA - 35,294\ kVA) \cdot \left(\frac{\$2}{kVA - Month}\right) \cdot (12\ Months/Year)$$

$$= \$112,941$$

7. The output of a variable frequency drive, as shown in the circuit below, is
157Sinωt. The VFD output is currently set at 60 Hz. This drive is connected
to a resistive load, capacitive reactance, inductive reactance, and a "black
box" load, Z_B, of $10\Omega\angle45°$. What should be the new frequency setting to
attain a power factor of 1, or 100%?

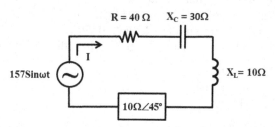

Solution:

Solution strategy: As explained earlier in this section, the power factor of
an AC circuit, consisting of inductive and capacitive reactance, peaks to the
maximum value of unity, or 100%, at resonance frequency, f_0. However, in this
case, because of the presence of the black box impedance of $Z_B = 10\Omega\angle45°$, we
cannot apply Eq. 5.8, directly, to compute f_0. We must convert impedance into
its rectangular form to derive the reactance component $X_{L,B}$, combine it with
the other inductive reactance in the circuit, derive the L and C values, and then
apply Eq. 5.8 to calculate the resonance frequency f_0.

$Z_B = 10\angle45° = 7.07\ \Omega + j7.07\ \Omega$. Therefore, $X_{L,B} = 7.07\ \Omega$.

Since the other inductive reactance is contributed by $X_L = 10\ \Omega$,

$X_{L,total} = X_L + X_{L,B} = 10\ \Omega + 7.07\ \Omega = 17.07\ \Omega$

∴ Total "equivalent L" in the circuit would be:

$X_{L,total} = 2\pi f L_{total} = 17.07\ \Omega$,

∴ $L_{total} = \dfrac{17.07\ \Omega}{2\pi f} = \dfrac{17.07\ \Omega}{2(3.14)(60\ Hz)}$, or, L = 0.0453 H

and,

$X_C = \dfrac{1}{2\pi f C} = \dfrac{1}{2(3.14)(60\ Hz)(C)} = 30\ \Omega$

∴ $C = \dfrac{1}{2(3.14)(60\ Hz)(30\ \Omega)} = 88.46\times10^{-6}\ F$

Then, according to Eq. 5.8,

$f_0 = \dfrac{1}{2\pi\sqrt{L\cdot C}} = \dfrac{1}{2(3.14)\sqrt{(0.0453)(88.46\times10^{-6})}} = 79.54\ Hz$

CHAPTER 6 – SOLUTIONS

1. The BMS at a truck assembly plant that operates 365 days a year is displaying following electrical power and energy consumption data:

Billing Days in the Current Month: 30
On-Peak Energy Consumption: 2,880,000 kWh
Off-Peak Energy Consumption: 11,520,000 kWh
Three highest 30-minute energy usages for the billing month are

   i.   12,500 kWh,
   ii. 12,300 kWh and
   iii. 12,290 kWh.

Assuming this facility is on OPT, Time of Use, contract with 30-minute demand interval, determine the following:
a. Average demand.
b. Peak demand.
c. The load factor for the current month.
d. Average annual demand

Solution:
a.   Average demand can be calculated by applying Eq. 6.1 as follows:

$$\text{Average Demand, in kW} = \frac{\text{Energy (kWh or MWh) consumed during the billing Month}}{\text{Total number of hours in the billing month}}$$

$$= \frac{\text{On Peak Energy} + \text{Off Peak Energy Consumption}}{\text{Total number of hours in the billing month}}$$

$$= \frac{2,880,000 \text{ kWh} + 11,520,000 \text{ kWh}}{(24 \text{ hours/Day}) \cdot (30 \text{ Days/Month})}$$

$$= 20,000 \text{ kW or } 20 \text{ MW}$$

b.   Peak demand can be calculated by applying Eq. 6.2 to the 30-minute interval during which the highest energy consumption is recorded. In this problem, three high 30-minute kWh consumptions are stated. However, we are interested in the "highest" kWh recorded and that is 12,500 kWh. Therefore, the peak demand would be as follows:

$$\text{Peak Demand} = \frac{\text{Peak Energy in kWh}}{\text{Time in hours}} = \frac{\text{Peak Energy in kWh}}{\frac{1}{2} \text{hour}}$$

$$= \frac{12,500 \text{ kWh}}{0.5 \text{ h}} = 25,000 \text{ kW or } 25 \text{ MW}$$

c. Load factor can be calculated by applying Eq. 6.3 as follows:

$$\text{Load Factor} = \frac{\text{Average Demand for the Month, in kW or MW}}{\text{Peak Demand for the Billing Month, in kW or MW}}$$

Using the Average Demand, calculated in part (a) as 20 MW, and the

Peak Demand calculated in part (b) as 25 MW:

$$\text{Load Factor} = \frac{20 \text{ MW}}{25 \text{ MW}} = 0.80 \text{ or } 80\%$$

2. A 200 kVA transformer has been tested by the manufacturer to safely and continuously sustain a load of 230 kVA. What service factor should the manufacturer include on the nameplate of this transformer?

Solution:

$$\text{Service Factor} = \frac{\text{Safe or Continuous Load, in kW, kVA or hp}}{\text{Nameplate rating of equipment, in kW, KVA or hp}}$$

$$= \frac{\text{Safe Operating Load}}{\text{Full Load Rating of the Transformer}} = \frac{230 \text{ kVA}}{200 \text{ kVA}} = 1.15$$

3. A 5 hp single-phase AC motor, rated at a service factor of 1.10, is being tested at maximum safe load, powered by 230 V_{AC} source. Determine the amount of current drawn by this motor from the power source if the motor efficiency is 90% and the power factor is 0.85.

Solution:
The 5 hp motor, with a service factor of 1.10, operating at its maximum safe load is, essentially, delivering energy or performing work at the rate of:

$$P = (5 \text{ hp}) \cdot (\text{Service Factor}) = (5 \text{ hp}) \cdot (1.10) = 5.5 \text{ hp}.$$

Also note that in most current computations the power, in hp, must be converted to power in watts or kWs.

$$\therefore \quad P_{\text{Motor-Watts}} = (5.5 \text{ hp}) \cdot (746 \text{ W/hp}) = 4,103 \text{ W}$$

As introduced in earlier chapters:

$$\text{Magnitude of apparent power } S = |S| = \frac{P_{\text{Motor-Watts}}}{\text{PF}} = |V||I|_{\text{Motor}}$$

$$\text{And magnitude of apparent power drawn from the source} = \frac{P_{\text{Motor-Watts}}}{(\text{PF}) \cdot (\text{Eff.})}$$

$$= |V||I|_{\text{Source}}$$

$$\therefore \ |I|_{Source} = \frac{P_{Motor\text{-}Watts}}{(PF)\cdot(Eff.)\cdot(|V|)} = \frac{4{,}103 \ W}{(0.85)\cdot(0.9)\cdot\left(230 \ V_{RMS}\right)} = 23.32 \ A$$

4. Consecutive electrical power meter readings at a home in Hawaii are listed below. Determine the total electrical power bill for the month of this residence if the flat $/kWh cost rate is ¢ 21/kWh. The renewable energy rider is $15 and the energy sales tax rate is 4%.

   Previous reading: 45,000
   Current or present reading: 46,000

Solution:
According to Eq. 6.6:

$$Baseline \ Charge = (46{,}000 - 45{,}000)\cdot(\$0.21 \ /kWh)$$

$$= (1{,}000)\cdot(\$0.21/kW) = \$210$$

$$Total \ Bill = (\$210 + \$15)\cdot(1 + 4\%) = (\$225)\cdot(1 + 0.04)$$

$$= (\$225)\cdot(1.04) = \$234.00$$

Note that the 4% sales tax is applied to the subtotal comprising of the baseline cost plus the renewable rider.

5. If the peak demand in Case Study 6.2 is reduced by 10% through implementation of peak-shaving measures, what would be the baseline cost for the demand portion of the bill?

Solution:
According to Table 6.1, the original peak demand for the month is 26,000 kW.

$$\therefore \ New, \ reduced, \ demand = (26{,}000 \ kW)\cdot(1 - 0.01)$$

$$= (26{,}000 \ kW)\cdot(0.9) = 23{,}400 \ kW$$

Then, revise the demand portion of Table 6.1 for the reduced peak demand of 23,400 kW as shown below:

Description	Demand and Energy Parameters	Rates	Line Item Charge
On-peak billing demand	23,400 kW		
On-peak billing demand charge			
For the first 2,000 kW	2,000 ×	$7.64/kW =	$15,280.00
For the next 3,000 kW	3,000 ×	$6.54/kW =	$19,620.00
For demand over 5,000 kW	**(23,400 − 5,000) ×**	**$5.43/kW =**	**$99,912**
Total demand cost for the billing month			$15,280.00 + $19,620.00 +$99,912 = **$134,812**

CHAPTER 7 – SOLUTIONS

1. A gas-powered prime mover is rotating the rotor of a single-phase alternator at a speed of 1,200 rpm. The alternator consists of six-pole construction.
   The effective diameter of the coil is 0.15 m and the length of the coil loop
   is 0.24 m. The coil consists of 20 turns. The magnetic flux density has been
   measured to be 1.2 T. Calculate the power delivered by this generator across
   a resistive load of 10 Ω.

   Solution:
   The RMS, effective or DC voltage produced through an alternator or generator
   can be computed by applying Eq. 7.6:

   $$V_{DC} = V_{RMS} = V_{eff} = \frac{V_P}{\sqrt{2}} = \frac{\pi n p\, NAB}{(60)\cdot\left(\sqrt{2}\right)}$$

   Given:
   n = 1,200 rpm
   p = 6
   N = 20
   B = 1.2 T
   A = (Eff. diameter of the coil conductor) × (Eff. length of the coil)
   = (0.24 m) × (0.15 m) = 0.036 m²

   $$V_{DC} = V_{RMS} = \frac{\pi n p\, NAB}{(60)\cdot\left(\sqrt{2}\right)}$$

   $$= \frac{(3.14)(1,200\ \text{rpm})(6)(20)(0.036\ \text{m}^2)(1.2\ \text{T})}{(60)\cdot\left(\sqrt{2}\right)} = 230.2\ \text{V}$$

   $$P = \frac{V^2}{R} = \frac{(230.2)^2}{10} = 5,299\ \text{W} = 5.3\ \text{kW}$$

2. A four-pole alternator/generator is producing electrical power at an electrical frequency of 50 Hz. (a) Determine the angular speed corresponding to
   the generated electrical frequency. (b) Determine the rotational (synchronous) speed of the armature/rotor. (c) Determine the angular velocity of the
   armature/rotor (rad/s).

   Solution:
   a. Angular speed, ω, corresponding to the generated electrical frequency,
      f, can be calculated using Eq. 7.8:

      $$\omega = 2\pi f = (2)\cdot(3.14)\cdot(50\ \text{Hz}) = 314\ \text{rad/s}$$

b. The rotational or synchronous speed of the armature/rotor is given by Eq. 7.7:

$$n_s = \frac{120f}{p} = \frac{(120)(50 \text{ Hz})}{4} = 1,500 \text{ rpm}$$

c. Angular velocity of the armature/rotor is simply the rotational speed, in rpm, converted into rad/s. Since there are 2π radians per revolution:

$$\omega_s = 1,500 \text{ rev./min} = \left(\frac{1,500 \text{ rev.}}{\text{min}}\right)\cdot\left(\frac{2\pi \text{ rad/rev}}{(60 \text{ s/min})}\right) = 157 \text{ rad/s}$$

3. A four-pole single-phase AC generator consists of windings constituting 80 series paths and is driven by a diesel engine. The effective or mean length of the armature is 18 cm and the cross-sectional radius of the armature is 5 cm. The armature is rotating at 1,800 rpm. Each armature pole is exposed to a magnetic flux of 1.0 T. The efficiency of this generator is 90% and it is rated 2 kW. Determine the following:

a. The maximum voltage generated.
b. The RMS voltage generated.
c. The horsepower rating of the generator.
d. The horsepower output of the prime mover.

Solution:
The maximum voltage, V_m, generated by this alternator is given by Eq. 7.7.

$$V_m = V_p = \left(\frac{\pi}{2}\right)\cdot\left(\frac{n}{30}\right)\cdot p\,NAB = \frac{\pi np\,NAB}{60} \qquad (7.7)$$

Given:
  $n = 1,800 \text{ rpm}$
  $p = 4$
  $N = \text{Number of series paths} = 80$
  $B = 1.0 \text{ T}$
  $A = (\text{Eff. diameter of the coil conductor}) \times (\text{Eff. length of the coil})$
  $= (2 \times 5 \text{ cm}) \times (18 \text{ cm}) = (0.1 \text{ m}) \times (0.18 \text{ m}) = 0.018 \text{ m}^2$

a. $V_m = \dfrac{\pi np\,NAB}{60} = \dfrac{(3.14)\cdot(1,800 \text{ rpm})(4)(80)(0.018 \text{ m}^2)(1.0 \text{ T})}{60}$

   $= 543 \text{ V}$

b. The RMS voltage could be calculated by using Eq. 7.6 or simply by dividing V_m, from part (a) by the square root of 2 as follows:

$$V_{RMS} = V_{eff} = \frac{V_m}{\sqrt{2}} = \frac{543}{\sqrt{2}} = 384 \text{ V}$$

c. The horsepower rating of this generator is the power **output** rating specified in hp, premised on the stated **output capacity** of 2.0 kW. Therefore, application of the 0.746 kW/hp conversion factor yields:

$$P_{hp} = \frac{P_{kW}}{0.746 \text{ kW/hp}} = \frac{2.0 \text{ kW}}{0.746 \text{ kW/hp}} = 2.68 \text{ hp}$$

d. The horsepower rating of the prime mover – or the propane fired engine – would need to offset the inefficiency of the AC generator. Therefore, based on the given 90% efficiency rating of the generator:

$$P_{hp\text{-Prime Mover}} = \frac{P_{hp\text{-Gen.}}}{\text{Eff.}_{Gen.}} = \frac{2.68 \text{ hp}}{0.90} = 2.98 \text{ hp}$$

4. A three-phase, four-pole, AC induction motor is rated 170 hp, is operating at full-load, 60 Hz, 460 V_{rms}, efficiency of 90%, power factor of 80%, and a slip of 4%. Determine the (a) motor shaft speed, in rpm, (b) torque developed, in ft-lbf, (c) line current drawn by the motor, and (d) the amount of reactive power, Q, sequestered in the motor under the described operating conditions.

Solution:
Given:

$P_{L,3-\phi}$ = Real power or rate of work performed by the motor = 170 hp
= (170 hp)·(746 W/hp) = 126,820 W
p = Four poles
$V_L = 460 \, V_{RMS}$
Pf = 80% or 0.80
Eff. = 90% or 0.90
n_s = Synchronous speed, in rpm = ?
Slip, s = 4%
f = Frequency of operation = 60 Hz

a. **Shaft or motor speed:** Rearrange and apply Eq. 7.10:

$$\text{Slip} = s = \left(\frac{n_s - n}{n_s} \right)$$

And, by rearrangement: $n = n_s (1 - s)$

Next, we must determine the synchronous speed of the motor by applying Eq. 7.9:

$$n_s = \text{Synchronous speed} = \frac{120f}{p} = \frac{(120) \cdot (60)}{4} = 1,800 \text{ rpm}$$

$$\therefore \, n = (1,800 \text{ rpm})(1 - 0.04) = 1,728 \text{ rpm}$$

b. **Torque developed, in ft-lbf**: There are multiple methods at our disposal for determining the torque developed. Formulas associated with two common methods are represented by Eqs. 7.12–7.14. Since the power is available in hp and the rotational speed in rpm, apply Eq. 7.12:

$$\text{Torque}_{(\text{ft-lbf})} = \frac{5,250 \times P_{(\text{horsepower})}}{n_{(\text{rpm})}} = \frac{5,250 \times 170 \text{ hp}}{1,728 \text{ rpm}} = 516 \text{ ft-lbf}$$

c. **Line current drawn by the motor:** Full-load line current is the current the motor draws from the power source or utility in order to sustain the full load. Application of Eq. 7.22 takes this into account in the computation of three-phase motor line current, after the motor hp rating (or actual load) is converted into watts:

$$\text{The 3-}\phi \text{ phase line curent, } |I_L| = \frac{|S_{L,3\text{-}\phi}|}{\sqrt{3}(|V_L|)} = \frac{P_{L,3\text{-}\phi} \text{ (in Watts)}}{\sqrt{3}(|V_L|)(\text{Pf}) \cdot (\text{Eff.})}$$

Three-phase (total) real power was converted into watts under "Given" as $P_{L,3\text{-}\phi} = 126,820 \text{ W}$

Therefore,

$$\text{The 3-}\phi \text{ phase line curent, } |I_L| = \frac{126,820 \text{ W}}{\sqrt{3}(460 \text{ V}_{\text{RMS}})(0.90) \cdot (0.80)} = 221 \text{ A}$$

d. **Reactive power, Q**: There are multiple approaches available to us for determination of reactive power Q for this motor application. We will utilize the power triangle or apparent power component vector method:

$$S = P + jQ, \text{ and } S^2 = P^2 + Q^2,$$

or,

$$Q^2 = S^2 - P^2$$

and

$$Q = \sqrt{S^2 - P^2}$$

Real power "P" was computed earlier as 126,820 W, and apparent power S can be assessed using Eq. 7.20:

$$|S_{L,3\text{-}\phi}| = \frac{P_{3\text{-}\phi} \text{ (in Watts)}}{(\text{Pf}) \cdot (\text{Eff.})} = \frac{126,820 \text{ W}}{(0.90) \cdot (0.80)} = 176,139 \text{ VA}$$

Therefore,

$$Q = \sqrt{S^2 - P^2} = \sqrt{176,139^2 - 126,820^2} = \sqrt{14,941,595,779}$$

$$= 122,236 \text{ VAR}$$

5. A three-phase, four-pole, AC induction motor is tested to deliver 200 hp, at 900 rpm. Determine the frequency at which this motor should be operated for the stated shaft speed. Assume the slip to be negligible.

Solution:
Given:

  p = Four poles
  n$_s$ = Synchronous speed, in rpm = 900 rpm = Shaft speed, since slip is zero
  f = Frequency of operation = ?
Rearrange and apply Eq. 7.9

$$n_s = \text{Synchronous speed} = \frac{120f}{p},$$

or,

$$f = \frac{(n_s) \cdot (p)}{120} = \frac{(900) \cdot (4)}{120} = 30 \text{ Hz}$$

6. A three-phase induction motor delivers 600 kW at a power factor of 80%. In lieu of installing power factor correction capacitors, a synchronous motor is being considered as a power fact correction measure. Determine the apparent power size of the synchronous motor – in kVA – that should be installed to carry a load of 300 hp and raise the (combined) power factor to 93%. The source voltage is 230 V$_{rms}$.

Solution:
Given:

  P$_I$ = Real power delivered by the 3-φ induction motor = 600 kW
  P$_S$ = Real power contributed by the synchronous motors = 300 hp = (300 hp) × (0.746 kW/hp) = 223.8 kW
  Pf$_i$ = Initial power factor = 80% = 0.80
  Pf$_f$ = Final power factor = 93% = 0.93
  V$_s$ = Source voltage = 230 V$_{rms}$. However, the voltage information is not needed to solve this problem.
This problem is similar to Example 7.6. So, the strategy for solving it is the same as the one employed for Example 7.6. Overall, we need to determine the reactive power, Q$_S$, contributed by the synchronous motor. Then, by applying the Pythagorean Theorem to Q$_S$ and P$_S$, we can derive the apparent power (kVA) of the synchronous motor.

$$\text{Since } S_T^2 = P_T^2 + Q_T^2, \quad Q_T = \sqrt{S_T^2 - P_T^2}$$

The combined real power of the induction motor and the synchronous motor:

$$P_T = 600 \text{ kW} + 223.8 \text{ kW} = 824 \text{ kW}$$

$$\text{Since } S_T \cos\theta = P_T, \ S_T = \frac{P_T}{\cos\theta},$$

And since θ_T, the final power factor angle $= \cos^{-1}(0.93) = 21.58°$,

$$S_T = \frac{P_T}{\cos\theta_T} = \frac{824 \text{ kW}}{\cos(21.58°)} = 886 \text{ kVA}$$

Therefore,

$$Q_T = \sqrt{S_T^2 - P_T^2} = \sqrt{886^2 - 824^2} = \sqrt{106,058} = 326 \text{ kVAR}$$

Now, in order to determine Q_S, the reactive power contributed by the synchronous motor, we must subtract the original reactive power, Q_O, from the final, total, reactive power, Q_T. However, Q_O is unknown and can be determined through the power triangle as follows:

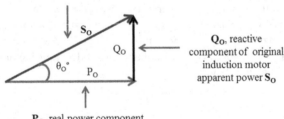

Since $\tan(\theta_O) = \dfrac{Q_O}{P_O}$, or, $Q_O = P_O \tan(\theta_O) = 600 \text{ kW} \tan(\theta_O)$

and

$$\theta_O = \cos^{-1}(\text{Pf}_O) = \cos^{-1}(0.80) = 36.87°$$

Therefore,

$$Q_O = 600 \text{ kW} \tan(36.89°) = 600 \text{ kW}(0.75) = 450 \text{ kVAR}$$

And, reactive power contributed by the synchronous motor would be:

$Q_S = Q_O - Q_T = 450 \text{ kVAR} - 326 \text{ kVAR} = 124 \text{ kVAR}$

Therefore,

$S_S = P_S + jQ_S$, and $S_S^2 = P_S^2 + Q_S^2$, or, $S_S = \sqrt{P_S^2 + Q_S^2}$

or,

$S_S = \sqrt{(224 \text{ kW})^2 + (124 \text{ kVAR})^2} = 256 \text{ kVA}$

CHAPTER 8 ANSWERS/SOLUTIONS

1. A substation in a manufacturing facility is being fed from a 13 kV transformer secondary. This switchgear in this substation would be categorized as:
   A. **Medium voltage**
   B. Low voltage
   C. High voltage
   D. None of the above

2. Power transmission lines would be categorized as:
   A. Medium voltage
   B. Low voltage
   C. **High voltage**
   D. None of the above

3. The breakers installed in residential breaker panels are:
   A. OCBs
   B. Thermal magnetic circuit breakers
   C. **Low voltage thermal magnetic circuit breakers**
   D. None of the above
   E. Both B and C

4. The vacuum circuit breakers tend to offer longer service spans between overhauls than do air circuit breakers.
   A. **True**
   B. False

5. The SF$_6$-type high-voltage circuit breakers are not preferred due to environmental concerns.
   A. True
   B. **False**

6. MCCs are not designed to accommodate PLCs and VFDs.
   A. True
   B. **False**

7. The bus bars in MCCs are commonly constructed out of:
   A. Aluminum
   B. Silver-plated copper
   C. Silver
   D. Iron

8. Pilot devices on MCCs:
   A. Control circuit breakers
   B. Indicate the status of MCC
   C. Indicate the status of motor/load
   D. Include "Start" and "Stop" controls
   E. Both (C) and (D)

9. Power control cubicles in MCCs are fixed and cannot be removed while the main fusible disconnect switch of the MCC is ON.
   A. True
   B. False

10. A control transformer in a given MCC compartment:

    A. Steps down the voltage for control circuit operation
    B. Provides power for MCC cabinet lighting
    C. Is seldom needed
    D. Serves as an isolation transformer
    E. Both (C) and (D)

CHAPTER 9 – SOLUTIONS

1. A given circuit is meant to carry a continuous lighting load of 16 A. In addition, four loads designed for permanent display stands are fastened in place and require 2 A each when operating. What is the rating of the overcurrent protective device (OCPD) on the branch circuit?

   Solution:
   From Article 210.20(a) of the NEC®, the OCPD must be rated at 100% of the non-continuous load plus 125% of the continuous load.

   $$\therefore \text{The Rating of OCPD} = (1.00)(2\text{ A} + 2\text{ A} + 2\text{ A} + 2\text{ A}) + (1.25)(16\text{ A})$$

   $$= 28\text{ A, minimum}$$

   A standard fixed-trip circuit breaker or a fuse rated at 30 A can be used (see Sec. 240.6).

2. A three-phase, four-wire feeder with a full-sized neutral carries 14 A continuous and 40 A non-continuous loads. The feeder uses an overcurrent device with a terminal or conductor rating of 60°C. What is the minimum copper conductor size? Assume no derating applies. Use Tables 9.1 and 9.2.

Solution:
Feeder conductor size, before derating, is based on 100% of the non-continuous load and 125% of the continuous load [Art. 215.2(a)].

$$\therefore \text{Amp load} = (1.00)(40\ A) + (1.25)(14\ A)$$

$$= 40\ A + 17.5\ A$$

$$= 57.5\ A \approx 58\ A$$

The total of 58 A is used since an ampacity of 0.5 A or greater is rounded up [Sec. 220.2(b)]. Using NEC® Table 310.15 – as presented in the form of Tables 9.1 and 9.2 in this text – TW, UF, **AWG 4 from the 60°C column should be selected.** The ampacity of AWG 4 is 70 A.
Note: AWG 6 (the size below AWG 4), with an ampacity of 55 A, would be undersized.

3. Electrical specifications for a brewery company call for a fusible disconnect switch enclosure that must be able withstand occasional splashing of water during periodic wash downs required by the local health codes. This design will be applied in breweries in the United States as well as Europe. The water flow from the 1-in wash down nozzles is expected to be less than 60 GPM from a distance of 11 ft for less than 4 minutes. (a) Determine the NEMA rating of enclosure for the US installations. (b) Determine the IP rating of enclosure for the European installations.

Solution:
a. Examination of the NEMA – IP rating table in this chapter shows that NEMA 4 enclosure is rated for:

  Watertight (weatherproof). Must exclude at least 65 GPM of water from 1-in. nozzle delivered from a distance not less than 10 ft for 5 minutes. Used outdoors on ship docks, in dairies, and in breweries.

  Therefore, a NEMA 4 enclosure would exceed the "1-in wash down nozzle, 60 GPM from a distance of 11 ft for less than 4 min" worst-case exposure. Hence, a **NEMA 4 enclosure** should be specified for the US installation.

b. Since the European installation would be exposed to the same worst-case conditions, the US NEMA 4's European counterpart, **IP 66 should be specified**.

  Note: If the health code, in the United States, were to require that chlorine solution be used for the wash down, since chlorine is corrosive, **NEMA 4X** should be specified.

4. **Overcurrent Protection and Conductor Ampacity:**
  Applicable Code/Codes: Articles 210.19 (A) (1), 210.20 (A), and 310.15. Table 310.15(B)(16) as presented in Tables 9.1 and 9.2 of this text. (Note: This is not the actual table and is only reproduced in this text for exercise and illustration purposes.)

TABLE 9.1
Partial NEC® Table 310.15(B)(16)

Partial Ampacity Information from NEC® Table 310.15(B)(16) – Allowable Ampacities of Insulated Conductors Rated Up to and Including 2,000V, 60°C through 90°C (140°F through 194°F), for Not More Than Three Current-Carrying Conductors in Raceway, Cable, or Earth (Directly Buried), Based on Ambient Temperature of 30°C (86°F)[a].

Size AWG or kcmil	Temperature Rating of Conductor; Ref: NEC® Table 310.104(A)						Size AWG or kcmil
	Copper			Aluminum or Copper-Clad Aluminum			
	60°C (140°F)	75°C (167°F)	90°C (194°F)	60°C (140°F)	75°C (167°F)	90°C (194°F)	
	Types TW, UF	Types RHW, THHW, THW, THWN, XHHW, USE, ZW	Types TBS, SA, SIS, FEP, FEPB, MI, RHH, RHW-2, THHN, THHW, THW-2, THWN-2, USE-2, XHH, XHHW, XHHW-2, ZW-2	Types TW, UF	Types RHW, THHW, THW, THWN, XHHW, USE	Types TBS, SA, SIS, THHN, THHW, THW-2, THWN-2, RHH, RHW-2, USE-2, XHH, XHHW, XHHW-2, ZW-2	
18	–	–	14	–	–	–	–
16	–	–	18	–	–	–	–
14b	15	20	25	–	–	–	–
12b	20	25	30	15	20	25	12b
10b	30	35	40	25	30	35	10b
8	40	50	55	35	40	45	8
6	55	65	75	40	50	55	6
4	70	85	95	55	65	75	4
3	85	100	115	65	75	85	3
2	95	115	130	75	90	100	2
1	110	130	145	85	100	115	1

(Continued)

TABLE 9.1 (Continued)
Partial NEC® Table 310.15(B)(16)

Partial Ampacity Information from NEC® Table 310.15(B)(16) – Allowable Ampacities of Insulated Conductors Rated Up to and Including 2,000V, 60°C through 90°C (140°F through 194°F), for Not More Than Three Current-Carrying Conductors in Raceway, Cable, or Earth (Directly Buried), Based on Ambient Temperature of 30°C (86°F)[a].

	Temperature Rating of Conductor; Ref: NEC® Table 310.104(A)			
Size AWG or kcmil	60°C (140°F) Types TW, UF	75°C (167°F) Types RHW, THHW, THW, THWN, XHHW, USE, ZW	90°C (194°F) Types TBS, SA, SIS, FEP, FEPB, MI, RHH, RHW-2, THHN, THHW, THW-2, THWN-2, USE-2, XHH, XHHW, XHHW-2, ZW-2	Size AWG or kcmil
	Copper			
1/0	125	150	170	1/0
2/0	145	175	195	2/0
3/0	165	200	225	3/0
4/0	195	230	260	4/0
250	215	255	290	250
300	240	285	320	300
350	260	310	350	350
	60°C (140°F) Types TW, UF	75°C (167°F) Types RHW, THHW, THW, THWN, XHHW, USE	90°C (194°F) Types TBS, SA, SIS, THHN, THHW, THW-2, THWN-2, RHH, RHW-2, USE-2, XHH, XHHW, XHHW-2, ZW-2	
	Aluminum or Copper-Clad Aluminum			
1/0	100	120	135	1/0
2/0	115	135	150	2/0
3/0	130	155	175	3/0
4/0	150	180	205	4/0
250	170	205	230	250
300	195	230	260	300
350	210	250	280	350

Note: Included for general illustration purposes only. Use actual NEC® tables for work.

[a] Refer to NEC® Table 310.15(B)(2)(a) for the ampacity correction factors where the ambient temperature is other than 30°C (86°F). Refer to NEC® Table 310.15(B)(3)(a) for more than three current-carrying conductors.

[b] Refer to NEC® Article 240.4(D) for conductor overcurrent protection limitations.

TABLE 9.2
Partial Table 310.15(B)(2)(a) Ambient Temperature Correction Factors Based on 30°C (86°F)

For Ambient Temperatures Other Than 30°C (86°F), Multiply the Allowable Ampacities Specified in the Ampacity Tables by the Appropriate Correction Factor Shown Below

Ambient Temperature (°C)	Temperature Rating of Conductor			Ambient Temperature (°F)
	60°C	75°C	90°C	
10 or less	1.29	1.2	1.15	50 or less
11–15	1.22	1.15	1.12	51–59
16–20	1.15	1.11	1.08	60–68
21–25	1.08	1.05	1.04	69–77
26–30	1	1	1	78–86
31–35	0.91	0.94	0.96	87–95
36–40	0.82	0.88	0.91	96–104
41–45	0.71	0.82	0.87	105–113
46–50	0.58	0.75	0.82	114–122
51–55	0.41	0.67	0.76	123–131

Note: For illustration purposes only. Use actual NEC® tables for work.

The branch circuit in the exhibit below consists of three continuous loads. Overcurrent protection in the branch circuit is provided through a 20 A circuit breaker. (a) Determine the size of copper conductor based on the ampacities given in Tables 9.1 and 9.2 assuming conductor temperature is at 75°C or less. Assume that 75°C operation and selection is allowed. (b) Verify the size/specifications of the circuit breaker. Assume no derating applies. (c) If the ambient temperature were to rise to 50°C, how would the conductor size be impacted?

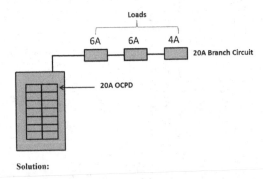

Solution:

Solution:
a. In accordance with Article 210.19 (A) (1): "Branch circuit conductors shall have an ampacity not less than maximum load to be served... (and) shall have an allowable ampacity not less than the non-continuous load plus 125% of the continuous load.

∴ The conductor ampacity for the given branch circuit

$$= 1.25 \times \text{Continuous Load} + 1.00 \times \text{Non-Continuous Load}$$

$$= 1.25 \times (16 \text{ A}) + 1.00 \times (0) = 20 \text{ A}.$$

According to Table 9.1 (in this text) for 75°C operation, with types RHW, THHW, THW, THWN, XHHW, USE, and ZW insulation, AWG 12 conductor carries an allowable ampacity of 25 A for conductors that are insulated, rated for 0–2,000 V operation, in situations with no more than three (3) current-carrying conductors in raceway, cable earth (directly buried); under ambient temperature (not exceeding) 30°C (85°F); with no required/applicable derating. With all conditions remaining the same, an AWG 14 could carry 20 A. However, the asterisk annotation on both AWG 14 and AWG 12, through the footnotes on Table 9.2, stipulates that if AWG 14 is selected it must be protected at no more than 15 A. Selection of 15 A protection for a continuous load could result in nuisance tripping or fuse clearing. Therefore, AWG 12, with the stated ampacity of 25 A, at 75°C operation should be selected. Also, selection of AWG 12 would maintain the existing 20 A breaker in compliance with the code.

b. In accordance with Article 210.20 (A):

Branch-circuit conductors and equipment shall be protected by overcurrent protective devices that have a rating or setting that complies with 210.20(A) through (D).

(A)   Continuous and Non-continuous Loads. Where a branch circuit supplies continuous loads or any combination of continuous and non-continuous loads, the rating of the overcurrent device shall not be less than the non-continuous load plus 125 percent of the continuous load.

(NEC® 2011)

The overcurrent protection should be rated = 1.25 × continuous load

$$= 1.25 \times (6 + 6 + 4) = 20 \text{ A}$$

∴ The 20 A circuit breaker as an overcurrent protection device is adequate for the given branch circuit provided AWG 12 conductor is used.

c. Ambient temperature rise and conductor size:

According to Article 310, Table 310.15(B)(16), as presented under Tables 9.1 and 9.2 of this text, when ambient deviates from 40°C, the derating/up-rating multipliers, listed at the bottom of Tables 310.15(B)(2)(b) – Table 9.2 in this text – must be applied to adjust the ampacity of the conductor.

As recognized above, the ampacity of an AWG #12, under 75°C operation, is 25 A. From Table 9.2, the multiplier for 50°C ambient, under 75°C terminal rating, is 0.75.

∴ The adjusted or derated ampacity of AWG #12 conductor, in this case, would be:

$$= 0.75 \times 25 \text{ A} = 18.75 \text{ A}$$

Since the derated ampacity of AWG #12, for this application, falls below the 20 A capacity mandated by 210.19 (A) (1), AWG #12 would no longer be adequate. Therefore, AWG #10, which is the next size above AWG#12, must be considered. According to Table 9.1, for 75°C operation, with types RHW, THHW, THW, THWN, XHHW, USE, and ZW insulation, AWG 10 conductor carries an allowable ampacity of 35 A. Then, if the 50°C adjustment rating of 0.75 is applied to 35 A ampacity of an AWG 10, the derated ampacity would be:

$$= 0.71 \times 35 \text{ A} = 26.25 \text{ A}$$

Since the 26.25 A derated ampacity of AWG #10 exceeds the 20 A requirement, it would meet the code. AWG #10 should be selected for this application at 50°C ambient temperature.

5. A US appliance manufacturer is planning to market a new appliance in Mexico. The most appropriate safety certification for this appliance would be:
   A. UL
   B. ULC
   C. ETL
   D. NOM: NOM Mark: (Norma Official Mexicana) NOM is a mark of product safety approval for virtually any type of product exported into Mexico.

6. Assume that the alarm switch in the control circuit depicted below is opened after being closed for a prolonged period of time. Which of the following conditions would best describe the status of the annunciating lights and the horn when the switch is opened?

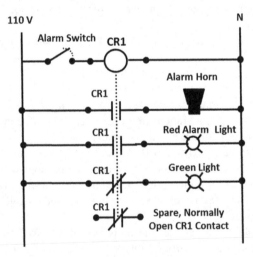

   A. Alarm horn will turn off
   B. Red light will turn off
   C. Green light will turn on
   D. All of the above

CHAPTER 10 – SOLUTIONS

1. Answer the following questions pertaining to the branch circuit shown in the schematic diagram below:

    a. What is the maximum voltage the power distribution system for this branch circuit rated for?

    b. How many wires and phases is the power distribution system for this motor branch circuit rated for?

    c. What would be the proper rating for the branch circuit disconnect switch?

    d. What should the solid-state overload device be set for at commissioning of this branch circuit?

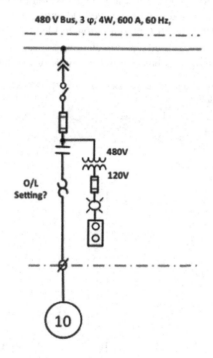

480 V Bus, 3 φ, 4W, 600 A, 60 Hz,

Solution:

a. The maximum voltage rating of the power distribution system for this branch circuit is included in the specification stated at the very top of the schematic diagram, as 480 V, within the caption:

    "**480 V Bus**, 3 φ, 4 W, 600 Amp, 60 Hz."

    Therefore, the answer is: **480 V.**

b. The number of wires and phases in this branch circuit can be assessed from the specification stated at the very top of the schematic diagram:

    "480 V Bus, **3 φ, 4 W**, 600 Amp, 60 Hz."

    Therefore, the answer is: **3 φ, 4 W.**

c. The proper rating for the branch circuit disconnect switch can be determined by using the NEC® or through the use of published tables such as the Buss® table introduced earlier in this chapter. As stated in the schematic diagram – in the circled motor symbol – the motor's full-load rating is **10 hp**. As specified at the top of schematic diagram, the motor is being powered by a **480 V, three-phase** source (480 V Bus, 3 φ, 4 W, 600 A, 60 Hz). Therefore, according to the Buss® table, and as highlighted (circled) below – under the "460 V(480V), 3-ph, section – the fusible disconnect switch size would be **3-φ or three-pole, 30 A, 480 V**. See the fourth column from the right.

d. Using the Buss® table above, for **three-phase, 10 hp** motor, operating at a **480 V,** the overload device should be set at **17.0 A**.

   Note: If NEC® Tables were used here, with the full-load current rating of 14 A (See the circled segment of the Buss® table below), the 115% multiplier by NEC® would result in an overload setting of 16 A. It is important to bear in mind that the NEC® requires that full-load Amp stated on the nameplate of the motor be used for computations associated with the code. In this text, the Buss® table is being used for simplicity and illustration purposes.

BUSS® SYSTEM 300 MOTOR PROTECTION GUIDE
"NO-DAMAGE" "TYPE 2" SHORT-CIRCUIT PROTECTION
OVERLOAD OR BACK-UP OVERLOAD PROTECTION

115VAC (120V), 1 ph. (LPN-RK-SP or LPJ-SP) / 200V (208V), 3 ph. (LPN-RK-SP or LPJ-SP)

HP	Motor Full Load Amps	Overload 1.15 S.F. Or Greater Or Rated Not Over 40°C	Overload All Other Motors	Back-up 1.15 S.F. Or Greater Or Rated Not Over 40°C	Back-up All Other Motors	Switch Or Fuseholder Size	Minimum NEMA Starter Size	Minimum Copper Conductor Size†	Minimum Conduit Size‡
1/6	4.4	5	5	5.6	5.6	30	00	14	1/2
1/4	5.8	7	6.25	7.5	7	30	00	14	1/2
1/3	7.2	9	9	9	9	30	0	14	1/2
1/2	9.8	12	10	15	12	30	0	14	1/2
3/4	13.8	15	15	17.5	17.5	30	0	14	1/2
1	16	20	17.5	20	20	30	0	14	1/2
1½	20	25	20	25	25	30	1	12	1/2
2	24	30	25	30	30	30	1	10	1/2
1/2	2.3	2.6	2.5	3	2.8	30	00	14	1/2
3/4	3.22	4	3.5	4.5	4	30	00	14	1/2
1	4.14	5	4.5	5.6	5	30	00	14	1/2
1½	5.98	7	6.25	7.5	7	30	00	14	1/2
2	7.82	9	8	10	9	30	0	14	1/2
3	11	12	12	15	15	30	0	14	1/2
5	17.5	20	20	25	25	30	1	12	1/2
7½	25.3	30	25	35	30	60	2	8	1/2
10	32.2	40	35	45	40	60	2	6	3/4
15	48.3	60	50	70	60	60	3	4	1
20	62.1	75	70	80	75	100	3	3	1
25	78.2	90	80	100	90	100	3	1	1¼
40	120	150	125	150	150	200	4	1/0	1¼
50	150	175	150	200	175	200	5	3/0	1½
60	177	200	200	225	225	400	5	4/0	2
75	221	250	250	300	300	400	5	300	2
100	285	350	300	400	350	400	6	500	3
125	359	400	400	450	450	600	6	2-4/0	2-2
150	414	500	450	600	500	600	6	2-300	2-2

230V (240V), 1 ph. (LPS-RK-SP or LPJ-SP) / 460V (480V), 3 ph. (LPS-RK-SP or LPJ-SP)

HP	Motor Full Load Amps	Overload 1.15 S.F. Or Greater Or Rated Not Over 40°C	Overload All Other Motors	Back-up 1.15 S.F. Or Greater Or Rated Not Over 40°C	Back-up All Other Motors	Switch Or Fuseholder Size	Minimum NEMA Starter Size	Minimum Copper Conductor Size†	Minimum Conduit Size‡
1/6	2.2	2.5	2.5	2.8	2.8	30	00	14	1/2
1/4	2.9	3.5	3.2	4	3.5	30	00	14	1/2
1/3	3.6	4.5	4	4.5	4.5	30	00	14	1/2
1/2	4.9	5.6	5.6	6.25	6	30	00	14	1/2
3/4	6.9	8	7.5	9	8	30	00	14	1/2
1	8	10	9	10	10	30	00	14	1/2
1½	10	12	10	15	12	30	0	14	1/2
2	12	15	12	15	15	30	0	14	1/2
3	17	20	17.5	25	20	30	1	12	1/2
5	28	35	30	35	35	60	2	8	1/2
7½	40	50	45	50	50	60	2	6	3/4
10	50	50	70"	60	60	3	4	3/4
1/2	1	1.25	1¼	1.25	1.25	30	00	14	1/2
3/4	1.4	1.6	1.6	1.8	1.8	30	00	14	1/2
1	1.8	2.25	2	2.25	2.25	30	00	14	1/2
1½	2.6	3.2	2.6	3.5	3	30	00	14	1/2
2	3.4	4	3.5	4.5	4	30	00	14	1/2
3	4.8	5.6	5	6	5.6	30	0	14	1/2
5	7.6	9	8	10	10	30	0	14	1/2
10	14	17.5	15	17.5	17.5	30	1	14	1/2
15				30		30			
20	27	30"	30"	35	35	60	2	8	1/2
25	34	40	35	45	40	60	2	6	3/4
30	40	50	45	50	50	60	3	6	3/4
40	52	60"	60"	70	60	100	3	4	1
50	65	80	70	90	75	100	3	3	1
60	77	90	80	100	90	100	4	1	1¼
75	96	110	110	125	125	200	4	1/0	1¼
100	124	150	150	175	150	200	4	2/0	1½
125	156	175	175	200	200	200	5	3/0	1½
150	180	225	200"	225	225	400	5	4/0	2
200	240	300	250	300	300	400	5	350	2½

* Fuse reducers required.   ** 100A switch required.
† THWN connected to 60°C terminations up to #1. AWG to 75°C terminations for 1/0 and larger. Consult equipment manufacturer for listed termination temperature rating. Higher equipment termination temperature ratings may allow smaller conductor and conduit size.
‡ Based on 3 conductors for 3 ph. circuits and 2 conductors for 1 ph. circuits.

MPG

2. Determine the sizes/specifications of the following components in the branch circuit shown below using the Bus® Table and information included in the questions:
   a. Conductor size.
   b. Conduit size.
   c. Overload setting based on the 115% NEC® stipulation.
   d. Disconnect switch size, safety and fusible.

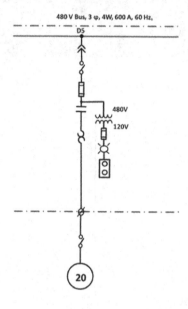

480 V Bus, 3 φ, 4W, 600 A, 60 Hz,

Solution:

   a. **Conductor size:** Since the load in the given branch circuit is a 20-hp motor, at 460/480 V, we will focus on the lower right section of the Bus® table. See the circled 20-hp row on the Bus® table below. The rightmost columns in the Bus® table represent NEC®-based switch, starter, conductor, and conduit sizes. The second column from the right, for the 20-hp motor load, shows that **AWG #8 conductor should be specified**.

   b. **Conduit size:** According to the rightmost column in the Bus® table, a ½″ conduit is permitted by the code for housing AWG #8 conductors associated with the 20-hp motor.

   c. **Overload setting based on the 115% NEC® stipulation**: First we must identify the FLA, Full-Load Amp, for the 20-hp motor, using the Bus® table. The FLA is listed in the column adjacent to the HP column. So, according to the Bus® table, FLA for a three-phase, 20-hp motor, operating at 460/480 V, would be 27 A. Then, as explained in this chapter, if the code requires that equipment be protected against overload at 115%, the overload should be set at:

   $$1.15 \times (\text{Motor Full Load Amp}) = 1.15 \times 27 \text{ A} = \textbf{31 A}$$

d. **Disconnect switch size, safety, and fusible:** In order to specify the sizes of the branch circuit fusible disconnect switch and the safety switch in the field, we will focus on the fourth column from the right in the Bus® table. In this column, we see that **60 A switch is required for a 20-hp motor.** This specification would apply to the branch circuit fusible disconnect switch and the safety switch in the field.

BUSS® SYSTEM 300

MOTOR PROTECTION GUIDE

"NO-DAMAGE" "TYPE 2" SHORT-CIRCUIT PROTECTION
OVERLOAD OR BACK-UP OVERLOAD PROTECTION

	HP	Motor Full Load Amps	Fuse Size And Type Protection: Overload — 1.15 S.F. Or Greater Or Rated Not Over 40°C	Overload — All Other Motors	Back-up — 1.15 S.F. Or Greater Or Rated Not Over 40°C	Back-up — All Other Motors	Switch Or Fuseholder Size	Minimum NEMA Starter Size	Minimum Copper Conductor Size†	Minimum Conduit Size‡
115VAC (120V), 1 ph. (LPN-RK-SP or LPJ-SP)	1/6	4.4	5	5	5.6	5.6	30	00	14	1/2
	1/4	5.8	7	6.25	7.5	7	30	00	14	1/2
	1/3	7.2	9	8	9	9	30	0	14	1/2
	1/2	9.8	12	10	15	12	30	0	14	1/2
	3/4	13.8	15	15	17.5	17.5	30	0	14	1/2
	1	16	20	17.5	20	20	30	0	14	1/2
	1½	20	25	20	25	25	30	1	12	1/2
	2	24	30	25	30	30	30	1	10	1/2
200V (208V), 3 ph. (LPN-RK-SP or LPJ-SP)	1/2	2.3	2.6	2.5	3	2.8	30	00	14	1/2
	3/4	3.22	4	3.5	4.5	4	30	00	14	1/2
	1	4.14	5	4.5	5.6	5	30	00	14	1/2
	1½	5.98	7	6.25	7.5	7	30	00	14	1/2
	2	7.82	9	8	10	9	30	0	14	1/2
	3	11	12	12	15	15	30	0	14	1/2
	5	17.5	20	20	25	25	30	1	12	1/2
	7½	25.3	30*	25*	35	30*	60	1	8	1/2
	10	32.2	40	35	45	40	60	2	6	3/4
	15	48.3	60	50	70**	60	60	3	4	1
	20	62.1	75	70	80	75	100	3	3	1
	25	78.2	90	80	100	90	100	3	1	1¼
	30	92	100	100*	125	110	200	4	1/0	1¼
	40	120	150	125	150	150	200	4	1/0	1¼
	50	150	175	150	200	175	200	5	3/0	1½
	60	177	200*	200*	225	225	400	5	4/0	2
	75	221	250	250	300	300	400	5	300	2
	100	285	350	300	400	350	400	6	500	3
	125	359	400*	400*	450	450	600	6	2-4/0	2-2
	150	414	500	450	600	500	600	6	2-300	2-2
230V (240V), 1 ph. (LPS-RK-SP or LPJ-SP)	1/6	2.2	2.5	2.5	2.8	2.8	30	00	14	1/2
	1/4	2.9	3.5	3.2	4	3.5	30	00	14	1/2
	1/3	3.6	4.5	4	4.5	4.5	30	00	14	1/2
	1/2	4.9	5.6	5.6	6.25	6	30	00	14	1/2
	3/4	6.9	8	7.5	9	8	30	00	14	1/2
	1	8	10	9	10	10	30	00	14	1/2
	1½	10	12	10	15	12	30	0	14	1/2
	2	12	15	12	15	15	30	0	14	1/2
	3	17	20	17.5	25	20	30	1	12	1/2
	5	28	35	30*	35	35	60	2	8	1/2
	7½	40	50	45	50	50	60	2	6	3/4
	10	50	60	50	70**	60	60	3	4	3/4
460V (480V), 3 ph. (LPS-RK-SP or LPJ-SP)	1/2	1	1.25	1⅛	1.25	1.25	30	00	14	1/2
	3/4	1.4	1.6	1.6	1.8	1.8	30	00	14	1/2
	1	1.8	2.25	2	2.25	2.25	30	00	14	1/2
	1½	2.6	3.2	2.6	3.5	3	30	00	14	1/2
	2	3.4	4	3.5	4.5	4	30	00	14	1/2
	3	4.8	5.6	5	6	5.6	30	0	14	1/2
	5	7.6	9	8	10	9	30	0	14	1/2
	7½	11	12	12	15	15	30	1	14	1/2
	10	14	17.5	15	17.5	17.5	30	1	14	1/2
	15	[illegible]	25	20	30	25	30	2	[illegible]	1/2
	20	27	30*	30*	35	35	60	2	8	1/2
	25	34	40	35	45	40	60	3	[illegible]	3/4
	30	40	50	45	50	50	60	3	6	3/4
	40	52	60*	60*	70	60*	100	3	4	1
	50	65	80	70	90	75	100	3	3	1
	60	77	90	80	100	90	100	4	1	1¼
	75	96	110	110	125	125	200	4	1/0	1¼
	100	124	150	125	175	150	200	4	2/0	1½
	125	156	175	175	200	200	200	5	3/0	1½
	150	180	225	200*	225	225	400	5	4/0	2
	200	240	300	250	300	300	400	5	350	2½

* Fuse reducers required. ** 100A switch required.
† THWN connected to 60°C terminations up to #1. AWG to 75°C terminations for 1/0 and larger. Consult equipment manufacturer for listed termination temperature rating. Higher equipment termination temperature ratings may allow smaller conductor and conduit size.
‡ Based on 3 conductors for 3 ph. circuits and 2 conductors for 1 ph. circuits.

MPG

3. Consider the wiring diagram for the 75-hp motor shown in Figure 10.6 and answer the following questions based on the control logic explained in this chapter:
   a. What would be the likely outcome if the START switch is depressed while the motor is operating?

      Answer: If the START switch is depressed while the motor is operating, the START switch contacts will provide a **redundant** 120V supply to the starter coil and the motor ON light. Therefore, the motor will continue to operate, with no change.

b. What would be the outcome if START and STOP switches are depressed simultaneously?

Answer: If the START and STOP switches are depressed simultaneously, the motor will stop. This is due to the fact that the 120 V supply to the motor starter coil is **interrupted** whenever the normally closed contacts of the STOP switch are opened.

c. What would be the likely outcome if the main disconnect switch for the MCC is opened?

Answer: If the main disconnect switch for the MCC is opened, the 480 V supply to the control transformer primary is removed. This de-energizes the secondary, turns off the 120 V supply for motor control circuit, and turns off the motor branch circuit.

d. Can the motor be stopped if the motor starter latching contact "welds" shut due to overheating?

Answer: When the spring-loaded STOP switch is depressed, the 120 V supply to the motor control circuit is removed and the motor shuts off. However, because the latching contact is sealed shut, the moment the STOP switch is released, the **motor control coil would re-energize and restart the motor.** This is due to the fact that with the coil latching contacts sealed closed, the START switch is **not needed** to start the motor.

4. Consider the logic associated with Timer P105.TD in Rung #11 of the PLC relay ladder logic program shown in Figure 10.10 and answer the following questions:

a. What would be status of the timer **Done** bit: "--||--" 105.TD_CASCADE_START. DN, if the XIC System E-Stop bit turns "**False**," 1 second after Timer P105.TD in Rung #11 is turned on?

Answer: The "--||--" 105.TD_ CASCADE_ START. DN bit is OFF when the timer begins the count. This bit remains OFF until the timer has timed out and accumulated time becomes equal to the preset time of 2,500 ms or 2.5 seconds. The XIC System E-Stop bit is representing the state of the System Emergency Stop switch. When this switch is depressed, the E-Stop bit turns "**False**" and the Timer P105.TD_ CASCADE_START rung (Rung #11) is "broken," and the timer stops. Since, at 1 second, or 1,000 ms, the accumulated time is not equal to the present time of 2,500 ms, the 105.TD_ CASCADE_ START. DN, **Done** bit will remain **OFF**.

b. Would the "P105 Conveyor _CMD" bit – commanding P105 Conveyor to turn ON – be **True** if and when the system E-Stop switch is engaged/pressed?

Answer: The answer is no. This is because the XIC System E-Stop bit, representing the state of the System Emergency Stop switch, will turn "**False**" and stop the timer before it times out and turns ON the

"P105.TD_ CASCADE_ START. DN" bit. From examination of Rung #11 we can see that if the P105.TD_CASCADE_START. DN, **Done** bit is **not True**, P105 Conveyor _CMD" coil will **not** energize and P105 Conveyor will remain **OFF**.

CHAPTER 11 – SOLUTIONS

1. HP, or hourly pricing, program is a standard feature in all OPT, or Time of Use, schedules.
   A. True
   B. False

2. The energy charge rate structure with electrical OPT schedule is:
   A. Flat, year round
   B. Tiered
   C. Exponential
   D. Is a function of time and season.

3. A large industrial electricity consumer set the peak demand in July's billing month at 40 MW. The demand rate structure is same as that included in Duke Energy® OPT-I, Time of Use, rate schedule, as shown in Table 11.2. Determine the demand cost for the month.
   A. $367,000
   B. $505,000
   C. $407,000
   D. $476,579

Solution:
As explained earlier in this chapter, the demand charges under OPT-I rate schedule are tiered. The tiered levels of demand charges are shown in Duke Energy's OPT-I rate schedule introduced in this chapter under Table 11.2.

As apparent from Table 11.2 excerpt below, July rates, under the OPT-I rate schedule falls under the summer months. Therefore, the shaded section of the excerpted table will be used for the tiered demand calculation.

II. Demand Charge	Summer Months	Winter Months
A. On-peak demand charge	June 1–September 30	October 1–May 31
For the first 2,000 kW of billing demand per month, per kW	$14.0767	$8.2981
For the next 3,000 kW of billing demand per month, per kW	$12.8972	$7.1075
For all over 5,000 kW of billing demand per month, per kW	$11.7067	$5.9064

Charge for the first tier, or $2,000 \text{ kW} = (2,000 \text{ kW})$

$$\cdot (\$14.0767/\text{kW})$$

$$= \$28,153$$

Charge for the second tier, or $3,000 \text{ kW} = (3,000 \text{ kW})$

$$\cdot (\$12.8972/\text{kW})$$

$$= \$38,691$$

Charge for the remaining demand $= (40,000 \text{ kW} - 5,000 \text{ kW})$

$$\cdot (\$11.7067/\text{kW})$$

$$= \$409,734$$

Total demand charge for the month $= \$28,153 + \$38,691 + \$409,734$

$$= \mathbf{\$476,579}$$

4. Calculate the **energy charge** for the month of July considered in Problem 3 if all of the energy is consumed during on-peak hours, for 10 hours per day. Assume that there are 30 days in the billing month and that the load factor is 1, or 100%.

A. $902,000
B. **$416,808**
C. $2,064,187
D. None of the above

Solution:
As explained earlier in this chapter, the energy charges under OPT-I rate schedule are classified into two categories: (a) On-peak energy charges per month, per kWh, and (b) off-peak energy charges per month, per kWh. The energy cost rates for on-peak and off-peak usage are shown in the shaded fields of Table 11.2 excerpt:

III. Energy Charges	All Months
A. All on-peak energy per month, per kWh	5.7847¢
B. All off-peak energy, per month, per kWh	3.4734¢

Since the total energy consumption is not given, we can derive it based on the parameters identified in the problem statement.

Since the load factor is unity, or "1," as discussed earlier in this text:

$$\text{Load Factor} = \frac{\text{Average Demand}}{\text{Peak Demand}}$$

Since Load Factor, in this case = 1,

$$1 = \frac{\text{Average Demand}}{\text{Peak Demand}}, \text{ or Average Demand} = \text{Peak Demand} = 40,000 \text{ kW}$$

The energy consumed during the billing month:

= (Number of hours of operation)·(average demand)
= (30 days/billing month × 10 hours of operation/day)·(average demand)
= (300 hours of operation/month)·(40,000kW)
= 12,000,000 kWh per month

Total energy cost for the month:

= (12,000,000 kWh)·(3.4734¢/kWh)·(100¢/$)
= **$416,808**

5. Which of the following statements describe the role of EPC and ESCO most accurately?
   A. The terms EPC and ESCO are synonymous
   B. **EPC is method for implementing energy projects and ESCOs are entities that offer this alternative**
   C. EPC is required by Department of Energy, ESCOs are not
   D. None of the above

CHAPTER 12 – SOLUTIONS

1. What type of battery would be most suitable for the following applications?
   a. UPS, Uninterruptible Power Supply system
   b. DC power for substation breaker operation
   c. Remote wind turbine
   d. Remote solar PV farm
   Answer: All four applications stated in this question would require large amounts of energy (Wh). Also, weight of the battery would not be a factor. While, a few years ago, with the cost of Li–ion batteries being substantially higher than the lead–acid batteries, the choice would have been, decisively, in favor of the lead–acid batteries; today, as evident through Table 12.1, **Li–ion or Li–Po batteries would be the best choice, from almost all points of view.**
2. In Example 12.1, if the volume of space reserved for the battery is 3L (3,000cm³) of less, and the mass constraint is 10kg, based on energy density considerations, which of the given battery designs would be suitable?

Answer: As mentioned in the energy density discussion in this chapter, and as shown in Figures 12.6 and 12.7, Li–Po has the best energy density, at 2.3 L (2,300 cc) per kWh. In addition, as stated in the solution for Example 12.1, the mass of 1 kWh Li–Po battery would not exceed 6 kg, which is well below the limit of 10 kg. Therefore, Li–Po would be the most suitable choice and would occupy the least volume.

Note: Based on charts in Figures 12.6 and 12.7, and other information provided in this chapter, a lead–acid battery system would occupy 11.8 L (11,800 cc) of space and a $LiPO_4$ battery would have a mass of 4.6 kg. The lead–acid battery would also be rejected due to the three-dimensional motion of the spacecraft.

3. It has been disclosed that the battery being selected for the mission to Mars, as outlined in Problem 2, must be capable of serving as an emergency source of power for 1 kW of ignition power without overheating. However, the mass of the battery must not exceed 3.5 kg. Based on the specific power data provided in this chapter, which of the three batteries would meet the specifications?

Answer: As mentioned in the specific power discussion in this chapter, and as shown in Figures 12.6, Li–ion battery would offer the best specific power (W/kg) among all battery choices discussed. Note that while capacitors might offer higher specific power than Li–ion batteries, they would not meet the specific energy criteria.

4. A $LiFePO_4$ cell battery is discharged at three times the nominal discharge rate. Will the energy density of the battery
   a. Remain 220 Wh/L?
   b. Be greater than 220 Wh/L?
   c. Be less than 220 Wh/L?
   d. None of the above?

Answer: Based on the discussion on energy density in this chapter, energy density of a battery will vary with abnormal demand, due to that fact that energy delivered by a battery, in Watt-hours, can vary based on the rate of discharge. Therefore, the energy density of a $LiFePO_4$ cell battery will decline when discharged at an excessive rate. Hence, the best answer is **(c)**.

5. A Li–Po, or lithium–polymer cell battery, is discharged at half the nominal discharge rate. Will the specific energy of the battery
   a. Remain 180 Wh/kg?
   b. Be greater than 180 Wh/kg?
   c. Be less than 180 Wh/kg?
   d. None of the above?

Answer: As implied in the discussion on specific energy in this chapter, specific energy of a battery will remain at or around the rated value when battery is discharged at, or less than, the nominal discharge rate. Therefore, the specific energy of a Li–Po battery will remain at, approximately, 180 kWh/kg. Hence, the best answer is **(a)**.

6. A double-layered capacitor is being considered as an energy storage device to in an electrical system. The amount of initial energy needed by the system is 10 Wh. Based on the specific energy and specific power chart in Figure 12.6, the approximate weight of the double-layered capacitor would be:
   a. 1 kg
   b. **10 kg**
   c. 100 kg
   d. 1,000 kg

   Answer: Based on the chart depicted in Figure 12.6, the specific energy of a double-layered capacitor is approximately 1 Wh/kg. Therefore, a double-layered capacitor capable of storing 10 Wh would have a mass of 10 kg. Correct answer: **(b)**

7. If, in Problem 6, the maximum weight allowed for the charge storage system is 1 kg, based on the specific energy and specific power chart in Figure 12.6, which charge storage system would be the most suitable?
   a. **Lead–acid battery**
   b. Double-layered capacitor
   c. Ultra-capacitor
   d. None of the above

   Answer: Based on the chart depicted in Figure 12.6, the specific energy of a lead–acid battery is 10 Wh/kg. Therefore, with the 10 Wh energy storage requirement, and the 1 kg limit on mass, lead–acid battery would be the best choice among the alternatives given in this problem. Correct answer: **(a)**

Appendix B
Common Units and Unit Conversion Factors

POWER

In the SI or Metric unit system, DC power or "real" **power** is traditionally measured in watts and:

kW = Kilowatt = 1,000 Watts
MW = Megawatt = 1,000,000 Watts = 10^6 W
GW = Gigawatt = 1,000,000,000 Watts = 10^9 W
TW = Terawatt = 10^{12} W
PW = Petawatt = 10^{15} W

Where k = 1,000, M = 1,000,000, G = 1 billion, and T = 1 trillion.

Some of the more common **power conversion factors** that are used to convert between SI system and US system of units are listed below:

1.055 kJ/s = 1.055 kW = 1 BTU/s
1–hp = One hp = 746 Watts

= 746 J/s
= 746 N-m/s
= 0.746 kW
= 550 ft-lbf/s

ENERGY

In the SI or Metric unit system, DC energy or "**real**" **energy** is traditionally measured in Wh, kWh, MWh, GWh, TWh (10^{12} Wh).

kWh = 1,000 Watt-hours
MWh = 1,000,000 Watt-hours = 10^6 Wh
GWh = 1,000,000,000 Watt-hours = 10^9 Wh
TWh = 10^{12} Wh
PWh = 10^{15} Wh

Some mainstream conversion factors that can be used to convert electrical energy units within the SI realm or between the SI and US realms are referenced below:

1,000 kW × 1h = 1 MWh
1 BTU = 1055 J = 1.055 kJ
1 BTU = 778 ft-lbf

ENERGY, WORK, AND HEAT CONVERSION FACTORS

Energy, Work, or Heat		
Btu	1.05435	kJ
Btu	0.251996	kcal
Calories (cal)	4.184	Joules (J)
ft-lbf	1.355818	J
ft-lbf	0.138255	kgf-m
hp-hr	2.6845	MJ
kWh	3.6	MJ
m-kgf	9.80665	J
N-m	1	J

POWER CONVERSION FACTORS

Power		
Btu/hr	0.292875	Watt (W)
ft - lbf/s	1.355818	W
Horsepower (hp)	745.6999	W
Horsepower	550.*	ft-lbf/s

TEMPERATURE CONVERSION FACTORS/FORMULAS

Temperature		
Fahrenheit	(°F − 32)/1.8	Celsius
Fahrenheit	°F + 459.67	Rankine
Celsius	°C + 273.16	Kelvin
Rankine	R/1.8	Kelvin

COMMON ELECTRICAL UNITS, THEIR COMPONENTS, AND NOMENCLATURE

Force	Newton	N	$kg\ m\ s^{-2}$
Energy	joule	J	$kg\ m^2\ s^{-2}$
Power	watt	W	$kg\ m^2\ s^{-3}$
Frequency	hertz	Hz	s^{-1}
Charge	coulomb	C	$A\ s$
Capacitance	farad	F	$C^2\ s^2\ kg^{-1}\ m^{-2}$
Magnetic induction	tesla	T	$kg\ A^{-1}\ s^{-2}$

COMMON UNIT PREFIXES

1.00E − 12	pico	p
1.00E − 09	nano	n
1.00E − 06	micro	μ
1.00E − 03	milli	m
1.00E + 03	kilo	k
1.00E + 06	mega	M
1.00E + 09	giga	G
1.00E + 12	tera	T

WIRE SIZE CONVERSIONS

A circular mil can be defined as a unit of area, equal to the area of a circle with a diameter of one mil (one-thousandth of an inch), depicted as:

1 circular mil is approximately equal to:

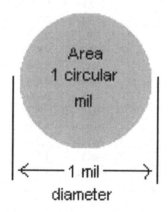

- 0.7854 square mils (1 square mil is about 1.273 circular mils)
- 7.854×10^{-7} square inches (1 square inch is about 1.273 million circular mils)
- 5.067×10^{-10} m²
- 506.7 μm²

1,000 circular mils = 1 MCM or 1 kcmil and is (approximately) equal to:

- 0.5067 mm², so 2 kcmil ≈ 1 mm²

AWG TO CIRCULAR MIL CONVERSION

The formula to calculate the circular mil for any given AWG (American Wire Gage) size is as follows:

A_n represents the circular mil area for the AWG size **n**.

$$A_n = \left(5 \times 92^{\frac{36-n}{39}}\right)^2$$

For example, an AWG number 12-gauge wire would use **n = 12**; and the calculated result would be 6,529.946789 circular mils.

CIRCULAR MIL TO MM² AND DIAMETER (MM OR IN) CONVERSION

kcmil or, MCM	mm²	Diameter in.	Diameter mm
250	126.7	0.5	12.7
300	152	0.548	13.91
350	177.3	0.592	15.03
400	202.7	0.632	16.06
500	253.4	0.707	17.96
600	304	0.775	19.67
700	354.7	0.837	21.25
750	380	0.866	22
800	405.4	0.894	22.72
900	456	0.949	24.1
1,000	506.7	1	25.4
1,250	633.4	1.118	28.4
1,500	760.1	1.225	31.11
1,750	886.7	1.323	33.6
2,000	1,013.4	1.414	35.92

Appendix C
Greek Symbols Commonly Used in Electrical Engineering

Greek Alphabet			
Aα	Alpha	Nν	Nu
Bβ	Beta	Ξξ	Xi
Γγ	Gamma	Oo	Omicron
Δδ	Delta	Ππ	Pi
Eε	Epsilon	Pρ	Rho
Zζ	Zeta	Σσς	Sigma
Hη	Eta	Tτ	Tau
Θθ	Theta	Υυ	Upsilon
Iι	Iota	Φφ	Phi
Kκ	Kappa	Xχ	Chi
Λλ	Lambda	Ψψ	Psi
Mμ	Mu	Ωω	Omega

Index

Printed in the United States
by Baker & Taylor Publisher Services